American Institute of Chemical Engineers

Process Plant Safety Symposium 2005

Held at the 2005 AIChE Spring National Meeting

April 10-14, 2005
Atlanta, Georgia, USA

Printed from e-media with permission by:

Curran Associates, Inc.
57 Morehouse Lane
Red Hook, NY 12571
www.proceedings.com

ISBN: 978-1-60423-472-5

Some format issues inherent in the e-media version may also appear in this print version.

American Institute of Chemical Engineers

Process Plant Safety Symposium
2005

TABLE OF CONTENTS

Session 92 - Case Histories and Lessons Learned - Joint Session (LPS and PPSS)
Chair: Henry L. Febo
Vice Chair: David G. Clark

1 **Insidious Explosion Hazards in the Mining Industry**
Steven J. Luzik and Derrick Tjernlund

21 **Two Large Losses: Refinery Fire and Vapor Cloud Explosion at a Natural Gas Processing Plant**
Jeffrey Yuill

39 **U.S. Chemical Safety Board Investigation: Positive Pressure Control Room Failure**
Lisa Long, Johnnie Banks, Mike Morris, John Murphy and Stephen Wallace

Session 93 - Human Factors Engineering and Ergonomic Engineering - Part I
Chair: Don Lorenzo
Vice Chair: Lawrence J. H. Schulze

46 **Rotational Force Capabilities of Males and Females Between the Ages of 30 and 55 Years**
Lawrence J. H. Schulze, Rebecca Ginsburg and Jehanzeb Quereshi

55 **Human Factors Analysis Techniques For Process Control Rooms and Control Systems**
Dennis Attwood

64 **Statistical Analysis of Human Factors Checklists**
Bruce K. Vaughen

75 **Methods for Improving Human Performance**
Don Lorenzo, Lee N. Vanden Heuvel and James J. Rooney

Session 94 - Human Factors Engineering and Ergonomic Engineering - Part II

88 **Human Reliability Analysis for Chlorine Bulk Unloading Procedures**
Joseph J. Chaback

Session 95 - Improving Safety Culture
Chair: M. Sam Mannan
Vice Chair: William J. Rogers

98 **Safety Climate: It"s not Just an Engineering Problem**
Stephanie C. Payne, Mindy E. Bergman, and Jaime B. Henning

111 **Operational Discipline at DuPont**
James A. Klein

126 **Values and Behavior: Building a Culture that Promotes Safety**
Terry E. McSween

136 **Integrated Management Systems and Safety Culture**
Robert Walter

148 **Essential Elements of a Sound Safety Culture**
Walt L. Frank

Session 96 - Risk Assessment
Chair: Philip M. Myers
Vice Chair: Michael Livingston

167 **Enhanced Layer of Protection Analysis (LOPA) Method**
Alvin Waller

183 **Selection and Evaluation of Release Scenarios for an LNG Import Terminal**
Jeffrey D. Marx and John B. Cornwell

193 **Criteria for the Escalation of Fires and Explosions**
Valerio Cozzani, Gianfilippo Gubinelli and Ernesto Salzano

212 **Quantitative Risk Analysis for Hazardous Materials Transportation**
Yuanhua Qiao, Nir Keren, William J. Rogers and M. Sam Mannan

224 **Statistical Analysis of Pressure Vessel Failure**
Thomas J. Mikschl and Ahmad Shafaghi

234 **Atlantis Case Study: Getting the Most from Your HAZOP**
Donnie J. Carter, Mary Gerschefski and Kevin Watson

248 **Operator Initiated Action as an Independent Protection Layer**
Angela Summers and Scott Sandler

Session 97 - Incident Investigations and Methods of Interpreting Evidence
Chair: Lisa M. Morrison
Vice Chair: Don Connolley

258 **The Transmitters Are Leaking in One Unit, But Are Working Fine in Another Unit, or Why Didn't the First Fixes Work?**
Arthur M. Dowell

259 **Incident Investigation of an Explosion at a Styrene Plant**
Dan Wiff

277 **Investigating Chemical Process Accidents: Examples of Good Practices**
D.R. Morrison, Russell A. Ogle, M.J. Viz, A.R Carpenter and Y. Su

290 **Investigation Challenges of the Kinston Dust Explosion Accident**
Quentin A. Baker, Robert C. Gombar James A. Lastowka, and Michael A. Polcyn

307 **Investigating Explosion Incidents – What Do You Look for and What Does It Tell You**
Darrell D. Barker

316 **Dust Explosion Case Study: How a "Wrong" Material of Construction Caused a Dust Explosion**
Dehong Kong

Session 98 - Inherently Safer Process Design and Operations
Chair: Jack Chosnek
Vice Chair: Vic Edwards

328 **Application of Inherently Safer Process Concepts in DuPont**
C. Curtis Clements

344 **Inherently Safe Process Plants: Practical Methods and Examples**
David G. Kehn

364 **Inherently Safer Technology: An Evolutionary Thing**
Tim Overton and George King

372 **Sub-Atmospheric Pressure Storage and Delivery For Gases**
Karl Olander and Peter Van Buskirk

382 **Implementing Inherently Safer Design in an Existing Plant**
Dennis Hendershot, Jonathan A. Sussman, Gerald E. Winkler, Jack Chosnek and G. Lee Dill

392 **Continuous Process Improvement Toward a Safer Ester Amidation Process**
Adrian David, Srinivas Tummala, Boguslaw M. Mudryk, Jing Liang, Lucius T. Rossano, David J. Kacsur, Karen L. Tenhuisen, Simon Leung and David Kacsur

Session 141 – Joint CCPS/LPS/PPSS Plenary Session

401 **Preventing Incidents At Newly Acquired Facilities – Implementation of Lessons Learned**
Brian Dunbobbin, C. Lorn Paxton, Glen A. Peters and Martin A. Dennehy

413 **Organizational Factors That Influence Safety: How NASA and Other Organizations are Changing Their Safety Culture**
Scott Stricoff and Patrick Killimet

Session 142 – CCPS/LPS/PPSS Conference Keynote

421 **Keynote Paper Holder (*paper not available*)**
Scott Berger

Session 154 - Case Histories - Joint Session (CCPS, LPS, and PPSS)
Chair: Henry L. Febo
Vice Chair: David G. Clark

422 **The 27 March 2003 Billy-Berclau Accident - A Technical and Organisational Investigation**
Jean-christophe LeCoze, Nicholas Dechy, Samantha Lim, Emmanuel LePrette, and Ruddy Branka

451 **When Risk Becomes Reality: Formosa Plastics' Response to a Plant Explosion.**
Robert Thibault

453 **Handling Chemicals in Small Containers**
Albert I. Ness, Richard Gibson

Insidious Explosion Hazards in the Mining Industry

Derrick M Tjernlund PE, CFEI
Steven J Luzik, PE, CFEI, CMSP
MSHA, Approval and Certification Center
Box 251, Industrial Park Rd. Triadelphia, WV 26059

Prepared for Presentation at
American Institute of Chemical Engineers
2005 Spring National Meeting
39th Annual Loss Prevention Symposium
Atlanta, Georgia
April 11-13, 2005

UNPUBLISHED

Insidious Explosion Hazards in the Mining Industry

Derrick M Tjernlund PE, CFEI
Steven J Luzik, PE, CFEI, CMSP
MSHA, Approval and Certification Center
Box 251, Industrial Park Rd. Triadelphia, WV 26059

Abstract

Explosions have occurred in industry throughout modern history. Most everyone, from those that either work in or manage industrial sites, to the general public, associate the word "explosion" with the large-scale catastrophic events that cause multiple fatalities and widespread property damage. These are the events that grab the front page headlines of newspapers or make the top stories of the evening television newscasts. Under the radar of these large-scale explosions are a number of small-scale events that occur on a frequent basis. These events often result in injury or loss of life without serious property damage, or represent near misses that few people removed from the accident ever hear or read about.

This paper discusses the Mine Safety and Health Administration's (MSHA) experience with these types of explosions that include incidents involving vehicle tire maintenance, industrial pumps, refining processes, and acetylene tanks. Identification of root causes and contributing factors are included, as well as a reemphasis of safety training, task evaluation, and outreach programs that the industry may consider to prevent future occurrences.

1. Introduction

On a cold February morning in 1999, just after midnight, a railroad tank car filled with liquid hazardous waste-derived fuel suddenly exploded at a cement plant in north central Indiana. The tank shell, weighing approximately 19 tons, was launched up and over two rotary cement kilns at a departure angle of approximately 60 degrees or greater. The shell finally came to rest about 800 feet from where it started. The explosion touched off a large fire in the waste fuel handling area of the plant that lasted for more than four hours and required the effort of numerous local fire departments and the crash-fire-rescue equipment from a nearby military base before it was brought under control. The emergency required evacuation of surrounding communities and resulted in the response from four federal agencies: the Mine Safety and Health Administration (MSHA), the Environmental Protection Agency (EPA), the Federal Railroad Administration (FRA), and the National Transportation Safety Board (NTSB). Needless-to-say, the story grabbed front page headlines for several days in regional newspapers. Yet, despite the news worthiness of the incident, there were miraculously no injuries or fatalities, due in large part, to the time of day of the event when few employees were on site.

In the early morning hours of July of the same year, another devastating explosion occurred at a bauxite ore processing plant in Louisiana when a number of very large pressure vessels containing superheated caustic slurry ruptured, destroying the central part of the facility. In this event, over 20 workers were injured, some seriously, with two receiving extensive, nearly fatal burns. The domed top of one pressure vessel, weighing more than 7,000 pounds, was thrown approximately 3,000 feet from where it started by the force of this BLEVE[1] event. Nevertheless, as with the railroad tank car, there were no fatalities in this event. A team of investigators, engineers, and mine emergency response specialists from MSHA, along with environmental specialists from the EPA, responded. The news worthiness of this event was also well documented.

In between these two events, on a warm sunny morning in Mississippi, a mine worker died when the impeller hub of the slurry pump he was working on exploded. He was struck by a metal fragment weighing less than one pound and died within minutes of receiving his injuries. Because this site too was a Mine Safety and Health Act regulated property, an MSHA investigation team responded to the accident. However, because of the limited nature of the event, only MSHA responded. Unfortunately, small, relatively un-newsworthy fire/explosion events such as the last one described occur far too frequently.

This paper presents a number of such events experienced by the mining industry in an effort to raise awareness about them. These fire/explosion hazards can be considered "insidious" because the hazards leading to these events often go unrecognized by the general work force (and not just the workforce within the mining industry). Industry itself may sometimes take too little notice of such events; hence, the lessons that could be learned never get delivered to those who could be exposed to similar hazards---hazards hidden in misunderstanding, inadequate knowledge, careless attitudes, or unsound practices.

2. Background

Fires and explosions have long been a deadly menace to those who work in industry. Nowhere was this truer than in the mining industry, where decades of history conveyed the deadly nature of unsafe coal mining practices which claimed thousands of miners' lives. During the Twentieth Century, miners, mine operators, and various government agencies worked diligently to halt this tragic loss of life. The U.S. Bureau of Mines, established in 1910 as a branch of the U.S. Department of the Interior, spent the better part of the century building a massive body of knowledge through research and investigation---a body of knowledge describing both the causes of and preventative measures needed to remove mining from the top of the list as the nation's most dangerous livelihood.

In 1969, Congress passed the Federal Coal Mine Health and Safety Act, which established significant binding safety and health standards for the coal mining industry. In 1977, Congress reorganized these efforts by passing the Mine Safety and Health Act which created the Mine Safety and Health Administration within the U.S. Department of Labor. MSHA's role is defined similar to that of OSHA[2], but with the authoritative mandate to specifically regulate and promote the health and safety of the nation's miners not only in coal mines, but also in metal and nonmetal mines.

---

[1] Boiling Liquid Expanding Vapor Explosion
[2] Occupational Safety and Health Administration

Over the years, the combined efforts of MSHA (and its predecessor agencies), along with efforts by NIOSH[3] (and its predecessor, the U.S. Bureau of Mines), have turned the nation's mines into some of the safest workplaces in the country. Yet despite these impressive accomplishments, MSHA has continued to observe a number of incidents that have resulted in single fatalities involving explosion events. In examining and investigating these accidents, one common theme has come to light; in all of these events, the hazards seemed hidden from the victims. Those involved (both victims and witnesses) often appeared to misunderstand what an explosion was, how it could happen, and how to prevent it.

Perhaps the first problem is that there are many different ideas as to what constitutes an explosion. The large scale, "headline grabbers" are easy to accept. A pump impeller blowing apart is less obvious.

NFPA[4] 921 defines an explosion as[1]:

> *The sudden conversion of potential energy (chemical or mechanical) into kinetic energy with the production and release of gases under pressure, or release of gas under pressure. These high pressure gases then do mechanical work such as moving, changing, or shattering nearby materials.*

While this definition is certainly a sound technical one, and a definition spoken in the language understood by those of us in the field of science and engineering, it is not necessarily a good one for the general work force.

The authors propose a more generic version for safety training and planning purposes that may better alert workers to examine their assigned work tasks <u>before</u> proceeding with the task. Our definition for an explosion is:

> *A violent release of pressure energy; often from a bursting container, vessel, or enclosure; causing, or capable of causing, death, injury, or damage to equipment or property.*

This definition is similar to a simple definition put forth by the U.S. Bureau of Mines as[2]:

> *A sudden release of pressure or energy.*

Our proposed definition utilizes the technical simplicity of the Bureau definition, while retaining emphasis on the potentially catastrophic nature of the event to cause death, injury, or damage, as implied by the NFPA definition.

MSHA classifies explosion-type accidents[5] into one of two categories[3]:

1. exploding vessels under pressure
2. ignition or explosion of gas or dust

---

[3] National Institute of Occupational Safety and Health

[4] National Fire Protection Association

[5] These classifications do not include accidents involving blasting activities and explosives.

What follows are examples of 11 mining accidents that fall into these two categories. These case histories resulted in fatalities and/or injuries. One important aspect in workplace understanding of explosion hazards is the ability to recognize the conditions under which pressure energy can be created. This energy can be generated by combustion; by other chemical reactions; by external application of energy such as flame or heat, by electrical energy, by mechanical energy generated by a rotating shaft; or by the structural failure of a vessel which is already normally under pressure, or thought to be under pressure.

3. Discussion of Accidents

Case History No. 1: Coal preparation plant; fatal exploding-vessel-under-pressure accident, Buchanan County, Virginia

Overview: A 54-year-old miner was fatally injured when a clean coal filter drain pump used for pumping a slurry of water and coal fines exploded at a coal preparation plant. The victim was standing approximately eight feet away at the pump's on/off switch while the pump was running, when the volute cover plate blew off and struck him. An examination of the pump after the accident revealed that the discharge pipe was plugged full of coal fines. This was not the first occurrence of such a pump failure at this facility as a similar failure had occurred 10 years earlier.

Findings and Conclusions: The pump became overheated when the water content in the slurry apparently decreased, causing a consequent rise in the specific gravity of the slurry and resultant decrease in pump performance. This permitted the suction and discharge lines to and from the pump to become clogged with coal fines, causing a total blockage of the pump. This caused the pump to become a closed pressure vessel hydraulically full of water/coal slurry. The slurry trapped within the pump continued to absorb energy from the pump as it continued to run for up to 35 minutes after the discharge became plugged. This transfer of energy into the entrapped slurry caused a significant rise in temperature and pressure of the contents, eventually resulting in explosive failure of the pump volute cover plate. The presence of welds at the access cover attachment points materially contributed to the mode and potentially the timing of the pump failure. The accident occurred because the discharge line to the pump became clogged while the pump continued to be operated. The confined coal slurry within the pump volute was consequently overheated, resulting in a pressure buildup to the point the access cover plate blew off.

Causal Factors:

- Company's failure to properly assess the hazards associated with overheated pumps.
- Company's failure to install pump casing temperature monitoring on all pumps of similar duty when the first pump failed ten years previous.
- Company's failure to establish a procedure for starting the pump from a safe location.
- A secondary causal factor was the presence of welds at the volute cover attachment points that materially contributed to the mode and potentially the timing of the pump failure.

Case History No. 2: Metal/Nonmetal sand and gravel; fatal exploding-vessel-under-pressure accident, Comal County, Texas

Overview: A 28-year-old maintenance worker was killed and two coworkers injured when a large off road tire exploded while the three workers were trying to remove the brake drum from the rim in an effort to replace the drum. The tire/rim assembly had been removed from the rear axle of a Caterpillar model 631G pan scraper and laid down on the ground with the drum side of the rim facing up. The tire had not been removed from the rim (nor depressurized) prior to commencing the repair work. The brake drum was bolted to the inside of the rim, but workers had trouble getting the drum to break loose even after removing all of the bolts and were using a sledge hammer to try and break it free. Approximately an hour before the accident, workers had used an oxyacetylene torch to cut the brake drum into two pieces in an effort to loosen it. When this failed, the workers took a break for lunch.

Within minutes after returning to the tire to continue the work, the tire exploded. One worker had just turned away from the tire to get a tool from his utility truck parked next to the tire when the explosion occurred. This worker did not witness the explosion but was knocked to the ground. The second injured worker was believed to have had a pry bar in his hands and was standing next to the tire. He was knocked unconscious and did not regain memory of what happened prior to the explosion. The worker killed was believed to have been standing directly next to, or on top of the tire. He had been holding a sledge hammer and may have begun hammering on the brake drum in an attempt to loosen it. He was killed instantly and both of his arms were severed at the shoulders by the force of the explosion. Significant shredding of the upper side wall of the tire had occurred as the steel belt cables were ripped from around the bead cable. The tire and rim had been propelled high enough into the air to allow the assembly to land up-side-down from its original orientation. The tire had also been turned inside out and pieces of its' sidewall had been propelled up to 165 feet from the original location.

Findings and Conclusions: Physical examination and Gas Chromatography/Mass Spectroscopy (GCMS) testing of the gas in the other rear tire from the scraper indicated it was pressurized with air to approximately 50 psig. Both tires were of the same make and model and had similar purchase and maintenance histories. An examination of the rim assembly indicated that the torch flame had directly impinged on the rim in the vicinity of the tire bead sufficiently to "nick" the rim during the attempt to cut the brake drum in half. It is believed that the use of the torch transferred sufficient heat through the rim steel to ignite the inner liner material in contact with the rim at this location. This resulted in a smoldering fire of the inner liner material. Examination of the tire's interior surfaces confirmed this.

It is known that smoldering tire rubber generates copious quantities of incomplete products of combustion that are, in and of themselves, combustible. Over the course of the delay of work during the lunch break, the tire likely filled with a sufficient quantity of combustibles to create an ignitable mixture within the tire. The recommencement of work after lunch may have dislodged a hot smoldering ember from the lining material that fell into the mixture, igniting it. The extensive shredding of the tire sidewall and the severe injuries to the worker killed were both indicative of a rapid rate of pressure rise and high peak pressure typically associated with a combustion explosion. The accident occurred due to the unsafe application of a cutting torch flame to the tire/rim assembly. Application of heat to a rim can only be done safely if the tire is first completely removed from the rim prior to conducting this work.

Figure 1. Case History 2–tire explosion

Figure 2. Case History 2–charred inner liner of tire

Causal Factors:

- Company's failure to establish maintenance procedures and training that mandated the need to always completely remove tires from rims before performing any cutting or welding to a rim assembly.
- Use of air instead of an inert gas such as nitrogen for pressurizing tires.
- Failure of the company to develop an emergency plan for immediately isolating any tire/rim assembly suspected of being heated. It is recommended that this isolation establish a no entry zone of at least 300 feet in all directions around the tire for at least 24 hours.

Case History No. 3: Metal/Nonmetal surface gypsum operation; fatal ignition or explosion of gas or dust, Sevier County, Utah

Overview: A 63-year-old quarry maintenance supervisor, with six years mining experience, was fatally injured while he and a coworker were replacing the seals on the rear suspension cylinder (shock absorber) of a haulage truck. The suspension cylinder assembly was mounted on the rear axle of a 35-ton, Model 769C Caterpillar haulage truck. While dismantling the suspension cylinder, a small threaded valve broke off. After unsuccessful attempts to remove the threaded portion of the broken valve, penetrating oil was sprayed into the valve port and heat was applied with a torch. An explosion occurred in the cylinder and the rod was forcefully ejected, striking the victim.

Findings and Conclusions: The cylinder rod measured approximately 25 inches long, seven inches in diameter, and weighed approximately 114 pounds. The upper mounting pin for the cylinder had been removed from the cylinder, allowing it to be rotated to a horizontal position. The nitrogen and hydraulic oil normally in the cylinder had been removed as part of the maintenance effort. As the nitrogen charging valve was removed, the lower threaded section broke off in the cylinder. The port within the broken portion of the valve tip was drilled out to a larger diameter to facilitate the use of an "easy-out" and compressed air from a 140 psig supply was used to blow the valve remnant free of material. An aerosol penetrating oil was applied to loosen the valve threads. An oxyacetylene torch with a welding/brazing tip was then used to heat the valve remnant.

Closed cup flash point testing conducted on samples of the hydraulic oil collected from the cylinder determined the flash point to be 374 degrees F. Similar tests conducted on a sample of the penetrating oil determined the flash point to be 175 degrees F. Tests were also conducted to determine if the use of compressed air would create enough internal pressure within the suspension cylinder to discharge the rod. No significant internal pressure was detected when 140 psi of air was supplied through the nitrogen charging valve. The compressed air escaped from several cylinder assembly ports, preventing any measurable increase of air pressure above atmospheric. These ports provided adequate venting for the applied compressed air, but not for the rapid rate of pressure rise due to ignition of the mixture of air and penetrating oil vapors.

Tests were conducted on the suspension cylinder assembly to recreate the accident by spraying the penetrating oil through the nitrogen valve, followed by the injection of compressed air, and then application of an oxyacetylene torch flame in a manner similar to what was done by the victim. During the tests, an ignition resulted within the suspension cylinder, resulting in the rod being forcefully ejected from the cylinder.

This accident occurred when an ignition source was applied to a combustible vapor contained with a vessel.

Figure 3 Case History 3–suspension cylinder explosion

Causal Factors:

- Company's failure to establish a job safety analysis involving tasks associated with components potentially containing combustible materials.
- Company's failure to provide adequate training in the service tasks the victim was assigned to perform.

Case History No. 4: Metal/Nonmetal sand and gravel; fatal exploding-vessel-under-pressure accident, Itawamba County, Mississippi

Overview: A 42 year old crusher operator with 2 years mining experience was fatally injured at a sand and gravel operation when he was hit by a fragment from an exploding slurry pump impeller. The victim and a coworker were attempting to remove the worn impeller from a pump in an effort to replace it. The impeller attached to the pump shaft by directly threading onto it. The end of the impeller was cast such as to form a cone-shaped hub that enclosed the end of the impeller shaft to protect the shaft from the abrasive nature of the slurry. The impeller metal was of a type that had a very high hardness to resist abrasion, but was also very brittle.

Efforts were made to unscrew the impeller from the shaft but were not successful. This may have been due to an incorrect assumption as to the direction of the threads. The two workers then hammered the sides of the impeller, breaking most of it off. However, the central threaded cone section of the impeller remained in place. The victim then proceeded to heat the impeller hub with the flame from an oxyacetylene torch. His coworker stated that this heating was applied for several minutes and that the surface of the hub turned a dull red. At that time, the victim turned off the torch and set it down while still leaning down and standing directly in front of the impeller hub. The hub suddenly exploded, blowing the cone tip off the hub and into the chest of the victim, causing fatal internal injuries.

Findings and Conclusions: An examination of the replacement impeller revealed that the cone hub section formed an enclosed cavity of approximately 3.7 cubic inches (60 milliliters) in volume. Examination of the damaged impeller revealed traces of a grease-like material on the threads. The pump manufacturer's literature indicated the recommended use of anti-seize compound on these threads. GCMS testing of small samples of grease recovered from the threads confirmed the presence of hydrocarbons.

It is believed that upon heating the impeller hub, some of the grease was vaporized and mixed with air trapped in the hub cavity. The auto-ignition temperature of the vapor/air mixture was likely reached by continued heating to the point of turning the impeller metal red. Normally, small cavities are not prone to development of significant pressures when internal combustible fuel/air mixtures are ignited because of the high surface area to volume ratios that exist. A high surface area to volume ratio encourages rapid heat transfer between the developing flame front and the sides of the container, resulting in a reduced flame temperature, and thus reduced pressure. However, in this case, the walls of the cavity were heated to a significant temperature, eliminating most of any heat sinking capability. The pump manufacturer's maintenance literature warned that the application of heat to the impeller created an explosion hazard. An examination of the pump frame also revealed an attached label warning against the application of heat to the impeller. The warning label had become covered with hardened rock dust over a period of time, completely obscuring it from sight. The accident occurred when a worker applied heat from a torch to an enclosed cavity containing an ignitable material.

Figure 4: Case History 4–slurry pump impeller explosion

Figure 5: Case History 4– replacement impeller, shown for reference.

Figure 6: Case History 4–cross sectional sketch of exploded impeller

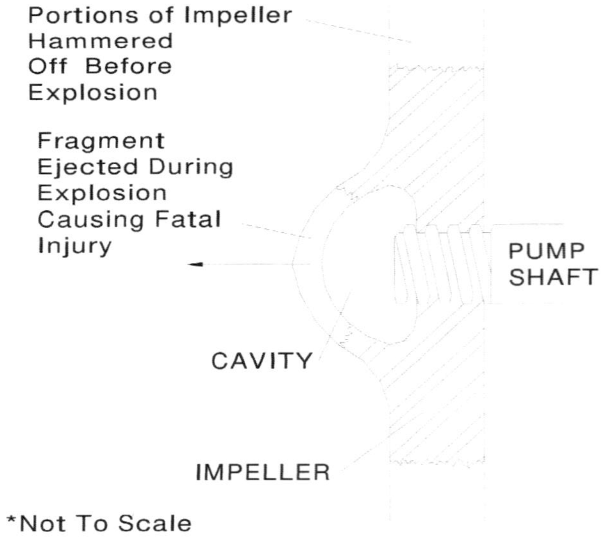

Figure 7: Case History 4–impeller pieces broken off using hammer before explosion.

Causal Factors:

- Company's failure to review the pump manufacturer's maintenance literature, which clearly warned against the application of heat to the impeller because it could create an explosion hazard.
- Company's failure to utilize the maintenance literature, which clearly identified the correct direction of rotation for removing the impeller, and identified use of a special tool needed to make removal of the impeller easier.

Case History No. 5: Metal/Nonmetal sand and gravel; fatal ignition or explosion of gas or dust accident, Fond du Lac County, Wisconsin

Overview: A 38-year-old foreman with 11 years experience was fatally injured at a limestone operation while trying to thaw out a frozen, in-ground truck scale with enclosed sump. The truck scale was scheduled to be returned to service after sitting during the winter. The March weather was still cold enough to prevent natural thawing of a 2-inch thick layer of ice in the bottom of the sump. The ice impeded the movement of the scale lever arms.

The foreman began his morning shift by putting tarps over the open edges of the scale in an apparent attempt to hold heat in the sump. He then placed a 20 pound propane cylinder down into the sump through an access hatch. The cylinder was equipped with a "weed burner" style torch head equipped only with a manual shutoff valve. The foreman placed the lit torch and cylinder down into the sump and left for most of the remainder of the shift. In the late afternoon, he returned to the scale with a 100 pound propane cylinder equipped with a second weed burner torch. One worker driving out of the quarry passed the victim in his truck as he headed toward the scale; however, no one was near the scale when the explosion occurred. Witnesses down in the quarry heard the explosion and saw several plates of the truck scale fly high enough into the air to be seen above the quarry spoil piles.

Findings and Conclusions: It is believed that the smaller propane torch self extinguished sometime during the day due to consumption of oxygen within the sump, before the cylinder was empty. The unlit torch continued to feed unburned propane gas into the sump space, and, along with the migration of fresh air into the sump after the flame extinguished, created an ignitable mixture. Upon returning with the second torch, the foreman apparently lit it and attempted to place it into the sump area. This ignited the gas/air mixture that had accumulated in the sump space, causing the explosion. The scale's six steel cover plates, weighing approximately 480 pounds each, were propelled up or away from the scale, with one plate landing approximately 100 feet from it. The victim was believed to be standing on the plate equipped with the access hatch and was killed instantly by the upward force of the plate. The investigation also revealed that the victim had used this same procedure on previous occasions when assigned the task of preparing truck scales after winter weather. The accident occurred when a worker applied a torch flame to an enclosed cavity containing an ignitable gas/air mixture.

Figure 8: Case History 5–truck scale explosion

Causal Factors:

- Company's failure to conduct a job safety analysis involving a procedure using a flammable gas.

Case History No. 6: Metal/Nonmetal cement plant; fatal exploding-vessel-under-pressure accident, Little River County, Arkansas

Overview: A 27-year-old waste fuel technician suffered fatal burns when a liquid waste fuel pumping system suffered a catastrophic overpressure within the system. The overpressure ruptured a casing on a waste fuel grinder, releasing substantial quantity of flammable fuel and covering the victim in the flammable liquid which was then ignited by an electrical arc from a power cable that was severed by the initial explosion. The accident occurred at a cement plant which burned liquid hazardous waste derived fuel in the plant's two rotary cement kilns.

Liquid fuel was received either by railcar or truck trailer and stored in vertical steel cone roof tanks within a large steel framed building. Two pumping trains took suction from either of two 100,000 gallon storage tanks and passed the fuel to the kiln burner floor for injection into either of the two kilns. Each pumping train consisted of nominal six inch steel piping, two variable-speed centrifugal pumps connected in series, followed by a pass-through vertical solid waste grinder, and then onto the kilns. The grinder was driven by a vertically mounted, direct coupled, 3 horsepower, 480-volt AC electric motor. The suction side of each pumping train had a number of pipe bends and low/high points capable of impeding flow and acting as air traps. When the pumps were started, a process computer took control of the pumps and normally ramped them up to full speed in several timed steps. The waste fuel technician was in close proximity to the pumps and grinder during the start up procedure. When it became apparent that the pumps were not moving any liquid, the technician began the process of opening bleed vents and may have opened a possibly closed discharge valve downstream of the grinder. During this effort, the explosion and fire occurred.

Findings and Conclusions: During a typical start up operation following a shutdown period, the pumps had occasionally failed to pick up a liquid prime from the storage tank because of entrapped vapors in the suction piping, resulting in zero pressure output. Small bleeder valves were installed on the downstream piping to allow for venting of trapped vapors within the piping during these occasions.

On this particular occasion, the pumps had already ramped up to full speed when the waste fuel technicians began the process of bleeding vapors from the system. A possibly closed supply valve downstream of the grinder may also have been opened during this process. At approximately that time, the first pump suddenly picked up the liquid supply from the tank and discharged it into the second pump. The second pump, in turn, suddenly boosted the pressure further, resulting in a large surge of liquid into the grinder which acted as a flow restriction. This caused a "water hammer like" pressure surge in the grinder. The severe overpressure spike catastrophically ruptured the grinder casing, forcing it upward and discharging flammable liquid throughout the area, including onto the waste fuel technician. The upward thrust of the ruptured grinder pulled the energized motor leads out of their conduit, causing a large electrical arc and igniting the liquid fuel. The fuel technician was engulfed in the ensuing fireball. A heat activated, deluge foam-water sprinkler system controlled the fire but could not respond fast enough to protect the technician from receiving second and third degree burns over 90 percent of his body.

Causal Factors:

- Company's failure to conduct an adequate chemical process safety management risk assessment of this operation.
- Company's failure to establish a safe location for workers during start up or during other transitional periods of operation.

Case History No.7: Coal surface operation; fatal ignition or explosion of gas or dust accident, Kanawha County, West Virginia

Overview: A 42-year-old contract mechanic, with 19 years mining experience, received fatal burns to the upper part of his body. He was preparing to use starting fluid to clean coal dust and oil from around the top of two motor mount bolts on a front-end loader. During this process, the aerosol can containing the starting fluid contacted an energized battery relay terminal and the frame of the machine, causing an electrical arc that burned a hole through the side of the can. The starting fluid ignited and sprayed through the hole and onto the victim, causing his clothes to become engulfed in flame. He ran around the front of the loader to a nearby service truck where another mechanic laid him on the ground and extinguished the flames. The victim was transported to a burn center where he later died as a result of his injuries.

Findings and Conclusions: Examination revealed that the cap (rubber insulating boot) provided to cover the energized battery relay terminal located inside the engine compartment was deteriorated (swollen from age and exposure to heat and oils) and was not covering the terminal. The accident occurred because a flammable fluid was being used near potential ignition sources. The ignition source, an energized battery relay terminal, was exposed because its insulating cap had deteriorated over time. When the side of the pressurized metal can containing the flammable liquid contacted both the energized terminal and the grounded mounting bracket, a high amperage electrical arc occurred that melted a hole through the side of the can. The starting fluid ignited and sprayed through the hole onto the victim, causing fatal burn injuries to his upper body.

Causal factors:

- Company's failure to maintain electrical equipment in adequate operating condition.
- The victim was using a highly flammable fluid near potential ignition sources.

Case History No. 8: Coal surface operation; fatal exploding-vessel-under-pressure accident, Buchanan County, Virginia

Overview: A 28-year-old utility man with 10 years of mining experience was fatally injured when unburned acetylene gas being discharged from a torch into a drum ignited. The fatally injured miner was holding an unlit oxygen/acetylene torch, through which he was delivering acetylene into an empty 55 gallon antifreeze drum when the drum exploded. Earlier in the day, the victim demonstrated the explosive nature of acetylene to a new mine employee by using a trail of starting fluid to ignite a bag filled with acetylene. At the time of the accident, he was intending to demonstrate a larger explosion. The drum was placed inside a large unattached Caterpillar, Model 992-G, front-end loader bucket, which concealed its view and shielded the blast from nearby buildings.

A mechanic who was standing next to the victim, and two other mine employees who were approaching the bucket at the time of the explosion, were also injured. The fatally injured victim received massive fatal head injuries when he was struck by debris from the explosion.

Findings and Conclusions: The accident occurred due to the explosion of an acetylene-oxygen-air mixture inside the 55 gallon drum. Forces and heat from the explosion injured four miners, one of whom died from his injuries. Ignition energy was most likely an electrostatic charge that built up on the torch tip from the rapid release of non-burning acetylene through the tip orifices. At the time of the accident, the fatally injured miner was using the unlit torch to intentionally fill an empty antifreeze drum with acetylene, and possibly oxygen, to demonstrate an explosion to a new employee. During this process, the torch tip neared the surface of the drum, which was grounded by contact with the front-end loader bucket, permitting an electrostatic arc that ignited the gas mixture inside the drum. Contributing to the accident was mine management's failure to establish, maintain, and enforce an effective safety program addressing horseplay and other hazards associated with mining activities, such as those related to the use of acetylene.

Figure 9: Case History 8 – empty 55-gallon drum explosion

Causal Factors:

- A safety program, adequate to deter horseplay and misuse of equipment, was not in place.
- The victim introducing acetylene/oxygen into the 55-gallon drum apparently without fear of reprisal on the company's part.

Case History No. 9: Surface of an Underground Coal Mine-; fatal exploding-vessel-under-pressure accident, Claiborne County, TN.

Overview: The victim was part of a two man crew performing repair work on a piece of mining equipment on the surface of an underground coal mine. The torch operator could not get the flame hot enough to perform the repair. The victim turned the acetylene regulator valve completely clockwise in an attempt to provide more gas to the torch. When this failed to solve the problem it was decided that the acetylene cylinder was depleted. The two man crew went to the supply building and retrieved a new cylinder of gas. The victim hooked up the new tank and his helper took his place onboard the mine scoop along with the torch. They were separated by about seven feet at that time. The torch operator tried to light the torch but was unsuccessful. The victim then stated that he may have backed the acetylene regulator out too far and was going to check it. It is unclear exactly what transpired after that but the torch operator stated that the victim went back over to the tank and performed an adjustment of some kind and told the torch operator to try to light the torch again. The torch operator could not say for certain if he struck his lighter but thought that he did. A violent explosion occurred at that time which dismembered the victim who was straddled over the Size B [40 cu. Ft.] acetylene tank. Examination of the remains of the tank suggested that a detonation had taken place.

Findings and Conclusions: It is believed that oxygen gas back-flowed into the acetylene supply hose and through the acetylene regulator as the acetylene cylinder became depleted. Absence of pressure gauges prevented the victim from recognizing the fact that the cylinder was nearing depletion. It also prevented him from realizing that he was dispensing the fuel gas at unsafe pressures above 30 psig. Absence of a check valve in the fuel line allowed oxygen to migrate back into the fuel line and regulator. Ignition is believed to have occurred from either the striking of the flint or adiabatic compression of trapped gases as the acetylene regulator was turned in clockwise by the victim. In either case, this primary ignition of the acetylene/oxygen mix in the fuel line is believed to have propagated into the acetylene cylinder and caused a much more violent secondary detonation within the cylinder. Testing and evaluation of the acetylene regulator confirmed that the flame path from the regulator into the tank would have exceeded the minimum quenching distances reported for acetylene gas in oxygen.

Causal Factors:

- Both the acetylene and oxygen regulators were being used without pressure gauges.
- Both the acetylene and oxygen pressure regulator diaphragm handles were turned almost completely clockwise allowing delivery of these gases at unsafe pressures.
- Check valves were not provided on either the fuel or oxygen lines between the torches and the hoses.
- Flashback arresters were not provided between the regulators and the delivery hoses.
- The acetylene and oxygen lines were not properly purged after changing the acetylene cylinder

Case History No. 10: Event: Nonfatal Fatal ignition or explosion of gas or dust accident, Gonzales County, TX.

Overview: An unplanned ignition of a fine organoclay dust (organically treated clay) occurred at a manufacturing plant while workers were in the process of troubleshooting a closed fluid bed batch dryer. The dryer was provided with a filter bag dust collection system which collected the fines. Periodically during the drying cycle the bags were shaken to allow collected dust to fall back into the drying bed. Hot air for drying was provided by a gas-piloted burner and air fan. The dryer was unexpectedly shutting down a few minutes into the drying cycle. The dryer was opened for troubleshooting purposes but the filter bags above the drying bed were not shaken prior to opening. Because of this, they were partially filled with collected dust. In order to facilitate troubleshooting, the bags were mechanically raised above the fluid bed. At that time an unexpected release of dust occurred from the bags. A flash fire ensued as a result of this ignition and five men suffered burns of varying degrees.

Findings and Conclusions: A leaking solenoid valve which allowed the gas pilot flame to remain on after shutdown of the dryer was found during the investigation. This provided a probable ignition source for the accumulations of dust within the combustion chamber, causing this material to smolder. Either this smoldering material, or the uncontrolled pilot flame itself, may have provided an ignition source for the dust cloud created when the workers failed to perform the shake cycles on the bags prior to opening of the dryer.

The unplanned ignition of organic clay dust occurred when a large quantity of the dust, used as a thickening agent in the manufacture of inks, dyes and paints, was released from the filter bags after the dryer was opened for troubleshooting purposes. It is highly probable that this cloud was ignited by a gas flame at the pilot burner that failed to extinguish after shut down of the burner due to a leaky gas pilot solenoid valve. The presence of both the fine dust cloud above the lower flammable limit and an ignition source of sufficient energy resulted in the ignition of the cloud and a flash fire. The construction of the dryer building and the geometry of the dryer allowed for adequate venting, preventing the build up of significant pressure.

Causal Factors:

- Company's failure to adequately maintain the combustion safeguards and fuel train appurtenances for the gas fired burner.
- Company's failure to establish a mandatory shakedown of the filter bags before any attempt to open the dryer under any conditions.

Case History No. 11: Metal/Nonmetal limestone hydration; nonfatal exploding-vessel-under-pressure accident, Clark County, Nevada.

Overview: A 45-year-old hydrator operator was seriously injured at a quicklime hydration plant when a piston-type slurry pump exploded while the operator was attempting to divert pump flow from a reaction tank to an exterior discharge bin. The valve used to divert flow was directly adjacent to the pump. The operator was struck by the pump cover plate when it blew off while he was attempting to operate the transfer valve.

Findings and Conclusions: The hydration reaction between the calcium oxide and water is substantially exothermic. In this operation, calcium oxide (quick lime) was mixed with water in an open chamber (mixing box) feeding a two cylinder piston pump. The pump discharged the reacting mixture through piping into a reactor vessel, where a vapor space could be used to control the pressure of the reaction. The suction side of each piston cylinder had a ball check valve that normally (but not always) prevented backflow out of the pump into the mixing box.

In this accident, the piping between the pump discharge and the transfer valve had plugged up, preventing flow out of the pump. If plugging of the pump discharge piping occurred, the operator received a pressure alarm and/or high amperage alarm on the pump motor, or heard the characteristic sound of the pump malfunctioning as flow backed up. Under those conditions, he had to manually change the position of a three-way valve on the pump discharge piping to redirect the back flow from the reactor vessel to the open bin. This was necessary to prevent the pressure and slurry in the reactor vessel from backing up into the pump, and out of the mixing box if the ball check valves failed. This event was not unusual since the mixture had a tendency to cause scaling and occlusion of the piping from time to time.

In this event, the reacting mixture was trapped in the pump. Heat from the reaction continued to be absorbed within the entrapped mixture and created increased pressure. During the accident, the pump was believed to have been hydraulically full. With no way to control the reacting mixture within the pump, dangerous internal pressure developed causing the pump lid to blow off, resulting in impact injury to the worker's arm and steam and thermo-chemical burns to his skin. A thermo-chemical analysis assuming isochoric and adiabatic conditions in the pump predicted that pressures in excess of 100,000 psig were possible under these conditions. This accident occurred because the pumping system was not capable of allowing for control of the chemical reaction rate within the entrapped mixture. Systems intended to handle chemically reacting materials must be designed using recognized chemical process safety management principles and technologies.

Causal Factors:

- Company's failure to conduct an adequate chemical process safety management risk assessment of this operation.
- Company's failure to establish a safe location for workers during start up or during other transitional periods of operation, such as plugged piping.

4. Methodology to Prevent Future Occurrences

The common thread in all of the above accidents is the failure of those doing the work to recognize the conditions created to permit an explosion to occur. The lack of recognition arises from a misunderstanding as to how pressure energy can be created or how it can be intensified beyond the structural design specifications of equipment. This misunderstanding, in turn, is often the result of poor procedures and inadequate training.

A dedicated effort in three distinct areas is needed to reduce the likelihood of repeating tragedies like those presented in this paper. These three areas are: 1) safety training, 2) job task evaluations, and 3) outreach programs. These three areas of effort can be effectively utilized as part of a safety culture approach taken by management.

4.1 Safety Training

Training is a fundamentally important component in any program to prevent deaths and injuries in the work place. Because job tasks and the hazards associated with them can be so diverse, it is difficult (if not impossible) to propose a simple, all-inclusive program for all those that are employed in any given operation.

One area that seems to need more work is educating personnel in the recognition of potential energy, and the role it can play in causing accidents. Certainly, such training will benefit the traditionally defined maintenance person. However, those charged with developing such programs within the work place must be cognizant of all those job positions which may expose the worker to potentially dangerous sources of potential energy while performing their assigned duties. In smaller operations in particular, care must be taken to identify jobs that may require a "jack-of-all-trades" role of the worker.

4.2 Job Task Evaluation

Formal safety analysis of all job tasks is slowly become more widely practiced within industry. An effective job safety analysis (JSA) coupled with training, can be a powerful tool in reducing tragic accidents often associated with traditionally unrecognized hazards. If not already completed, workplace safety officials should conduct reviews of all job assignments and tasks to prioritize them in terms of conducting JSA's. Any tasks that involve potential energy such as the application of high temperature heat; electricity; use of chemicals; the removal, repair, or installation of equipment, or is associated with pressurized fluids are all candidates for a red flag---a red flag that implies the need to immediately conduct a JSA.

4.3 Outreach Programs

Experience can often be an unforgiving teacher. Hence, industry must set a priority to communicate within itself to alert the workforce to unforeseen hazards. MSHA, for example, uses both its web site (www.MSHA.gov) and direct mailings to stakeholders in an effort to alert miners, mine operators, manufacturers, and others involved with the mining industry to accidents occurring within the industry. However, the workforce and those charged with safety management and training within industry must encourage the use of, and provide input to, outreach programs if the message is to truly get to those who need to hear it.

5. Conclusion

An aggressive, effective safety program utilizing a variety of methodologies is needed to eliminate the unseen insidious explosion hazards that can endanger those in the workplace. Training, job safety analysis, and communications are some of these important methodologies. Job safety analysis of all industrial work activities is particularly necessary in order to identify hidden potential explosion hazards. Outreach programs must be delivered to the grassroots personnel, and must be designed to affect the attitude of personnel toward working safer.

Emphasis should be placed on each worker, supervisor, manager, and trainer to develop and exercise a "questioning attitude" in approaching all work tasks. A questioning attitude is a mindset that must replace the all-too-often mentality of "I didn't think, I didn't see, I didn't know." It is a skill only sharpened with use. This will serve to enhance the awareness of workers who may otherwise face the insidious explosion hazards that still claim the lives of the nation's most precious industrial resource, its workers.

6. References

[1] NFPA 921 – 2001, Guide for Fire and Explosion Investigations, National Fire Protection Association, One Batterymarch Park, Quincy, MA, 02169

[2] Kuchta, Joseph M, "Investigation of Fire and Explosion Accidents in the Chemical, Mining, and Fuel-Related Industries – A Manual", Bulletin 680, U.S. Bureau of Mines, 1985

[3] MSHA Handbook Series, Handbook PH00-I-5, Accident/Illness Investigations Procedures, (http://www.msha.gov/readroom/handbook/handbook.htm)

Two Large Losses: Refinery Fire And Vapor Cloud Explosion At A Natural Gas Processing Plant

Jeffrey Yuill, P.Eng.
Starr Technical Risks Agency, Inc.
145 Wellington Street West,
Toronto, Ontario, CANADA M5J 1H8
Jeff.Yuill@starr-tech.com

Prepared for Presentation at
American Institute of Chemical Engineers
2005 Spring National Meeting
39th Annual Loss Prevention Symposium
Atlanta, Georgia
April 11-13, 2005

December 1, 2004

UNPUBLISHED

Two Large Losses: Refinery Fire And Vapor Cloud Explosion At A Natural Gas Processing Plant

Jeffrey Yuill, P.Eng.
Starr Technical Risks Agency, Inc.
145 Wellington Street West,
Toronto, Ontario, CANADA M5J 1H8
Jeff.Yuill@starr-tech.com

Abstract

This paper discusses two large losses.

The first loss involves a fire at a lube oil complex which forms part of a larger refinery site. The fire was the result of a failure of a portion of the flare header. The fire resulted in one death and $10 million dollars worth of physical damage.

The second loss involves a vapor cloud explosion at a natural gas processing facility. A release of a large amount of ethane and heavier hydrocarbons occurred for eight to 10 minutes before the cloud was ignited. The explosion resulted in eight deaths and $136 million in physical damages.

Details for much of this paper is based on reports provided by internal reports from adjusters and other specialists working on our behalf, references are not being submitted to ensure the anonymity of the losses.

1. Fire At A Refinery

South America, December 15, 1995

1.1 Background

The oil refinery where the incident took place is owned by the principal oil company operating in that country. Through its subsidiaries the company is involved in all phases of the oil industry including exploration, transmission, refining and retail sales. The company has three wholly-owned refineries, the largest being the subject of this case study and rated at 180,000 bbls/day.

The refinery was originally constructed in 1925 with a crude processing capacity of 11,340 bbls/day. Over the years, the capacity was increased with the addition of a topping unit, a FCC and delayed coker, a petrochemical complex and a lube oil complex.

The lube oil complex where the incident took place consists of the following units:

- Lube vacuum distillation - 26,566 bbls/day. Column bottoms are directed to the propane de-asphalting unit. Side draws from the column are directed to the furfural unit and then the MEK dewaxing unit for additional processing.
- Propane de-asphalting (PDA) - 10,064 bbls/day. LP gasses have the ability to precipitate asphaltic and resinous materials from crude residues while the lubricating oil components remain in solution. All liquified hydrocarbon gasses have this property to a marked extent, but propane is used due to its relatively low cost and its ease of separation from lubricating oils.
- Furfural unit- 10,600 bbls/day
- Solvent/MEK dewaxing.

The units were constructed in 1969 by a European engineering firm. All units are located in a common process block measuring 203 feet by 362 feet to the battery limits.

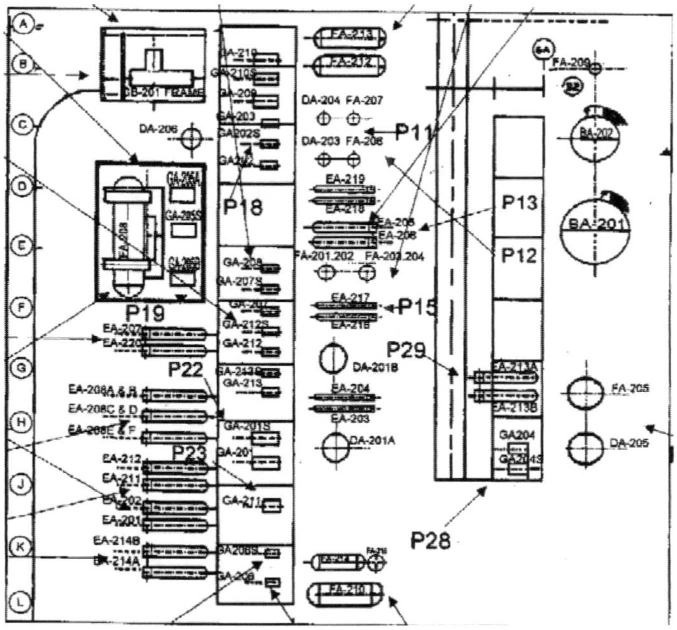

Figure 1: Plan of the PDA unit

1.2 Events

A chronological sequence of events is as follows. All times are local.

December 15, 1995:

- 4:40 PM- A loud hissing noise is heard by outside operations staff followed quickly by a white-yellowish cloud in the vicinity of flash drums FA-201/FA-202. A fire ensues almost immediately followed by an explosion. The fire is immediately reported to the control room who initiate a shutdown of the entire refinery site.

- ~4:43 PM- The refinery's own fire department arrives, responding with two fire engines. Soon afterwards a total of 13 pumper trucks arrive from the surrounding towns to fight the fire. The incident consisted of a main fire and 3 localized secondary fires. The fires not only involved the PDA unit but spread to the furfural and lube vacuum distillation units as well, due to inability of the sewer system to remove the volume of water and hydrocarbons. The fires are fought with a combination of water and foam.
- 5:25 PM- Fire declared under control.
- 6:10 PM- All fires extinguished except for propane being released from towers DA-201 A/B (1st and 2nd stage extractors) which is allowed to intentionally burn to avoid any vapor cloud explosions.

December 17, 1995

- 9:15 AM- Propane which was being purposely burned is exhausted.

1.3 Outcome

Two employees were seriously injured, one subsequently died.

Figure 2: View Of Damaged Piperack Between Pipe Columns G And H

Equipment damage is as follows:

- **Heaters** - The asphalt heater (BA-201) and gas oil heater (BA-202) reveal little damage. Melting of non-metallic components such as sight glasses, plastic tubing and wiring is not apparent.

- **Vessels** - Damage to vessels and heat exchangers is limited to loss of insulation and cladding as well as the loss of instrumentation cables.
- **Piping** - Piping between column lines C and K receive the high degree of damage. Some spalling of the fireproofing over structural steel supports occurs. The heat caused several other hydrocarbon lines (between 3 and 6 inches in diameter) to rupture increasing the combustible load present.
- **Pumps** - Pumps, associated electrics, and instrumentation are all severely damaged between column lines E & H. Limited damage occurs to the concrete pump bases.
- **Compressor** - The compressor (GB-201) located on the south-west corner of the process block is undamaged.

Figure 3: Overall View Of Fire Damaged Area With The Arrow Showing The Location Of The Ruptured Flare Line

1.4 Cause

The fuel that fed the fire came from a ruptured section of flare line inside the PDA unit. The rupture started at a small imperfection adjacent to a weld in the reducing section where the flare line increased from 14 inches in diameter to 20 inches in diameter. The rupture progressed into both the 20 inch and 14 inch portions of the line.

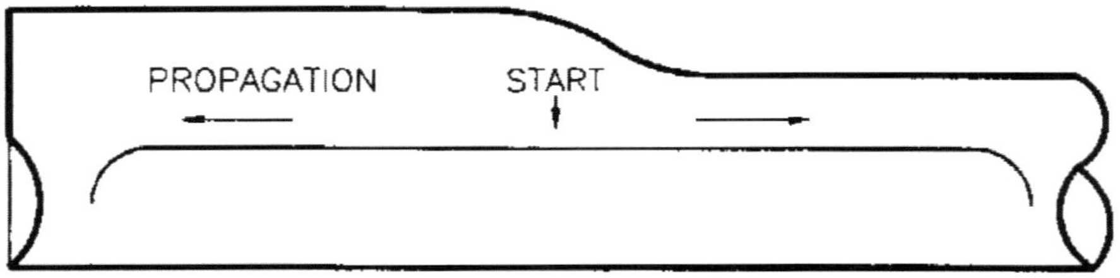

Figure 4: Flare Line

The reducer was not a forged fitting, rather was fabricated from a 20 inch diameter pipe and then slotted, trimmed and compressed to 14 inches and joined together by several longitudinal welds.

The metal composition is analyzed after the incident and determines:

Material	% Carbon	%Manganese	%Sulfur
Main pipe	0.19	0.33	0.07
Reducer	0.20	0.63	0.05

The flare line was refrigerated by mixed phase propane that was vented from pump GA-205B into the flare line.

At the same time, an apparent failure of the distributive control system allowed the high flow of heating oil to pass through several heat exchangers (EA-217, EA-218 and EA-219). The increased flow resulted in a higher temperature on the process side of the exchangers with an increase in the system pressure.

The pressure resulting from this temperature excursion caused various relief valves to open, venting into the flare header that was also receiving the mixed phase propane vented from pump GA-205B.

When the hot hydrocarbons from the relief valves hit the cold propane, they caused a very rapid increase in vapor load in the flare header.

Figure 5: Photograph Of The Flare Line After It Was Removed From The Field

The flare line contained about 15 m^3 of liquid which was lying at the bottom of the pipe, allowing the material being vented from the pump to pass over the liquid with little resistance. However, when the relief valves vented hot material into the line, the velocity lifted the liquid resulting in plug flow and because of the inertia of this liquid resulted in a high pressure spike.

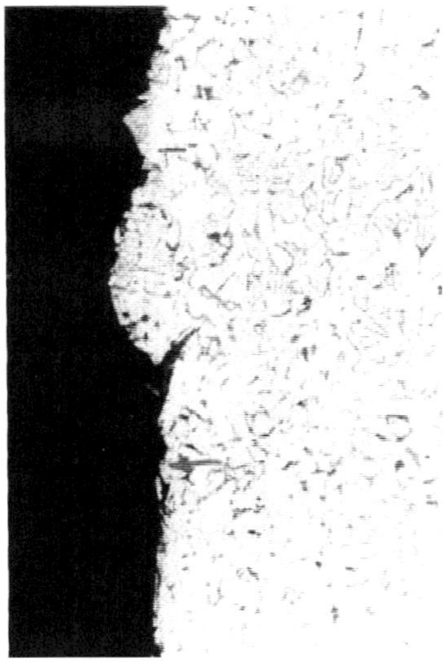

Figure 6: Micrograph Of Interior Of Pipe At Weld (X 200)

With the reducer cooled below the brittle zone, the pressure spike caused the metal in the reducer to fail by rapid fracture. After progressing a short distance, the crack entered a more ductile zone and finished the failure by ductile failure.

With both propane and bright stock emanating from the failed pipe, this mixed phase caused a large static charge which probably ignited the mixture.

In addition the storm water sewer system was not large enough to remove the fire protection water and spilled hydrocarbons which resulted in the carry over of burning hydrocarbons into the vacuum unit.

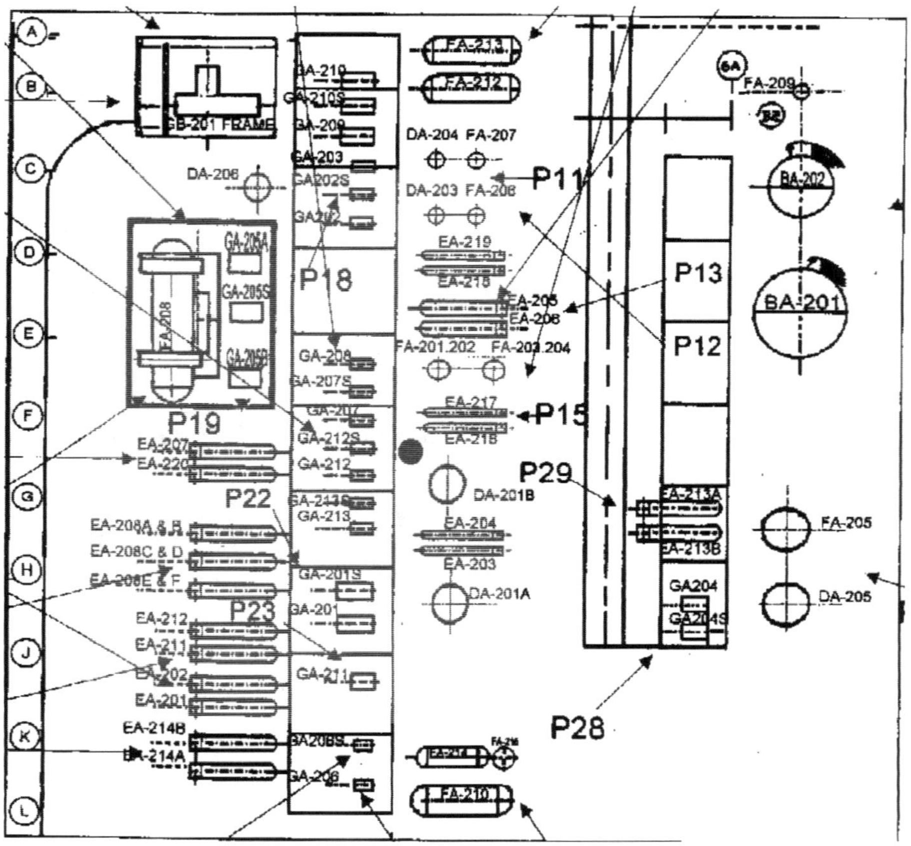

Figure 7: View Of PDA Unit Shaded Area - Piping & Steel Replaced Red Dot Indicates Approximate Position Of Flare Line Rupture

1.5 Damages

Physical damages were in the amount of US$10 million in 1995 dollars. A breakdown of some of the damages include:

Civil/mechanical repairs	$5.7 million
Instrument repairs	$1.6 million

Electrical repairs $512,000

The loss of profits associated with the downtime was not insured (it was assumed by the company directly). Total downtime is unknown but repairs did not start until April 1996.

1.6 Lessons Learned

The following major lessons can be drawn from the loss

- **Drainage -** The need to safely remove fire water and hydrocarbons to limit damages. This can be accomplished via drainage and/or an arrangement of diversions/swales and sloping to safely remove the material.
- **Engineering Design -** The original design of the flare header should have considered the auto-refrigeration effect of the propane and the metallurgy modified accordingly. Additionally, a proper transition piece should have been utilized instead of a shop assembled unit.

2. Vapor Cloud Explosion at a Natural Gas Processing Plant

Central America- July 1996.

2.1 Background

The facility is owned by the national oil and gas company. Through its subsidiaries the company is involved in all phases of the oil industry including exploration, oil refining, gas processing and petrochemicals. The natural gas production unit of the company had eight gas processing operations.

The site where the incident took place is occupied for the processing of sour natural gas and sour condensates. The gas plant is located on 700 acres and was originally constructed in 1974. The plant is rated as follows:

- Sour gas treating- 1350 million scf/d
- Sour condensate treating- 48,000 barrels/day
- Cryogenic separation- 1450 million scf/d
- Liquid fractionation- 105,000 barrels/day

Products from the site are sweet, dry natural gas, ethane, propane, butane, light natural gasoline (naphtha), sweet stabilized condensate and sulfur.

Processing is conducted using 10 desulfurization units to remove hydrogen sulfide and 5 cryogenic units which lowered the gas temperature to –130°F. Cryogenic plants #3, #4, #5 and #6 were constructed in 1978 and were each rated at 150 million scf/d. Cryogenic unit #3 was relocated off-site several years before the loss. Cryogenic plants #1 and #2 were constructed in 1981 and are each rated at 500 million scf/d.

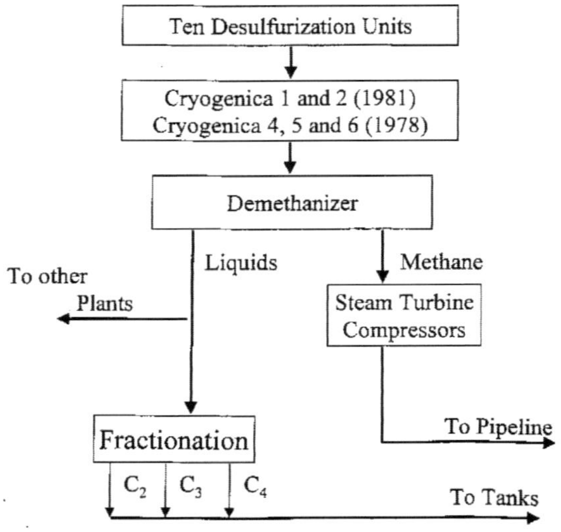

Figure 8: Process Block Flow Diagram

All of cryogenic plant #1 sat up to 15 feet below the perimeter of the road surrounding the unit.

The pump station where the incident occurred consists of a concrete slab sitting about 16 inches above grade. A concrete roof structure supported by 6 concrete columns covered the pumps. There are no walls. The pump station is located in the south-east corner of the Cryogenic #1 process unit.

The pump station consists of two, 10 stage centrifugal pumps (BA-110A and BA-110B) each driven by a 500 HP electric motor mounted vertically above. Each pump is used alternatively to increase the line pressure from 400 psi to 700 psi. Both the suction and discharge piping was 16 inches in diameter.

The pumps receive their input from Vessel FA-108 located about 200 feet from the pumps. The vessel holds a hydrocarbon mix of 42% ethane and 58% light hydrocarbons (propane to gasoline) at 90°F and 400 psi.

A number of means of isolation was provided both upstream and downstream of the pumps. Upstream, a manually operated valve was provided at the discharge of vessel FA-108. A motorized plug type valve was provided just prior to each of the booster pumps (BA-110A & B), this valve could be operated from the control room and from the valve itself. These motorized actuators were installed in 1995 as part of a process hazard analysis. Each of the actuators featured a selection dial with:

- Off (allows for local manual operation only)

- Remote
- Local

The selection dial could be locked-in-place at the actuator.

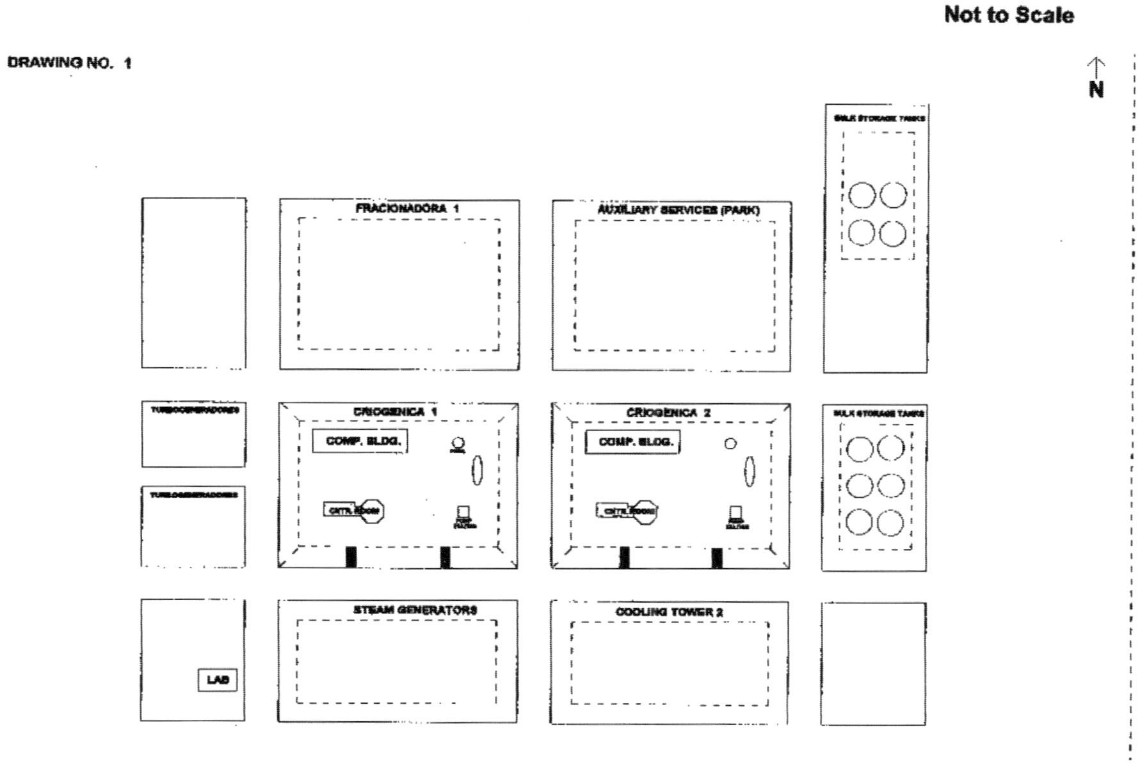

Figure 9: Simplified site plan

Downstream of the pumps, an in-line check valve followed by a manually operated valve is provided. The isolation valves that operated from the control room were not coupled to the DCS system. Only four staff on shift had authorization to operate any of the valves (even during emergencies).

There were two different types of safe work permits in use at the site:

- General
- Potentially dangerous- This type of work order required the review of several safety oriented prerequisites and ultimately the signature of an Operations Engineer

Figure 10: Valve Actuator Similar To That Involved In The Loss (Finger Points At The Locking Mechanism)

Some of the fixed fire protection and detection systems consist of:

- Four electric motor driven and two diesel engine drive fire pumps capable of delivering 14,000 gpm. Manual start.
- A remote, manually actuated deluge system over some of the process pumps (one head each for pumps BA-110A & B).
- Monitor nozzles spaced around the perimeter of each process unit.
- Combustible gas detection throughout the process units. Five gas detectors are provided within cryogenic plant #1, gas detection was specifically provided over pumps BA-110.

Fire protection consists of 14 dedicated firefighters per shift. Mobile firefighting equipment consists of three pumper trucks and a site inventory of 7200 gallons of protein foam.

2.2 Events

A chronological sequence of events is as follows. All times are local.

- **May 1-** Pump BA-110B in cryogenic plant #1 started as part of the normal practice of rotating pumps every four months.

- **July 25-** Pump BA-110B trips electrically. Further investigation finds a mechanical seal located between the vertically oriented motor and the pump beneath leaking. A decision is made to fix the seal and then determine why the motor tripped. Pump BA-110A is started until repairs are completed.

- **July 26**

  9:00 AM- 'Potentially dangerous' type of work order is issued. Separate work orders are issued for the seal replacement, blind flange installation and electrical repair.

  11:00 AM- After safety department attendance, mechanical seal replacement initiated by disconnecting the suction & discharge lines and then inserting blind flanges. Between 8 and 13 people are in attendance during this procedure.

  1:00 PM- Reattached suction & discharge lines. Shaft coupling disconnected.

  6:00 PM- Mechanical seal replaced. Electrician(s) investigate cause of motor failure and replace an electrical switch in one of the substations.

  7:00 PM- Suction & discharge lines disconnected. Blind flanges removed and replaced with 'flex-steel' seals. The shaft coupling had not as yet been reconnected.

  7:20 PM- All of the discharge line bolts have been replaced. The 20 suction-line bolts were in the process of being replaced. Flanges did not line up correctly. All but 3 bolts had been replaced (albeit only hand-tightened) when the **SUCTION VALVE OPENS**. It should be noted that there were no operation staff on hand at the repair site at this time.

  ~7:28 PM to 7:30 PM- Easterly wind drifts vapor cloud toward cryogenic plant #2

  Cloud covers eastern third of cryogenic plant #1 and western two-thirds of cryogenic plant #2. Cloud was most likely ignited by a process furnace in cryogenic plant #1 (located in the north-east quadrant of the unit). Prior to the explosion, personnel try to disperse vapor with monitor nozzles, however, the flow is weak. It should also be noted that plant personnel are told to stand-by to shutdown entire plant site only. **VAPOR CLOUD IGNITES**.

  ~7:32 PM- Vessels EA-1109 & EA-1110 in cryogenic plant #2 BLEVE.

  10:30 PM- Ensuing fire under control. In total 2100 to 2600 gallons of foam is used

2.3 Outcome

The explosion results in the death of eight employees and injury to 40 others. Damages were extensive and are summarized as follows:

- **Cryogenic plant #1-** Much of the plant suffers heavy blast damage. The control room is destroyed. The east end of the compressor building is damaged by fire. Many product pumps are damaged or destroyed.

- **Cryogenic plant #2-** This unit is the center of the blast and receives the brunt of the damage. The entire process unit is essentially destroyed. The entire compressor building and its contents are destroyed when the roof collapses onto the compressor deck and the floor collapses onto the condensers. The control room is totally destroyed.

* **Utilities-** Heavy blast and fire damage occurs to several electrical substations. Even underground high-voltage electrical cable are damaged to the point of requiring replacement. There is blast damage to the duct work associated with four of the process boilers. Two cooling tower receive blast damage but are repairable. A portion of the water treatment plant also receives damage.

Figure 11: Aerial photograph
Arrow 1- Pump BA-110B
Arrow 2 - Cryogenic unit #2

Figure 12: Aerial photograph
Arrow #1- Cryogenic unit #1
Arrow #2- Cryogenic unit #2

2.4 Cause

Following examination of the scene, coupled with statements provided by witnesses, the explosion and ensuing fire were the result of the opening of the BA-110-B suction line valve while the downstream 16 inch pipeline was open for repairs.

Due to the death of several key witnesses (control room board operator, shift engineer and chief engineer) as well as the degree of damage, the reason why the valve opened could not be determined with certainty. Three possibilities exist:

- **Human error-** Someone inadvertently, accidentally or through miscommunication, either locally or remotely, opened the suction valve.
- **Electrical malfunction-** Due to an electrical malfunction, the valve opened without specific instructions from anyone.

- **Mechanical stress-** Due to ongoing repairs, the handling of the components in the repair area resulted in the activation of the motorized valve's relay resulting in its opening.

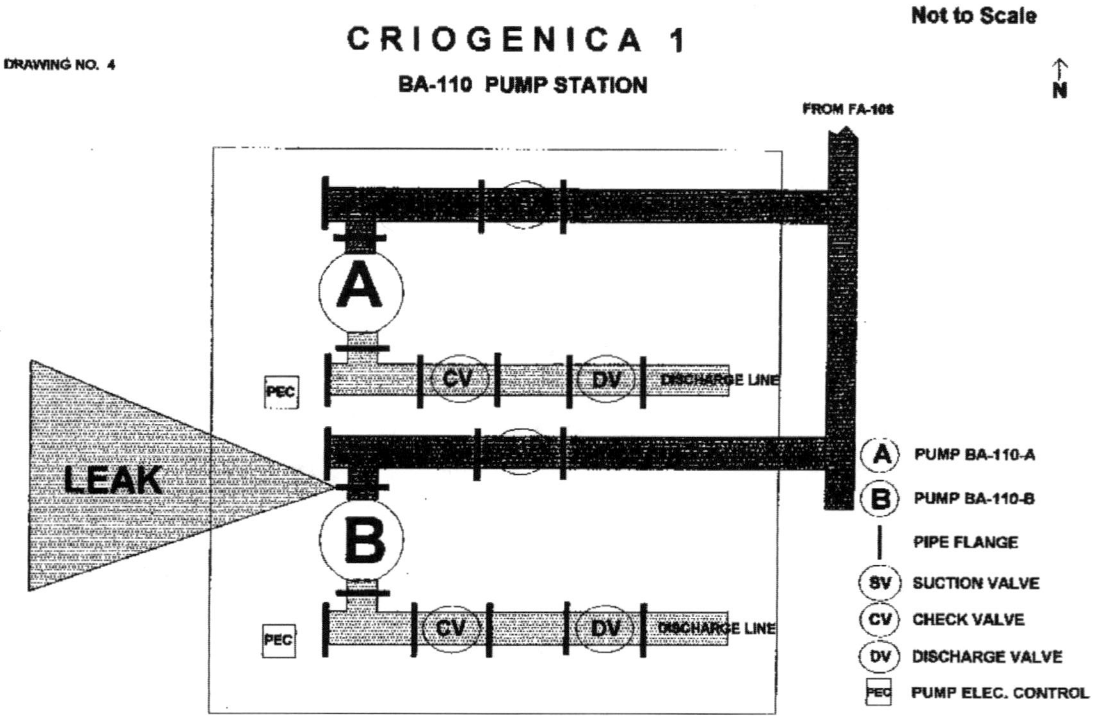

Figure 13: BA-110 Pump station

2.5 Damages

Physical damages were in the amount of US$137.6 million in 1996 dollars. A breakdown of some of the damages include:

Cryogenic unit #1	$27.5 million
Cryogenic unit #2	$75.7 million
Auxiliary services	$23.1 million
Contracts, debris removal etc	$11.4 million

The numbers stated above are gross amounts and do not include deductibles or salvage.

The loss of profits associated with the downtime was not insured (it was assumed by the company directly). It did however, require 19 months to rebuild cryogenic plant #2 which was built at another site.

Figure 14: View of Suction side of pump BA-110B

Figure 15: Suction valve for pump BA-110B

2.6 Lessons Learned

The following major lessons can be drawn from the loss

- **Maintenance/Safe Work Procedures -** Maintenance practices should be augmented to include to locking out of critical control valves when hydrocarbon or high energy steam lines are opened. Many maintenance staff were unaware of the status of the motorized or manual valves during their activities.
- **Training of Maintenance Staff -** All maintenance staff need to understand the process and specifically the location of alternative isolation valves. Only senior staff and operators were aware of the location of the manual isolation valve near FA-108.
- **Procedures -** All staff need to be able to operate valves in the event of an emergency. This item ties back to training as well.
- **Emergency Procedures -** Emergency procedures should be revised. Plant staff had no recollection of the plant alarm being sounded during the 8 to 10 minute period prior to the explosion. Additionally, operators were told to only 'standby' to shut down the operation during the same period.
- **Control Rooms -** The control rooms for each unit should be relocated to an area not subject to damaging overpressures or reinforced to sustain the impulse of an overpressure. The majority of deaths occurred to staff occupying the control rooms.
- **Fire pumps -** The fire pumps should be automated to start on pressure drop. Several of the operators and maintenance staff attempted to operate monitors prior to the explosion but only a 'trickle' of water was available.

Paper TG006 - 3

U.S. Chemical Safety Board Investigation: Positive Pressure Control Room Failure

Lisa Long – Speaker
Johnnie Banks
Mike Morris
John Murphy, P.E.
Stephen Wallace, P.E., CSP
U.S. Chemical Safety Board
2175 K St, NW
Washington, DC 20037

Prepared for Presentation at
American Institute of Chemical Engineers
2005 Spring National Meeting
39th Annual Loss Prevention Symposium
Atlanta, Georgia
April 11-13, 2005

UNPUBLISHED

U.S. Chemical Safety Board Investigation: Positive Pressure Control Room Failure

Lisa Long – Speaker
Johnnie Banks
Mike Morris
John Murphy, P.E.
Stephen Wallace, P.E., CSP
U.S. Chemical Safety Board
2175 K St, NW
Washington, DC 20037

Abstract

On July 20, 2003, a release of chlorine gas from the Honeywell chemical plant in Baton Rouge, Louisiana, resulted in the hospitalization of eight plant workers and required residents within a half-mile radius to shelter in their homes. The chlorine was released when a cooler failed allowing chlorine to enter a coolant system not designed for chlorine. Plant operators noticed the release and attempted to stop it. Response efforts were halted when the chlorine entered the unit's positive pressure control room. Operators in the control room were overcome by the chlorine, and had to evacuate the plant prior to determining and stopping the source of the leak or shutting the process down. The U.S. Chemical Safety Board investigated the incident. This paper presents details of the positive pressure control room failure and its consequences, discusses available guidance and standards for positive pressure control rooms, and details the inadequacies in this particular positive pressure system.

1. Chlorine Release

At 3:10 am on July 20, 2003, operators inside the control room noticed a chlorine odor. One operator went outside to investigate and observed a leak near the coolant system pumps. Plant personnel outside the control room quickly realized the need for additional help and protection to stop the leak. However, before they could take any action, the chlorine concentration both inside and outside of the control room became overwhelming. Operators experienced difficulty breathing and were forced to evacuate the area before they could determine why chlorine would be leaking from the coolant pumps. The process was still running when they were forced to leave the area.

2. Incident Response

By 3:25 am, plant personnel had evacuated to the main plant gate. A level II incident (i.e., effects inside the plant, with potential for offsite effects) was reported to local authorities through the East Baton Rouge I-notification system. By 3:29 am, the incident was raised to level III (i.e., effects outside the plant).

Several operators from the unit were exposed to chlorine during the release; they were given oxygen in the first-aid building. The site water deluge towers[1] were turned on to suppress chlorine vapor, and additional off-duty plant personnel were called into the plant to help stop the release.

Plant emergency response personnel activated the incident command system and began attempts to stop the leak. However, the chlorine that entered the control room had corroded the process control system and rendered it inoperable. Response personnel had to manually shut down the process. Once it was shut down, they realized that the chlorine was still being released, and the chlorine railcar (the source of the chlorine feed) would have to be isolated to stop the leak.

The chlorine leak was finally stopped at approximately 6:46 am, when the emergency response team manually closed the railcar valve. As a result of this incident, eight employees were transported to the hospital for medical evaluation and treatment.

The exact amount of chlorine released could not be determined. Honeywell reported a maximum estimated release of 14,400 pounds to the Louisiana Department of Public Safety and Corrections, Office of the State Police. This estimate was calculated based on pre- and post-incident estimates of chlorine railcar weight. The release may have been less if some of the chlorine dissolved in the coolant or remained in the system piping. Additionally, the amount of chlorine that drifted offsite may have been less than the amount released because of mitigation by the deluge tower water spray.

3. Community Impact

At 4:00 am, the local fire department issued a shelter-in-place advisory for all residents within a 0.5-mile radius of the plant. Community sirens were activated, and the auto dialer was used to notify these residents. The advisory remained in effect for 3 hours. At 7:09 am, the fire department downgraded the incident to level I (i.e., effects inside the plant only).

At the time of the release, the predominate wind speed was 3 to 5 miles per hour (mph) from the southwest. Chlorine concentrations in air were not measured until 5:30 am—at which time, Honeywell recorded a concentration of 1.5 parts per million (ppm) at the main gate. At 7:13 am, LADEQ arrived and sampled air near the facility; these measurements showed less than 0.35-ppm chlorine outside the plant fenceline.[2]

CSB interviewed emergency responders, who reported noticing a strong chlorine odor over 1.5 miles away, in the vicinity of Plank and Prescott Roads. As they proceeded west on Chippewa Street toward the site, they stated that the odor was much less pronounced. A few citizens also noticed a strong chlorine odor on Interstate 110, north of the Honeywell facility. Ten members of the general public sought treatment at the local emergency room—three initially and seven

[1] Water deluge towers spray water from a nozzle. Studies indicate that the water spray can suppress chlorine vapors, reducing the concentration of chlorine outside the area of release (The Chlorine Institute, Inc., 1990).

[2] These measurements are indicators of concentrations at a specific location and time and do not necessarily represent concentrations at other locations and time.

throughout the day. They complained of headaches and sore throats. They were either treated or observed and then released from the emergency room.

4. CSB Investigation

CSB conducted a full investigation of this incident. A detailed report with root causes and recommendations will be available on CSB's website (www.csb.gov) once the investigation has been voted on and approved. This paper presents details of the positive pressure control room failure and its consequences, discusses available guidance and standards for positive pressure control rooms, and details the inadequacies in this particular positive pressure system.

4.1 Design and Maintenance of Positive Pressure Control Rooms

The control room had been converted into a positive pressure control room in 1998. Plant personnel had complained of odors getting into the control room; and an engineering study had identified the potential infiltration of hydrogen fluoride, a highly hazardous substance used in the plant.

A divisional engineering group managed the positive pressure control room conversion project. The control room was designed to provide short-term protection for personnel during a chemical release so that they could safely shut down the process prior to evacuating. Escape respirators, located in the control room, were to be used for operators to evacuate after shutting the process down.

The design of the positive pressure control room included the following features:

- Air intake piped to pull air from the highest point in the plant.
- Sealing of the room to allow the heating, ventilating, and air conditioning (HVAC) system to maintain a positive pressure.
- Manual shutoff of the HVAC system intake.
- Audible alarms to indicate low pressure.

During the July 20 release, chlorine entered the control room. CSB investigators found that it was drawn into the building through holes and gaps in the HVAC intake ducts located on the roof. Some holes appear to have been drilled in the duct and not plugged, while others resulted from gaps in joints. Duct tape had been used to seal some of the gaps, but it became dry and brittle over time. Figures 1 and 2 show some of this deterioration. Figure 3 shows the location of the coolant system chlorine leak.

Figure 1: Condition Of Duct Tape Used To Seal Gaps.

Figure 2. Hole In Ductwork.

Figure 3. HVAC intake system on control roof (foreground). Arrow points to location of coolant system chlorine leak.

The positive pressure control room system did not protect personnel or equipment during the July 20 chlorine release. When interviewed by CSB, engineers stated that they were unaware of any standards for design and maintenance of control rooms to withstand toxic gas releases.

The following deficiencies in the positive pressure control room system contributed to its failure:

- There was no maintenance program for the control room, including HVAC ductwork.[3]
- Duct tape was used to seal some joints in HVAC ductwork. This tape eventually became brittle and exposed gaps in the joints.
- There were numerous entrances and exits from the control room—none of which had a double door air lock system.
- There were no toxic gas alarms on the HVAC intake system. Alarms could have warned the operators to use the escape respirators and exit the control room.
- Although the positive pressure control room system was listed as a safeguard in the hazard analysis, it was not designated as critical equipment or maintained as such.[4]

[3] When asked, Honeywell indicated that there was no preventive maintenance on the positive pressure control room system, including ductwork.

CSB researched available industry standards and guidance for control rooms, and found that existing standards aimed at the chemical manufacturing industry do not adequately address design and maintenance for protection against toxic chemical releases. Current guidance is summarized below:

- ***American Petroleum Institute (API) Recommended Practices***: API develops standards for petroleum refineries. However, API Recommended Practices 550, 551, and 752 provides direction in the area of positive pressure control rooms that can also be helpful to the chemical industry (API, 1977; 1995; 1999).
- ***National Fire Protection Association (NFPA)***: NFPA 496, Standard for Purges and Pressurized Enclosures for Electrical Equipment (1998), contains only a few recommendations for pressure and air velocities. It was not intended to address the protection of personnel.
- ***Instrument, Systems, and Automation Society (ISA)***: Several ISA standards include design recommendations to protect process measurement and control systems; however, none of the standards address protection of personnel.
- ***U.S. Nuclear Regulatory Commission (NRC)***: Although NRC regulatory guides are not intended for the chemical industry, they may provide relevant information in the area of positive pressure control room design and maintenance (USNRC, 2003a; 2003b; 2001). NRC staff use the guides to implement specific regulations, to evaluate specific problems or postulated accidents, and to review applications for permits and licenses.

The United Kingdom Chemical Industries Association provides some guidance on control room design and maintenance, including recommendations for inspecting and maintaining HVAC systems and for entrance and exit design. In *Loss Prevention in the Process Industries*, Lees (1996) briefly covers topics such as the need for an airtight design, positive pressure requirements, and possible use of toxic gas monitors.

As outlined above, U.S. guidance and standards for designing controls rooms in chemical manufacturing facilities to protect against toxic releases are limited. If the positive pressure control room system involved in this incident had been better designed, documented, and maintained, it is likely that the operators would have been better protected and able to more expeditiously stop the July 20 chlorine leak.

At the time of this writing, the Chemical Safety Board has not yet voted this on investigation report. More information regarding CSB's final report and recommendations will be available during this paper's presentation at the 2005 AIChE Loss Prevention Symposium, and can be found on CSB's website (www.csb.gov).

This paper has not been independently approved by the Board and is published for general informational purposes only. Any material in the paper that did not originate in a Board-approved report is solely the responsibility of the authors and does not represent an official finding, conclusion, or position of the Board.

---

[4] In its citations, OSHA considered the positive pressure control room to be covered by the PSM Standard.

Rotational Force Capabilities of Males and Females Between the Ages of 30 and 55 Years

Lawrence J. H. Schulze[1], PhD, PE, CPE, Rebecca Ginsburg[2], and Jehanzeb Quereshi[1]
[1]Department of Industrial Engineering, University of Houston, Houston, Texas, [2]Shell Lubricants, Houston, Texas

Prepared for Presentation at
American Institute of Chemical Engineers
2005 Spring National Meeting
Process Plant Safety Symposium
Atlanta, GA
April 11 – 13, 2005

ABSTRACT

Opening and closing valves in process industries is done by the use of handwheels. Significant force may be necessary to operate the valve handwheels. In extreme cases, a metal bar or pipe wrench often called a 'cheater' may be used to increase the moment (lever) arm of the handwheel, which may result in physical injury to the worker and/or damage to the valve. The objective of this study was to find the rotational force capabilities of males and females ranging from 30 to 55 years of age. Each participant was screened for pre-existing medical conditions. After initial screening, the eligible participants were asked to make 3 consecutive maximum isometric exertions and one isotonic effort for each treatment condition using a LIDO® Workset rotational force transducer. The data obtained was analyzed using the non parametric Kruskal–Wallis test for the main effects. The Spearman–Rho test was conducted to access associations between the dependent and independent variables. In this study, the results indicated that the main effect of handwheel size had an impact on torque production capabilities of male participants. The height of the handwheel significantly affected the force generation capabilities of male participants, as well. The angular orientation of the handwheel had an affect on both the production of torque and force. For the female participants, the size of the handwheel, the height of the handwheel and the angular orientation of the handwheel all had a significant impact on both torque and force generation capabilities. The distance the handwheel was located away from the coronal plane of the participants did not have an impact on the torque nor force generation of either the female or the male participants. These results indicate that torque and force production capabilities appear to be more independent of the physiological characteristics of the male participants while females, traditionally non-socialized for these types of activities, may be impacted by their physical characteristics. In addition, these findings will lead to the development of a database of rotational force capabilities that can be used to help prevent occupational injuries related to the manual operation of valve handwheels through application to good engineering design.

Introduction and Background

Valves are the components in a fluid flow or pressure system, which regulate either the flow or the pressure of the fluid. This duty may involve stopping and starting flow, controlling flow rate, diverting flow, preventing back flow, controlling pressure, and/or relieving pressure (Zappe, 1987). Adjusting the position of the valve, either manually or automatically, enables these functions; manual operation is commonly conducted by the use of a handwheel.

Handwheels are a part of a large group of control devices known as continuous adjustment rotary controls. They have a wide range of industrial applications, which require workers to exert varying amounts of force to operate them (Wood, Schulze, Chen and Cleveland 1998). The term 'cracking' or breaking force is used to denote the force required to turn the valve handwheel from a locked position to an unlocked position for operation; it ends at the initial movement of the handwheel (Amell, 2001).

Process plants in the petroleum, chemical and waste and municipal water systems, as well as power generation facilities, utilize handwheels actuated by human operators to regulate the flow of material within a process system (Schulze, 1997a; Wood and Schulze, 1998). Handwheels are also used in the railway industry as a means of regulating the movement of rail cars (Woldstad, McMulkin and Bussi, 1995). Handwheels are commonly found as control devices for valves, and it has been estimated that 50% of all valves currently in use are operated manually (Shih, 1997).

According to Parks and Schulze (1998), 56.5 % of back injuries (20% upper back and 64% lower back) and 75% of head, neck and face injuries over a three-year period reported by the process operators at a large corporation were attributed to industrial valve handwheel actuation. Further, the amount of force generated by the person turning a handwheel can vary according to several factors such as: individual strength, wheel size, wheel position and environmental conditions (i.e. traction). When a person is presented with a wheel turning task, such as opening a valve, these factors will either enhance or detract from the person's ability to perform the task.

In theory, a handwheel should be part of an integrated design, which results in a device that can be turned with minimal effort. In reality, the design of a system using handwheels may not consider the human factors/ergonomics, resulting in awkward body positions and the use of 'cheaters' (Wood and Schulze, 1998), as shown in Figures 1 and 2. The only effective solution to overcome these hazards and to protect the process operator is to integrate the proper design of handwheel into the system (Wood and Schulze, 1998).

Figure1. Process Plant Worker Operating a Vertically Oriented Valve Handwheel in an Awkward Position (Schulze, Stanton, Patel, Cheli, 1997a).

Figure 2. Awkward Location of a Handwheel (Lakehead Inc.).

Van Cott & Kinkade (1972, 1997) and Rogers (1983) provided recommended design values for parameters such as the outer diameter, the rim diameter, the clearance around the wheel, and the resistance. In relation to the placement of the valve in the subsystem, several criteria are defined. First, appropriate placement allows the operator to actuate the handwheel with ease, and avoids overreaching (Van Cott and KinKade, 1972). A second criterion indicates that the wheel should be placed where a person can see and use it from a normal working position without excessive shifting of the head or body (Bailey, 1989). Rogers (1983) provides guidelines with respect to vertical distances for vertically and horizontally oriented handwheels. The force, speed, accuracy, and range of body movement required should never exceed the abilities of the least capable operator and the handwheel should be placed far enough away from adjacent controls and/or structures to permit adequate grasp and manipulation through the entire range of motion (Bailey, 1989). It should be recognized that these recommendations are based on the NASA 1024 Anthropometric Standard for data collected between 1957 and 1964 (NASA, 1978). Therefore, it is important to determine the torque production capabilities of operators who will be required to manually operator such valves using data from populations that represent those working in the processing industries.

Methodology

Participants

56 female and male volunteers between the ages of 30 and 55 years (average age of 37 years) participated in this study. Recruitment was done through personal contact and participant search by other co-workers and advisor. It was made sure that the participants did not have any pre-existing medical conditions that could aggravate due to physical exertion.

Apparatus

The test apparatus used was a biomechanical testing machine LIDO® Workset (Loredan Biomedical Inc., West Sacramento, California) consisting of a rotational dynamometer, an analog to digital converter, a computer workstation and a printer. The dynamometer provided a voltage output in response to a rotational force input; the A/D board converted the voltage output to a digital torque output, which was displayed on the computer screen. The height, pitch angle and distance from the operator were adjusted. The test apparatus was chosen because it offers isometric, isotonic and isokinetic capabilities with concentric, eccentric and continuous passive motion modes. For the purpose of this experiment, isometric and isotonic testing was used.

The isometric testing was used to determine the participant's peak trial. The isotonic testing was used in the breakaway mode and set at the participant's peak torque to confirm that the participant was at or near their peak. These modes helped in determining the maximum capabilities of a participant. If the participant was able to successfully crack the handwheel, the torque was increased by 5ft-lb (6.8 Nm) until he was no longer able to perform the motion. Dynamic torque could be set up as high as 162N-m and static torque could be set at 202.5 N-m, to allow practical testing of maximal strength. The Work Set could also be tested for minimum measurements as low as 28 N-m and 14 N-m.

Additional support equipment included: An anthropometry kit to measure the anthropometric measurements of the test participant; A sphygmomanometer and heart rate monitor to record the heart rates, and blood pressures of each participant before, during and after participation; and a Hand Dynamometer used to measure the grip strength of the test participants.

Procedure

Prior to performing the tests, each potential participant was screened for any pre-existing medical conditions that might affect their performance. Participants with previous heart problems, severe musculoskeletal injuries or any physical disability were excluded. Screening was required so that any participant with illness or injury could be prevented from risking their conditions.

Each qualified test participant was given a verbal description of the study procedures and their role in participation by a member of the research team. An opportunity to ask questions was then provided. Following this, a witnessed, written informed consent was reviewed and signed by each participant. Prior to testing, the study protocol was reviewed and approved by the University of Houston Committee for the Protection of Human Subjects.

The following anthropometric and physiological data was measured prior to the start of testing: Stature, Shoulder Height, Elbow Height, Knee Height, Trochantric Height, Ankle Height, Functional Reach, Upper Arm Length, Forearm Length, Hand Length, Hand Breadth, Wrist Circumference, Grip Strength, Shoulder Breadth, Chest Circumference, Forearm Circumference, Upper Arm Circumference (Bicep), Weight, Heart Rate, and Resting Blood Pressure (Sitting).

On the day of the test, each participant was briefed again on the procedures, given a demonstration of the dynamometer, and then allowed time to familiarize themselves with the test equipment (including performing trial run). At the start of testing, each participant was instructed to perform three consecutive, maximum-efforts, counter clockwise isometric exertions at the first test position. The maximum and average amount of torque produced during the rotations was measured and recorded.

After each test position, the person was given a 3-minute rest break (timed with a stopwatch). This rest break allowed the participant's heart rate and blood pressure to return to pre-test levels, which was verified by measuring the heart rate and blood pressure at the beginning and end of the rest break. During the rest break, grip strength of the participant was measured twice, once at the start of the break and once at the completion. In addition, the position of the dynamometer was changed to the next setting in the test matrix (Table 1).

Few constraints were placed on the position of participant's feet and legs, except that they be able to squat into position for the 60.96 cm wheel height setting; the remaining test settings were performed while standing. The data gathered from the test participants was stored in a computer Microsoft Excel spreadsheet. This data was analyzed using the Statistical Analysis Software (SAS Institute, Cary, North Carolina). Data was first tested for normality using the Kolmagorov-Smirnoff Goodness of fit and the Shapiro-Wilk test. Since the data was not normal, therefore a non-parametric Kruskal-Wallis test and Spearman-Rho test was conducted for the main effects and correlation respectively.

Results and Discussion

Tables 1 and 2 provide the results of the torque production assessments for the female participants and Tables 3 and 4 present the torque production results for the male participants. The results for the female participants are presented first, followed by the results for the male participants.

Females

As can be seen from a review of Tables 1 and 2, the main effects of Size (Torque), Height (force) and Angle (both torque and force) were statistically significant. Force was determined by dividing torque be the radius of the handwheel to eliminate the affect of the length of the lever arm.

Table 1. Kruskal-Wallis Test of Main Effects for the Female Participants

	Size (df = 2)	Height (df = 1)	Angle (df = 1)	Distance (df = 1)
Peak Torque	$\chi^2 = 323.3502$ ($p < 0.0001$)	$\chi^2 = 10.3083$ ($p < 0.0013$)	$\chi^2 = 66.8082$ ($p < 0.0001$)	$\chi^2 = 0.0778$ ($p < 0.7803$)
Peak Force	$\chi^2 = 13.2985$ ($p < 0.0013$)	$\chi^2 = 16.9120$ ($p < 0.0001$)	$\chi^2 = 124.5324$ ($p < 0.0001$)	$\chi^2 = 0.0074$ ($p < 0.9314$)
Average Peak Torque	$\chi^2 = 332.7652$ ($p < 0.0001$)	$\chi^2 = 11.1913$ ($p < 0.0001$)	$\chi^2 = 62.6323$ ($p < 0.0001$)	$\chi^2 = 0.0590$ ($p < 0.8081$)
Average Peak Force	$\chi^2 = 16.4266$ ($p < 0.0003$)	$\chi^2 = 20.2919$ ($p < 0.0001$)	$\chi^2 = 112.8366$ ($p < 0.0001$)	$\chi^2 = 0.0298$ ($p < 0.8628$)

Table 2. Summary Statistics for the Female Participants by Independent Variable and Level

		Size			Height		Angle	
		8 in.	12 in.	16 in.	2 ft	5 ft	0°	90°
Peak Torque	Mean	21.11	33.43	40.89	30.61	33.01	27.71	35.91
	Median	21	33	39	28	32	26.0	35
	SD	5.71	9.34	12.33	13.29	11.62	10.20	13.29
	Range	5-45	10-61	14–79	5-79	11-75	5-60	11-79
Peak Force	Mean	31.67	33.43	30.62	30.74	33.07	27.97	35.84
	Median	31.5	33	29.25	30	33	27.75	34.5
	SD	8.56	9.34	9.23	10.24	7.67	7.57	8.83
	Range	7.5-67.5	10-61	10.5-59.25	7.5-67.5	15-60	7.5-54	15-67.5
Average Peak Torque	Mean	19.66	31.48	38.34	28.57	31.09	26.15	33.51
	Median	20	31	37	26	31	24	33
	SD	5.21	8.55	11.26	12.17	10.93	9.77	12.16
	Range	5-40	10-56	13-70	5-70	10-65	5-58	10-70
Average Peak Force	Mean	28.98	31.89	29.12	28.84	31.15	26.35	33.64
	Median	28.5	31	27.75	27	31	26.25	32.63
	SD	27	9.10	8.96	10.13	7.00	7.10	8.79
	Range	7.5-60	10-57	9.75-63	7.5-63	14.25-58.5	7.5-48	11.25-63

Size. Size was significant for peak torque and the average peak torque as well as peak force and average peak force. Quite logically, greater torque was generated by using larger size handwheels (longer lever arm). This was true even when the length of the lever arm was subtracted from torque (torque divided by radius of handwheel). This would indicate that individual personal characteristics (human anthropometrics) are important factors in rotational force production capabilities for the female participants.

Height. The height of the handwheel also had an influence over the peak and average peak torque as well as the peak force and average peak force. These results may also indicate that torque and force generation of an individual female participant is dependent on her lower body muscles.

Angle. The angle of the handwheel played a significant role in determining the peak torque, peak force, average peak torque and the average peak force production capacities of the female participants. Observations of the participants indicated that that keeping the arms straight to operate a handwheel oriented horizontally would assist in the production of more torque and force, in comparison, to the vertical orientation when the participant bends his arm at the elbow to produce the required amount of torque and force.

Distance. Distance did not impact either the torque or force production capabilities of the female participants.

Table 3. Kruskal-Wallis Test of Main Effects for the Male Participants

	Handwheel Size (df = 2)	**Height (df = 1)**	**Angle (df = 1)**	**Distance (df = 1)**
Peak Torque	$\chi^2 = 329.6392$ ($p < 0.0001$)	$\chi^2 = 3.0491$ ($p < 0.0808$)	$\chi^2 = 62.7445$ ($p < 0.0001$)	$\chi^2 = 0.1085$ ($p < 0.7419$)
Peak Force	$\chi^2 = 2.9232$ ($p < 0.2319$)	$\chi^2 = 6.6686$ ($p < 0.0098$)	$\chi^2 = 125.1628$ ($p < 0.0001$)	$\chi^2 = 0.0016$ ($p < 0.9680$)
Average Peak Torque	$\chi^2 = 337.1762$ ($p < 0.0001$)	$\chi^2 = 2.5635$ ($p < 0.1094$)	$\chi^2 = 58.4295$ ($p < 0.0001$)	$\chi^2 = 0.0117$ ($p < 0.9137$)
Average Peak Force	$\chi^2 = 4.4204$ ($p < 0.1097$)	$\chi^2 = 5.9314$ ($p < 0.0149$)	$\chi^2 = 117.4083$ ($p < 0.0001$)	$\chi^2 = 0.1845$ ($p < 0.6676$)

Table 4. Summary of Statistics for the Male Participants by Independent Variable and Level

		Handwheel Size			**Height**		**Angle**	
		8 in.	12 in.	16 in.	2 ft	5 ft	0°	90°
Peak Torque	Mean	34.81	52.4	66.91	50.21	52.54	45.33	57.42
	Median	33	51	65	46	49	42	55
	SD	8.91	13.8	20.1	19.75	20.03	17.22	20.6
Peak Force	Mean	104.42	104.8	100.37	101.02	105.4	91.48	114.91
	Median	99	102	97.5	97.75	102	87	112
	SD	26.74	27.57	30.14	28.9	27.41	25.08	26.28
Average Peak Torque	Mean	32.6	49.2	62.57	47.18	49.05	42.78	53.45
	Median	32	48	60.5	43.5	46	39	52
	SD	8.2	12.42	18.69	18.5	18.4	16.28	18.96
Average Peak Force	Mean	97.78	98.36	93.86	94.86	98.48	86.26	107.08
	Median	96	96	90.75	92	96	82	103.5
	SD	24.58	24.85	28.05	26.6	25.12	23.43	24.06

Males

As can be seen from a review of Tables 3 and 4, the main effects of Size (Torque), Height (force) and Angle (both torque and force) were statistically significant. Force was determined by dividing torque be the radius of the handwheel to eliminate the affect of the length of the lever arm.

Size. Size was significant for both the peak torque and the average peak torque. Quite logically, greater torque was generated by using larger size handwheels (longer lever arm). When the length of the lever arm was subtracted from torque (torque divided by radius of handwheel) there was no significant difference between handwheel sizes. This would indicate that individual personal characteristics (human anthropometrics) are not the limiting factors in rotational force production capabilities.

Height. The height of the handwheel had a slight influence over the peak force and average peak force, while it did not have any effect on the peak torque and the average peak torque. These results may indicate that force generation of an individual is dependent on his lower body muscles. Furthermore, if the individual squats for the 2-feet high handwheel, his force generation tends to be better than standing up for a 5-feet high handwheel.

Angle. The angle of the handwheel played a significant role in determining the peak torque, peak force, average peak torque and the average peak force. These results would denote that keeping the arms straight to operate a handwheel oriented horizontally would produce more torque and force, in comparison, to the vertical orientation when the participant bends his arm at the elbow to produce the required amount of torque and force.

Conclusions and Recommendations

It appears that torque generation ability is independent of body size and shape indicating that optimal handwheel placement for a wide range of personnel is most important. Further, by optimizing handwheel placement, it would possible to allow a wide range of personnel to generate the necessary amount of torque to operate the handwheel in safer, less risky positions, especially without the use of some type of cheater.

Logically, larger handwheels will produce greater torque, but in reality, at times it may not be feasible to implement larger handwheels due to limited space in industry. Therefore, it can be concluded that how strong a worker is has less importance than where and how the handwheel is located.

This is clearly a design issue, and therefore, system designers should consider valve operation characteristics to compensate for limited handwheel size, including horizontal handwheel orientation and not the human variable. Designers should also consider force aiding applications such as gear reduction, packing change and type of valve (e.g., 1/4 turn ball valve vs. gate valve)

This implies that facilities designers must consider operator characteristics and limitations when designing and specifying valves that will be operated manually. These recommendations are made independent of gender.

References

Amell, T.K., Kumar, S., (2001). *Industrial hand wheel actuation and the human operator: a review*. International Journal of Industrial Ergonomics, 28, 291-302.

Bailey, R.W., (1989). Human Performance Engineering. Prentice Hall, New Jersey.

Lakehead Incorporate, Hudson, Michigan. Retrieved April 8th, 2004 http://lakeheadinc.com/images/valvewheelwrenches1.jpg

NASA, (1978). Anthropometric Source Book. NASA Defense Publication, 1024 (US National Aeronautics and Space Administration).

Parks, S.C. and Schulze, L.J.H., (1998). *The effects of valve wheel size, operation position and in-line pressures on required torque for gate valves*. Process Safety Progress, 17(4), 263-271.

Rogers, S. A. (1983). Ergonomic Design for People at Work. 1, Van Nostrand Reinhold, New York.

Schulze, L.J.H., Stanton, E., Patel, A., Cheli, M., (1997a). *Torque Production using hand wheels of different size during a simulated valve operation task*. International Journal of Occupational Safety and Ergonomics, 3(3-4), 109-118.

Shih, Y.C., Wang, M.J.J., (1997). *Evaluating the effects of interface factors on the torque exertion capabilities of operating hand wheels*. Applied Ergonomics, 28(5/6), 375-382.

Van Cott, H.P., Kinkade, R.G., (editors, 1972). Human Engineering Guide to Equipment Design. US Government Printing Office, Washington, D.C.

Wood, K.K., Schulze, L.J.H., Chen, J.C., Cleveland, T.G., (1999/2000). *The Effects of handwheel position on torque production capability of operators*. Occupational Ergonomics, 2(1), 53-65.

Woldstad, J.C., McMulkin, M.L., Bussi, C.A., (1995). *Forces applied to large hand wheels*. Applied Ergonomics, 26(1), 55-60.

Zappe, R. W., 1987. Valve selection handbook. 2nd ed., Gulf Publishing Company, Houston.

HUMAN FACTORS ANALYSIS TECHNIQUES FOR PROCESS CONTROL ROOMS AND CONTROL SYSTEMS

By

Dennis Attwood, Ph.D, CPE
Principal Human Factors Engineer
RRS Engineering
2525 South Shore Blvd, Suite 206
League City, TX 77573

ABSTRACT

Increasing integration of plant systems into fewer, and more complex DCS systems, combined with separate plant optimization systems and a variety of supporting equipment such as independent gas alarm systems have significantly increased the workload of control room operators. Increased workload can result in errors in information processing and decision making thus increasing the risk of a significant plant incident. The design of a control room environment to accepted standards, the proper design and layout of control room task areas, equipment and facilities and the user-centered design of process control systems are critical to the safe and error free operation of the process control system.

Human Factors / Ergonomics is the systematic process of designing systems to match the capabilities, limitations and needs of human operators. Mismatches lead to the potential for an increase in errors and in the number and severity of injuries. This paper addresses the tools that have been developed to systematically identify the issues associated with the design and operation of process control rooms and process control systems, to analyze the issues and to provide the most cost-beneficial solutions.

Introduction

The following is reproduced from Attwood et al., 2003. It is a hypothetical account of an incident that could easily happen in today's process control environment.

> *At 03:00 hours of his fifth consecutive night shift, a process control operator in a chemical plant received a "GROUP 1" alarm on the visual display monitor of the plant's Distributed Control System (DCS). He was not alert at this time of the morning. He recalled that during the shift change meeting, his colleague reported that this same alarm had occurred several times during the previous shift. Each time the problem was traced to a faulty transducer on a fin-fan. So, our operator acknowledged the alarm without checking further. If he had checked, he would have found that this time, the "GROUP" alarm was notifying him that the power had been lost to the DCS and that the entire system was now on battery. In four hours, when the batteries discharged, the DCS failed and the control valves on the furnaces went fully open. The ensuing temperature increase burst the tubes in the furnace and started a major fire. Without power on the DCS, the control valves could not be closed. The damage was $30 million dollars. Fortunately, nobody was hurt.*

Human Factors/ Ergonomics is defined as the systematic process of designing for human use through the application of our knowledge of human beings to the equipment they use, the environments they operate in, the tasks they perform and the management systems that guide the safe and efficient operations of refineries, chemical plants, upstream operations and distribution terminals. Neglecting any of the above elements could lead to the failure of the entire system. System failure could take many forms including injury or the loss of property, hardware or information.

The principles of Human Factors can be applied in any operation where humans interact with their working environment. None is more important for the proper use of Human Factors than the process control area.

Areas of Human Factors Implementation

For the purposes of this report, Human Factors implementation in the control space can be divided into three separate areas: The control room, the control room operator and the process control system. In each area, the systematic application of Human Factors requires analyses that are focused on the unique issues that each presents.

The Control Room

The design and layout of the control room should meet several objectives. The control room should:

- Properly route traffic around the control space and through the building use the control room space efficiently
- Locate areas for optimal interaction
- Have entries and exits properly located
- Ensure that personnel are secure, and
- Meet ergonomic and environmental standards

The analyses that have been developed to meet these objectives include:

Human Factors Control Room Survey: A set of questions that capture operator opinions about the layout of the space and the design of the equipment. The answers are analyzed by a Human Factors Specialist and form the basis for the Ergonomics/Environmental Assessments.

Task Area Needs Assessment: Determines the type and number of specific task areas and the personnel, equipment, storage, work station and space needs for each area (Table 6-5, Attwood et al, 2003).

Affinity Analysis of Task Areas and Equipment: Identifies relationships between the various items (e.g., task areas, equipment, entrances and exits) within the control room and determines how each of these items should be located relative to each other. The analysis starts by identifying each of the items in the control room that could have a relationship with each other. Then, the subjective strengths of the relationships between each item are estimated. The result is used to decide how the items can be relocated for maximum effectiveness.

Ergonomics Assessment: An assessment of the physical design of the existing workstations, control panels and equipment layout from an ergonomic perspective. The data collected in the assessment is compared against accepted industry standards for furniture, equipment design and spacing.

The Process Control Operator

Control room operators are the key to the effective operation of the process control system. It is essential that systems are designed to ensure that they:

- Are properly trained
- Have effective communication in the control space and that their
- Workload is properly understood and balanced

Three analyses are proposed to test that the above requirements are met.

Communications analysis: The communications analysis consists of a series of questions conducted in an interview format that attempts to understand and describe the following:

- What are the communications systems that are used for normal operations?
- What are the communications systems that are used in Emergency conditions?
- Are the systems usable under all operating conditions?
- What types of usage does each system receive?
- What are their strengths and weaknesses?
- How could they be improved?

Training Analysis: In many cases, process control operators are not properly trained for their task at the control console. On one hand, they may not first receive the proper orientation to the plant by not first learning the outside posts that they will be controlling from inside. On the other hand, experienced outside technicians may not receive the proper training on the control console before being given the responsibility to run it. This analysis is conducted as an interview with the following objectives:

- Describe initial training to the plant and on the control console
  - Course content
  - Instructor capability
  - Prerequisite knowledge
- What are the plant's adequacy ratings?
- Describe your refresher training
  - How often
  - How thorough
- What performance criteria have been established?

Workload Analysis

- ***Task Analysis and Workload Review:*** Identifies the operator tasks at each of the existing control consoles that could lead to excessive workload levels, lack of control, and task ambiguity in the consolidated work space. RRS proposes that the workload assessment be divided into three activities:
  - *Identify* the tasks performed at each of the existing control consoles.
  - *Conduct* a workshop to identify the physical and mental workload issues associated with each task, the cumulative time required to perform each task during the work shift the potential for improving each task and the potential interventions that could be applied to improve each task.
  - *Perform* a subjective workload assessment and prioritization of each task using the NASA, Task Load Index (TLX).

- ***Activity Analysis:*** The purpose of this analysis is to determine which activities are performed over a selected time period. Before beginning the analysis, it necessary to know what activities are performed over the time period of interest. The activities are listed, then the work is observed at predetermined intervals and the activity being performed at each interval is recorded.

For example, in one study, it was necessary to know how the activities performed by control room operators were distributed over their shifts. Five categories of activities were identified prior to the analysis:

a. Use of the phone or radio
b. Report preparation, paper work
c. Breaks, social discussions
d. Passive monitoring
e. Active monitoring

The control room operator was observed at sample intervals of 2-minutes, the activity was noted and the observer waited for another 2-minutes to take the next observation. Over a 12-hour shift, 360 samples (observations) were taken.

- ***NASA Task Load Index:*** The NASA-TLX is a subjective workload assessment tool that is used to identify those activities that have the most effect on the workload of the operators.

Operators begin the assessment by identifying each activity that they perform in a work shift. They then rate each activity on six workload dimensions:

- Physical Demand (PD): How much physical activity was required (e.g., pushing, pulling, turning, controlling, activating, etc.)?
- Mental Demand (MD): How much mental and perceptual activity was required (e.g., thinking, deciding, calculating, remembering, looking, searching, etc.)?
- Performance (OP): How successful do you think you were in accomplishing the goals of the task/scenario?
- Temporal Demand (TD): How much time pressure did you feel due to the rate or pace at which the tasks or task elements occurred?
- Frustration (FR): How insecure, irritated, stressed and annoyed versus secure gratified, content, relaxed and complacent did you feel during the task/scenario?
- Effort (EF): How hard did you have to work (mentally and physically) to accomplish your level of performance?

First, the operators are asked to decide which of the dimensions are most applicable to each task. They do this with a paired-comparison task. They are presented with 15 pairs of dimensions, covering all possible combinations of the six dimensions, e.g. PD/EF, and are asked to circle the dimension in each pair that would provide the most significant source of workload for the task. The number of times that each dimension is chosen ($N_{DIM\ X}$) is recorded for each of the six dimensions.

Second, they are asked to estimate the magnitude of each workload dimension (Mag_X) present in each task. Continuous scales anchored with LOW/HIGH extremes are provided as follows:

|---|---|
LOW HIGH

Finally, they estimate the percent of time that they spend on each task during the shift.

The Overall Workload Level (OWL) is calculated for each task from Equation (1)

$$OWL_Y = [\Sigma(N_{DIM\ X}) \times (Mag_x)] \qquad (1)$$

Where X = Dimension 1, 6

OWL depends only on subjective workload estimates and not on the time spent performing the task. Workload estimates based on OWL accurately reflect the true effort that the task requires from the operator.

The Weighted workload level is the Overall Workload Level (OWL) multiplied by the percentage of time spent on the task. Based on the three indices from above, overall weighted workload is estimated for task (Y) by Equation (2):

$$\text{Weighted } OWL_Y = [\Sigma(N_{DIM\,X}) \times (Mag_X)] \times [\%\ \text{time}] \qquad (2)$$
$$\text{Where } X = \text{Dimension } 1, 6$$

The results from Equation (2) are highly influenced by the time that an operator spends performing the tasks. If strategies could be put in place to reduce the percentage of time devoted to each task, the rankings would be much different.

The Process Control System

The design and layout of the process control system should meet the following objectives:

- Provides the operator with the proper information for process status and alarms
- Provides the information in an appropriate format

The analyses that have been developed to meet these objectives include:

Critical Task Analysis**:** An analysis of the major upset, abnormal or difficult tasks that the operators perform. The tasks are divided into different categories and prioritized using a Risk Matrix. The tasks with the highest risk are then analyzed at the control console. The analysis conducted is a modified task analysis. The analysis begins by identifying the response of the process control system when the upset takes place. This is usually an alarm, but could be the result of communication form the field or another process control room operator. The analysis continues by identifying the operator's response, then the return response of the system and so on. At each step the operator is questioned on potential alternative responses or alternative information available. The result of the analysis is to define the operator information requirements to reduce the probability of operator error.

Alarm Function Analysis: This analysis is conducted in an interview format. It evaluates the capability of the alarm system to provide the proper warnings to the control room operator of processes that are exceeding their control boundaries. Some alarms are more important than others, so the analysis looks especially at alarm priority settings. Finally, nuisance alarms and alarm flooding, which are both important in today's operations are investigated.

- Alarm priority
- Coding Conventions (e.g. time/date, nature/severity etc)
- Priority coding (visual/auditory)
- Conventions for presentation format
- Expert systems to prevent alarm flooding
- Hierarchical alarm summaries
- Status notification for uncleared alarms
- Alarm test capability
- Importance categories

Process control display interface: The optimal process control display interface uses the latest technology, helps the operator effectively manage normal and abnormal operations, and yet does not overload the operator during abnormal operations (Chapter 5, Attwood et al. 2004). The operator interface consists of process display screens, keypads and/or keyboards, pointing devices and panels that contain controls and non-computer displays.

Analysis of the process control system interface consists of comparing the design of the displays and controls against internationally accepted guidelines. Two most recently developed guidelines are from the Abnormal Situation Management consortium in the United States (ASM, 2002) and the International Standards Organization (ISO, 1999).

Summary

In summary, the systematic Human Factors analysis of the process control rooms and control systems requires the application of techniques that target three areas of control room operation: the control room, the control room operator and the process control system.

References

Attwood, D.A., Deeb, J.M and Danz-Reece, M.E. (2003) *"Ergonomic Solutions for the Process Industries"* Butterworth Professional Publications (Elsevier), Boston MA.

ASM (2002) *"ASM Consortium Guidelines: Effective Operator Display Design"* Abnormal Situation Management Consortium, Displays Subcommittee, Version 4.00, 19 June 02

ISO (1999), *"Ergonomic Requirements for the Design of Displays and Control Actuators. Part I: Human Interactions with displays and control actuators, Part II: Displays"* ISO 9355, International Standards Organization, Geneva.

Title: Statistical Analysis of Human Factors Checklists

Presenter: Bruce K. Vaughen, Ph.D., P.E.
DuPontTeijinFilms

Prepared for presentation at the AIChE 2005 Spring National Meeting
April 10-14, 2005
Process Plant Safety Symposium (Topical Session TH)
Human Factors Engineering and Ergonomic Engineering

Unpublished

1. Abstract

Two Human Factors checklists modeled after CCPS checklists were used and statistically analyzed during a revalidation Process Hazards Analysis (PHA). The first, simple checklist was used during the Failure Mode Effect Analysis (FMEA), with particular emphasis placed on the "human factors" elements affecting the equipment's failure mode and the lines of defense. The second checklist was used to survey the "Okay As Is" to "Needs Work" responses from ten personnel from operations, maintenance, engineering and management. The second checklist is more comprehensive, consisting of 106 questions. Statistical analysis of the survey results, a standard bell curve, was used to pinpoint questions for discussion. The questions that showed general agreement had either high or low relative scores with low standard deviations; the questions that showed either diverse interpretations or opinions had high standard deviations. These human factors assessments led to five of the PHA Team's recommendations. These recommendations were then proposed to and accepted by the site leadership.

2. Introduction

This statistical analysis of Human Factors checklists was a part of a revalidation Process Hazards Analysis (PHA). By definition, a "revalidation PHA" reviews, updates as necessary, and confirms the work performed in prior PHAs. The reviews include incidents and modifications since the last PHA. Reviews of the incidents showed that almost half of the incidents identified "operating procedures" or "human factors" as contributors to the incident. The most significant incident occurred during the flammable solvent loading operation. Reviews of the modifications showed that the "management of change – technology" documentation had not been completed for the new equipment added to the flammable solvent loading operation (after the incident). Reviews of the operating procedures showed that some were not accurate and up-to-date. This included lack of documentation for changes to the original procedure for new equipment at the solvent loading operation. These discussions led the PHA Team to include a detailed human factors assessment in its revalidation PHA.

The first human factors assessment incorporated a simple one-page human factors checklist which was used during the analysis of each process block of a Failure Mode and Effect Analysis (FMEA). The second assessment consisted of a personnel survey using a comprehensive 106-question Human Factors checklist. Both checklists were modeled after the Center for Chemical Process Safety (CCPS) checklists. Statistical analysis, using Six Sigma tools, was used on these survey responses to efficiently select a few key questions for discussion in the PHA Team meeting.

3. Human Factors – Definition

Human factors is primarily concerned with the interactions between people and the equipment, systems, and information in their work environment. A human factors analysis focuses on the identification and avoidance of potential error-likely situations in the operation of the process and in the maintenance of the associated equipment and systems.

4. Human Factors – FMEA Survey

The first human factors approach used the simple checklist shown in Table 1. There are five general categories associated with this checklist: 1) operator / process interface; 2) task design and job organization; 3) work place and work environment; 4) training; and 5) procedures. Human factors-specific items were noted as each "process block" was discussed during the Failure Mode and Effect Analysis (FMEA). The number of human factors items noted during the FMEA are summarized in Table 2. Of the twenty-nine items noted for each category relative to a process block, the "operator / process interface" and the "procedures" questions generated the most discussion items. All of these human factors categories generated discussions for the flammable solvent loading "process block," as well. After the FMEA was completed, the PHA Team decided that this simple assessment was not sufficient, and chose the more comprehensive human factors assessment described below.

5. Human Factors – Comprehensive Checklist Survey

The second human factors approach used a more comprehensive 106-question checklist to survey ten personnel independently (those on the PHA Team and additional operators from the area). There are six general categories associated with this checklist: 1) management; 2) procedures; 3) training; 4) task design and organization; 5) ergonomics; and 6) control systems. Each question was rated numerically and independently, from "1" for "Okay As Is" to "5" for "Needs Work." The results from these ten responses for each question were combined and statistically analyzed, with each question providing an average numerical score and standard deviation.

Histograms of the average and standard deviation results from this survey are plotted in Figures 1 and 2, respectively. The "outlier questions" on this normal distribution curve shown in Figure 1 were selected from the list and used for a focused and more detailed discussion. These "outlier" questions were separated into the two main categories shown in Table 3, "Okay As Is," with an average less than two; and in Table 4, "Needs Work, " with an average greater than three. The questions with low and high standard deviations represented Team "agreement" or "disagreement," respectively, and are listed in Table 5. The questions representing disagreement were discussed in further detail to answer this: "Why such a broad range in the response?"

A summary of these results is shown in Table 6. The "training" and "control systems" categories produced the most discussion items, with training (procedure) and calibration effort questions noted for each category, respectively. Questions that showed up in BOTH the "needs work" and "disagreement" categories were discussed in detail. Consistent with other observations during the team meetings, these questions addressed concerns for the flammable solvent loading operation. Questions in the "procedures" category reinforced prior discussions that some of the procedures needed improvement, including their thoroughness, control of and ensuring that they are up-to-date.

6. Conclusion

The combined human factors assessments, focusing on the flammable solvent loading operation, reinforced the findings from the incident, modification, and operating procedure reviews (all parts of the revalidation PHA protocol). Five of the PHA Team's recommendations were created with support from the results of these human factors assessments.

In general terms, the first recommendation addressed improving the operator / process interface (including ergonomics): "Review potential upgrades to the operator data gathering protocols." The next three recommendations addressed improving the operating procedures: "Update the operating procedure for the solvent loading operation;" "Write and implement the solvent loading system calibration procedure;" and "Establish an automatic operating procedures review frequency to ensure that they are current and up-to-date." The final human factors-related recommendation addressed improving the training: "Review and upgrade the training protocol."

These five recommendations, with more specific actions, responsibilities, and due dates, were submitted as a part of all of the proposed PHA recommendations to the site leadership for their review and discussion. Site leadership approved of all of the recommendations.

7. References

Guidelines for Technical Management of Chemical Process Safety, CCPS, AIChE

Guidelines for Preventing Human Error in Process Safety, CCPS, AIChE

DuPont Corporate SHE Standard on Process Hazards Analysis (S25A)

MINITAB® Release 14.12.0; Six Sigma Academy Module Release 14.12.0

Table 1. Simplified FMEA-based Human Factors Checklist

1) Operator/Process Interface

Are display and control panels clearly identified?
Are computer displays clearly formatted?
Is operator information clear and accessible?
Are the implications of actions and their effects clear?
Are indicators and codes clearly defined?
Does instrument layout minimize chances of error?

2) Task Design and Job Organization

Is operator performance monitored and measured?
Is operator assistance readily available?
Is there team responsibility?
Are the operator's individual responsibilities clearly defined?
Is there incident follow-up?

3) Work Place and Working Environment

Have posture, movement, and accessibility been considered?
Have environmental conditions been considered?
(e.g., illumination, noise, temperature, and /or humidity)

4) Training

Are training requirements defined?
Are training methods developed?

5) Procedures

Are procedures concise?
Are mandatory procedures clearly identified as such?
Are supporting procedures developed?
Are operational procedures correct?

Table 2. Summary of Results from the FMEA-Based Human Factors Assessment

	Human Factors Categories					
	1) Operator/Process Interface	2) Task Design and Job Organization	3) Work Place and Work Environment	4) Training	5) Procedures	Total
FMEA "Process Blocks" Noted With Discussion Items	11	2	2	2	12	29

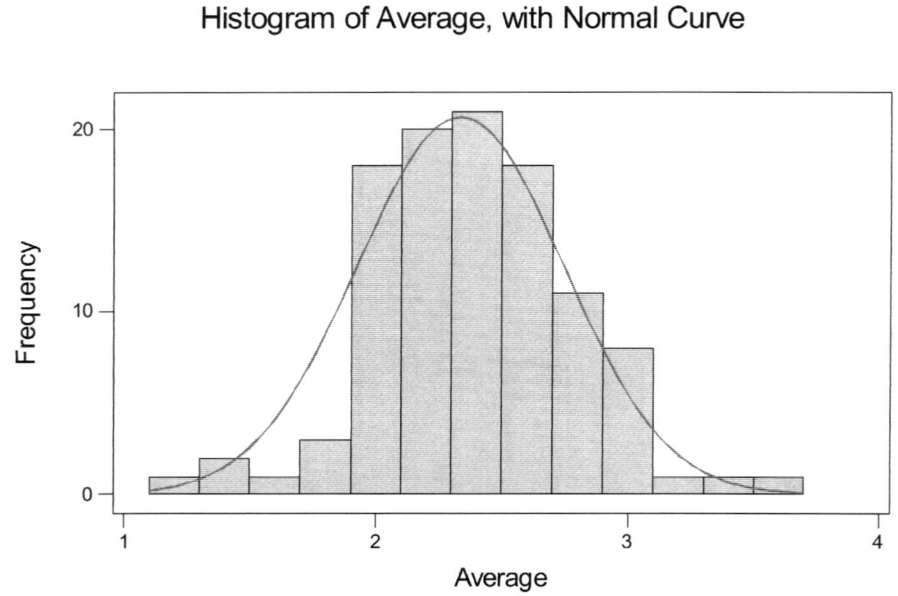

Figure 1. Results from the Comprehensive Human Factors Checklist Survey

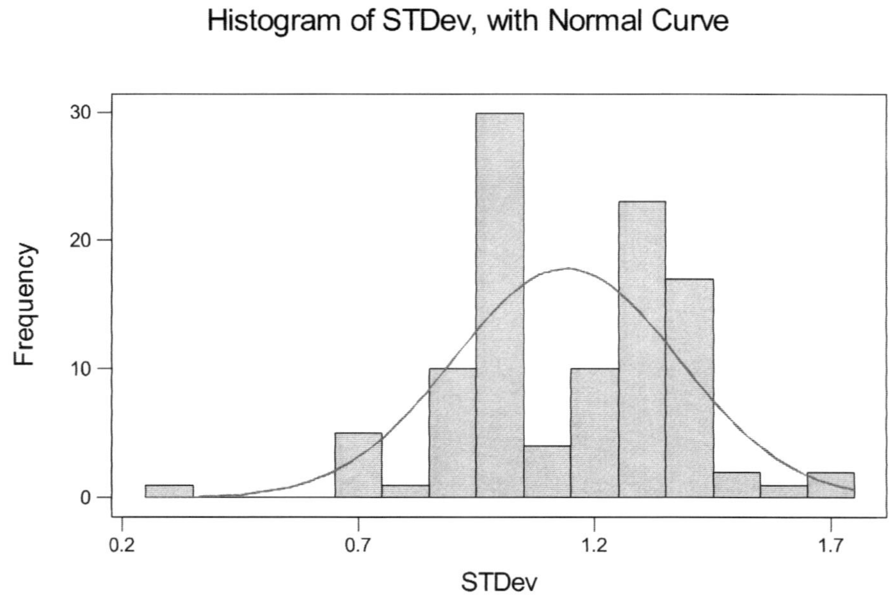

Figure 2. Results from the Comprehensive Human Factors Checklist Survey

Table 3. Results from the Comprehensive Human Factors Assessment

Questions Rated "Okay As Is" (Averages Less Than 2)

	Checklist Question	Average	Standard Deviation
1.02	Is the work environment, including, for example, general housekeeping and congestion, maintained in an acceptable manner, or is there a pattern of tolerance until conditions become unbearable? Is there a deteriorating trend evident?	1.1	0.32
1.06	Is there an effective means to detect and correct substance abuse?	1.3	0.67
1.11	Is the personnel turnover rate maintained at a tolerable level? Is there a policy to manage personnel change?	1.3	0.73
6.25 e	Does the user have time to determine the source of the alarm, or will the alarm scroll off the screen too quickly?	1.5	0.85
5.08	Is it easy to work at or move past control panels without accidentally altering settings? (For example, are there cover plates over emergency shutdown buttons? If so, make sure the cover plates don't restrict access in case of emergency.)	1.8	0.92
4.01	Are operators' job descriptions well defined (e.g., do overlaps or gaps in responsibility exist, having the potential to result in important tasks being left undone due to confusion regarding responsibility)?	1.8	1.00
1.05	Are safety rules understood by all? Are personnel held accountable to them? Have personnel been fully apprised of safety priorities and disciplinary practices for failure to follow procedures?	1.8	1.40
5.02	Does successful use of emergency equipment require physical skills or training beyond the capability of some of the people likely to respond?	1.9	0.99
5.05	Would area congestion play a role in tasks where speed is of the essence (e.g., emergency access to shutdown devices, evacuation routes, etc.)?	1.9	0.99
6.10	Are the criteria clear and unambiguous for taking over manual control from automatic? Is it understood when a system can be placed into automatic or supervisory control?	1.9	0.99
6.13	Can instrument (or video display terminal) lag/update time be slow enough to create a potential for operator overcontrol?	1.9	0.99
6.19	Are systems designed to avoid overly sensitive controls? In other words, is there a reasonable range of control movement? (For example, error would be likely if attempting to regulate flow at 50 GPM with a control knob that changed flow 1000 for each half-turn.)	1.9	0.99
6.26 a	Are there proper checks and balances to prevent computer programming errors?	1.9	0.99
6.08	Are there area process control/alarm color or sound conventions used that run counter to conventions used elsewhere on the site?	1.9	1.05
1.04	Are the various elements of process safety management audited? How well is the business responding to recommendations made?	1.9	1.36

Table 4. Results from the Comprehensive Human Factors Assessment

Questions Rated for "Needs Work" (Averages Greater Than 3)

	Checklist Question	Average	Standard Deviation
3.05	Are operating teams trained together in the transfer of information and responsibility?	3.0	0.71
6.03	Is the control system labeling terminology consistent and compatible with the comprehension level of the user? (For example, does "0% valve loading" always mean the valve is closed?)	3.0	0.94
6.20	Are instruments calibrated or checked regularly?	3.0	1.70
4.08	For manual ingredient preparation tasks (e.g., for charging a reactor), is there a means to prevent adding the wrong amount or multiple charges?	3.1	0.99
3.10	Have people been assigned to jobs based upon a demonstration of required skills?	3.3	1.49
4.02	Are there parts of the process where split/shared ownership can lead to a lack of accountability?	3.6	1.26

Table 5. Results from the Comprehensive Human Factors Assessment

Question Standard Deviations (Team Agreement or Disagreement)

	Checklist Question	Average	Standard Deviation
	Greatest Alignment --- Strong Agreement (Standard Deviation Less Than 0.7)		
1.02	Is the work environment, including, for example, general housekeeping and congestion, maintained in an acceptable manner, or is there a pattern of tolerance until conditions become unbearable? Is there a deteriorating trend evident?	1.1	0.32
3.12	Are critical maintenance procedures identified, understandable, and accurate?	2.9	0.67
1.06	Is there an effective means to detect and correct substance abuse?	1.3	0.67
3.09	Do the training requirements (or skill demonstrations) for a job accurately reflect the requirements for both routine and nonroutine operations?	2.6	0.70
	Greatest Difference --- Strong Disagreement (Standard Deviation Greater Than 1.4)		
2.04	Are the procedures kept up-to-date? Are they routinely checked, compared with user actions, and revised as appropriate?	2.7	1.49
3.10	Have people been assigned to jobs based upon a demonstration of required skills?	3.3	1.49
6.26 f	Are software safety interlocks well-documented, including a written description of how they work?	2.7	1.64
6.24 b	Color coded as needed? Color coded consistently? (Is colorblindness an issue?)	2.4	1.65
6.20	Are instruments calibrated or checked regularly?	3.0	1.70

Table 6. Summary of the Results from the Comprehensive Human Factors Assessment

	1) Management	2) Procedures	3) Training	4) Task Design	5) Ergonomics	6) Control Systems	Total Number of Questions
"Okay As Is"							
Questions	5			1	3	6	15
"Needs Work"							
Questions			2	2		2	6
Greatest Alignment (Standard Deviation < 0.7)							
Questions	2		2				4
Greatest Difference (Standard Deviation > 1.4)							
Questions		1	1			3	5
Total "Need Work" and "Difference"	0	1	3	2	0	5	11

Methods for Improving Human Performance

Donald K. Lorenzo
ABS Consulting, Risk Consulting Division

Lee N. Vanden Heuvel
ABS Consulting, Risk Consulting Division

James J. Rooney
ABS Consulting, Risk Consulting Division

Introduction

Historically, human errors have been found to be significant factors in almost every accident, equipment shutdown, or quality problem in industrial and manufacturing facilities. In fact, a study of accidents in refining and petrochemical plants[1] identified the following causes: equipment and design failures (41%), operator and maintenance errors (41%), inadequate or improper procedures (11%), inadequate or improper inspection (5%), and miscellaneous causes (2%). Some managers assume that virtually all human errors are the consequence of lazy, stupid, or inattentive workers, so simply finding and punishing the guilty will deter others from making the same or similar mistakes.

However, this philosophy blinds managers to the error-likely situations that underlie most human errors and usually does more harm than good. Instead of blaming human error on ill-trained or unmotivated workers, systems must be established to investigate and analyze near misses so that root causes are determined and corrective actions are implemented to prevent recurrence. To assist in this process, the American Institute of Chemical Engineers' Center for Chemical Process Safety (CCPS) has written a new book, *Human Factors Methods for Improving Performance*. This book describes a number of practical tools that can be used to identify and correct error-likely situations in the workplace. This paper describes five key elements that can be used by any facility to effectively reduce human error.

Performance Shaping Factors

Humans are an essential element of any process system. To minimize human errors, managers must ensure that the worker interfaces, which include interactions with other workers as well as with the equipment and environment, are compatible with the capabilities, limitations, and needs of the worker. A *performance shaping factor* (PSF) is anything that affects a worker's performance of a task within the system. PSFs can be divided into three general classes: (1) internal PSFs that act within an individual, (2) external PSFs that act on an individual, and (3) stressors.[2] Table 1 lists some internal PSFs, which are the individual skills, abilities, attitudes, and other characteristics that a worker brings to any job. Some of these, such as training, can be improved by managers; others, such as a short-term emotional upset triggered by a family crisis, are beyond any practical management control (however, a manager's style can influence workers' mental/emotional states, as can counseling programs). Note that the PSFs managers often focus on, such as motivation and work attitude, are generally the more difficult to address. Telling someone to improve his or her attitude usually has nothing more than a temporary (and sometimes opposite) effect if the underlying PSFs are not addressed.

Table 1 Internal PSFs (based on Table 3-2 in Reference 2)

Training/Skill	Emotional State
Practice/Experience	Gender
Knowledge of Required Performance Standards	Physical Condition/Health
Stress (Mental or Bodily Tension)	Influences of Family and Other Outside Persons or Agencies
Intelligence	Group Identifications
Motivation/Work Attitude	Culture
Personality	

Table 2 lists external PSFs that influence the environment in which tasks are performed. External PSFs are divided into two groups: (1) situational characteristics and (2) task and equipment characteristics. Situational characteristics include general PSFs that may affect many different jobs in the plant. Situation characteristics generally describe what it is like to have a specific job at a facility, such as an outside operator. Task and equipment characteristics are pertinent to a specific job or a specific task within a job. These characteristics describe what it is like to do a specific task, such as charging a batch of material. Job and task instructions are a particularly important part of the task characteristics because they have such a large effect on human performance. By emphasizing the importance of preparing and maintaining clear, accurate task instructions, managers can significantly reduce the likelihood of human errors.

The interaction between internal and external PSFs creates stress in the individual performing the task. Mismatches between internal and external PSFs result in disruptive stress that degrades job performance. If too little stress is present, a worker will not remain sufficiently alert or motivated to do a good job. For example, a control room worker who repetitively fills batch tickets may not be alert enough to notice that an ingredient was omitted. On the other hand, too much stress will quickly overburden a worker and degrade job performance. In such situations, workers tend to (1) focus on the largest or most noticeable signals and ignore some information entirely, (2) omit or delay some responses, (3) process information incorrectly and reject information that conflicts with their diagnosis or decision, or (4) mentally and/or physically withdraw. Examples of disruptive psychological and physiological stressors are listed in Table 3.

Although stress usually has a negative connotation, some stress is actually necessary for humans to function at optimum performance, as illustrated in Figure 1. Facilitative stress is anything that arouses us, alerts us, prods us to action, thrills us, or makes us eager. When a positive balance exists between internal and external PSFs, workers experience facilitative stress and their job performance is at its best.

Managers must recognize that most PSFs (including many internal PSFs) are within their control. By (1) designing work situations that are compatible with human needs, capabilities, and limitations, (2) carefully matching workers with job requirements, and (3) rewarding positive behaviors, managers can create conditions that optimize worker performance and minimize human errors.

Table 2 External PSFs (based on Table 3-2 in Reference 2)

SITUATIONAL CHARACTERISTICS (Job Level)	TASK, EQUIPMENT, AND PROCEDURAL CHARACTERISTICS (Task Level)
Architectural Features Environment (Temperature, Humidity, Air Quality, Lighting, Noise, Vibration, General Cleanliness, etc.) Work Hours/Work Breaks Shift Rotation Availability/Adequacy of Special Equipment, Tools, and Supplies Staffing Levels Organizational Structure (Authority, Responsibility, Communication Channels, etc.) Actions by Supervisors, Coworkers, and Accreditation and Regulatory Personnel Facility Policies General Rewards/Punishments	Procedures (Written or Not Written) Written or Oral Communications Cautions and Warnings Work Methods/Practices Dynamic vs. Step-by-step Activities Team Structure and Communication Perceptual Requirements Physical Requirements (Speed, Strength, etc.) Anticipatory Requirements Interpretation/Decision Making Complexity (Information Load) Long- and Short-term Memory Load Calculational Requirements Feedback (Knowledge of Results) Hardware Interface Factors (Design of Control Equipment, Test Equipment, Process Equipment, Job Aids, Tools, etc.) Control-display Relationships Task Criticality Frequency/Repetitiveness Mapping Error-proofing Conformance with Populational Stereotypes Visual Systems Implementation Equipment Functionality (maintenance issues)

General Approaches for Reducing Human Error

When contemplating ways to improve human performance, managers must address two basic types of errors: (1) errors whose primary causal factors are individual human characteristics unrelated to the work situation and (2) errors whose primary causal factors are related to the design of the work situation. Employing appropriate hiring and job assignment policies is an important means for managers to reduce the causes of the first type of error. But, on any given day, a worker could be emotionally upset, fatigued, and so forth, and commit an error. However, human factors specialists estimate that only 15 to 20%[3] of workplace errors are caused primarily by such internal human characteristics.

Table 3 Stressor PSFs (based on Table 3-2 in Reference 2)

Psychological Stressors	Physiological Stressors
Suddenness of Onset	Long Duration of Stress
High Task Speed	Fatigue
Heavy Task Load	Pain or Discomfort
High Jeopardy Risk	Hunger or Thirst
Threats (of Failure, of Loss of Job, etc.)	Temperature Extremes
Monotonous, Degrading, or Meaningless Work	Radiation
Long, Uneventful Vigilance Periods	Exposure to Diseases
Conflicting Motives About Job Performance	Vibration
Negative Reinforcement	Movement Constriction
Sensory Deprivation	Movement Repetition
Distractions (Noise, Glare, Movement, etc.)	Lack of Physical Exercise
Inconsistent Cueing	Disruption of Circadian Rhythm
Lack of Rewards, Recognition, Benefits	

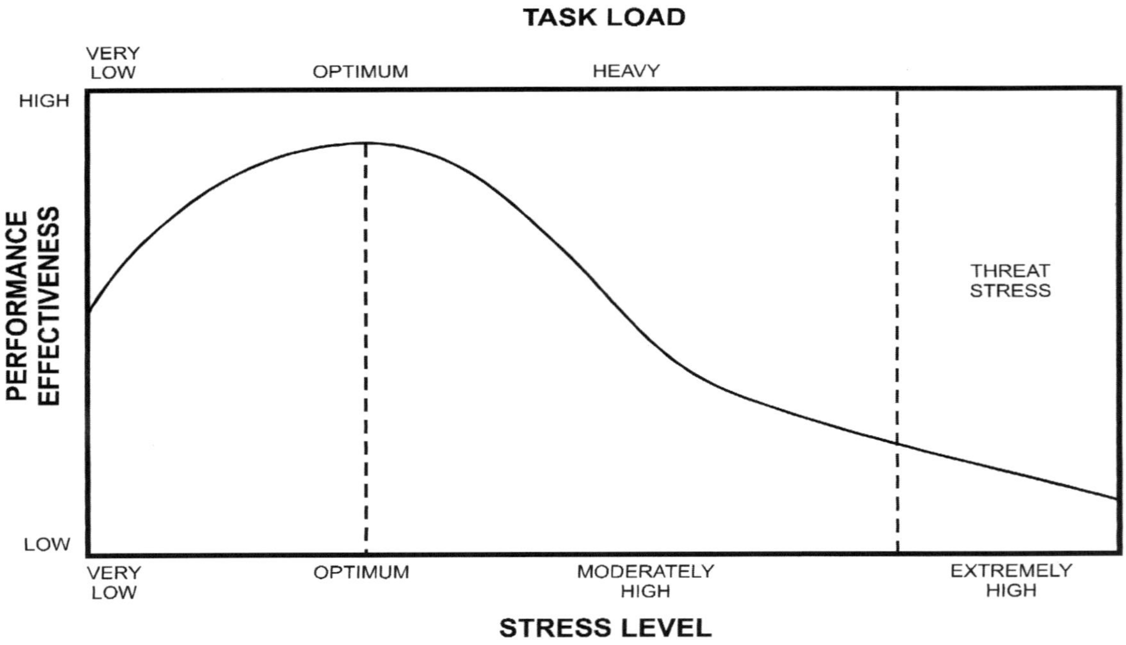

Figure 1

The vast majority (80 to 85%[3]) of human errors result primarily from the design of the work situation (the tasks, equipment, and environment), which managers can directly control. A work situation in which the PSFs are not compatible with the capabilities, limitations, or needs of an employee to perform a task correctly is called an *error-likely situation*. In a sense, an error-likely situation is a situation in which a person has unintentionally been "set up" to make a mistake. Error-likely situations can result from a variety of causes, such as:

- Deficient procedures (e.g., incorrect/incomplete/nonexistent, poor format, lack of warnings, inappropriate language, conflicting requirements, inconsistent with user needs/requirements)
- Poor communication between workers (e.g., no shift overlap, vague oral instructions, incomplete instructions, inconsistent instructions)
- Inadequately trained workers (e.g., poor mental model[s] for diagnosing process upsets, no/rare practice)
- Conflicting interests of workers (e.g., productivity vs. safety, workers vs. management, operations vs. maintenance, practice vs. policy)
- Inadequately labeled equipment (e.g., parts, raw materials, motor numbers, locations, valves)
- Equipment poorly designed for human use (e.g., inaccessible/inconvenient, violates populational stereotypes, excessive/inappropriate automation, user needs not addressed)

By providing the resources necessary to identify and eliminate error-likely situations, managers can improve the PSFs and dramatically reduce the frequency of human errors. Instead of setting up workers for error, we set them up for success! This strategy is called the *work-situation approach*, which involves the five elements described below. To maximize the benefits of such a strategy, managers should solicit workers' input into this strategy at every opportunity. After all, the workers can best identify factors that hinder their performance, and they likely will enthusiastically support such a strategy if they are not penalized for telling the truth. In addition, human factors specialists can be employed to provide expert assistance in implementing any changes.

Element 1: Implementing Good Human Factors Engineering of Equipment, Control Systems, and the Work Environment

Many human errors are caused by equipment and work environments that were not initially designed with an emphasis on how humans would interact with them (i.e., operability, maintainability, and human factors engineering principles). Therefore, a new system design plan should account for all necessary operational and maintenance behaviors through proper labeling and visibility; accessibility for use, repair, removal, and replacement; proper inspection and testing; and availability of spare parts and tools. Examples include (1) handles that can be grasped and turned with reasonable force, (2) labels that can be read from a reasonable distance, and (3) critical and/or unique tools permanently located at the job site. Where possible, error-proofing principles should be used to eliminate the potential for error. For example, unique hose fittings that can only be connected to the correct tank; only data in the valid range can be accepted by the computer.

Identification and elimination of error-likely situations early in the system- design phase is obviously more desirable than the frustrating and costly task of retrofitting. Because most systems are designed based on a similar system currently in service, evaluating the existing system to help determine future requirements would be prudent. Reviewing old drawings, performing walkthroughs, and interviewing operators, maintainers, and other experienced personnel can be effective ways of evaluating an existing system to help determine the requirements and identify error-likely situations for a new facility. Including these personnel in the design reviews, or including them as permanent members of the design team, is an excellent way of involving the users (and possible future owners) of the new process in the design phase and using their knowledge and experience to help identify potential error-likely situations. Having these resources available during design reviews has other advantages, such as (1) providing cost-effective and practical ways of resolving the problem areas that are identified, (2) providing an excellent opportunity for training on the new process, and (3) developing a sense of ownership of the new process among workers.

Many resources, such as checklists to systematically walk the user through different human factors considerations, are available to help maximize the benefits of design reviews. Pages 384 through 388 of Reference 4, and Appendix 1 in Reference 5, provide two examples.

Element 2: Providing Clear, Accurate Procedures, Instructions, and Other Job Aids

Many human errors can be prevented by ensuring that clear, accurate procedures and job aids are available and appropriately used by all workers. In addition to reducing workers' reliance solely on skill and memory to perform a task, procedures and job aids assist the worker in decision making and help ensure that a given task is performed consistently. Written procedures provide step-by-step directions describing how and when to perform the various portions of the task. Job aids (e.g., flowcharts, decision tables, and checklists) can be used to (1) provide concise organization of information needed to perform problem diagnosis and planning tasks and (2) aid workers when performing tasks that involve numerous steps.

A procedure or job aid's effectiveness in reducing human errors is increased by incorporating the following key principles:

- ***Select a procedure style or format that is usable and familiar and that best communicates the information to the worker.*** A consistent style and format will assist workers in finding relevant information.

- ***Ensure that the procedure is accurate and complete to maintain credibility and to facilitate continued use.*** If procedures contradict the way tasks are performed in practice, workers will soon lose faith in the procedures and will not use them. Involving experienced workers in procedure development and establishing a proper frequency for updating procedures (e.g., after any relevant equipment modifications) will help create and maintain accurate procedures. To further ensure that procedures remain accurate, a periodic review by experienced workers should be conducted.

- ***Include the appropriate level of detail in each procedure***. Procedures that contain too little detail will make the procedure unusable by the inexperienced worker, and too much detail may discourage the experienced worker from using it. The appropriate level of detail depends on the level of worker expertise and the criticality of the task. An easy check is to ask a "qualified" person who does not normally perform the task (e.g., someone who might temporarily perform the job if the regular worker were sick or absent) to follow the procedure. Inclusion of warnings, cautions, and other critical parameters is appropriate for all levels of worker expertise. Some procedure formats, such as an outline format, allow experienced workers to identify the highlights, yet still provide sufficient details for the inexperienced worker.

- ***Use language that is understandable to the worker to reduce the potential for errors, especially in stressful situations (hurried to complete a task, etc.).*** Procedures can be made more understandable by (1) including only one action in each procedure step, (2) using simple, clear language, (3) using the active voice for the actions (e.g., open the coolant valve), and (4) using short sentences and positive phrases.

In addition, procedures and job aids must be readily available to workers to encourage use. Ideally, the procedures should be located where they will be used. For example, the directions on how to use a sample station should be located on the front of the access door. Otherwise, a single set of procedures should be kept in a place that is centrally located but readily accessible to all workers. However, if procedures are stored electronically and only printed when needed, sufficient work stations and printers must be provided to give the workers convenient access. The use of personal copies of procedures should not be allowed unless a system has been implemented to control and update all the copies.

Element 3: Providing Job-relevant Training and Practice

Training ensures that workers possess the basic skills necessary to effectively perform their functions; proper training of workers is critical to reducing human errors. Several types of training have proven most effective in reducing human errors, including (1) initial skill training, (2) refresher training, and (3) management systems training.

Classroom knowledge-based training provides the basic expertise, technical background, and understanding of the procedures. Initial skills training is a bridge between the knowledge-based training and on-the-job performance. Initial skills training should be viewed as the means for preparing workers for real experiences they will routinely encounter, as well as those they will infrequently encounter. If skill training does not include the critical, infrequent events or situations, the likelihood of successfully handling such situations will depend solely on the problem-solving and decision-making skills of the worker (which may not be sufficient). Initial skills training is generally conducted in the classroom and supplemented with on-the-job experience. On-the-job training supplements the classroom training by developing necessary skills and providing a minimum level of experience for performing required skills. A combination of knowledge, skill, and on-the-job training is needed.

In addition to the initial training, refresher training on nonroutine or modified tasks will minimize worker errors and reduce the potential for a worker's skills to deteriorate. A refresher training (or practice) program that regularly challenges the worker's knowledge and provides opportunities to practice skills is needed to assist workers in developing and maintaining a high skill level. Not only will such a program address a worker's loss of skills; it will also enhance skills beyond the initial training level. The sawtooth curve in Figure 2 presents the potential benefits of a refresher training (or practice) program on worker responses versus performing only initial training. The curve illustrates the conversion of *knowledge* to *skill* through repeated training and practice.

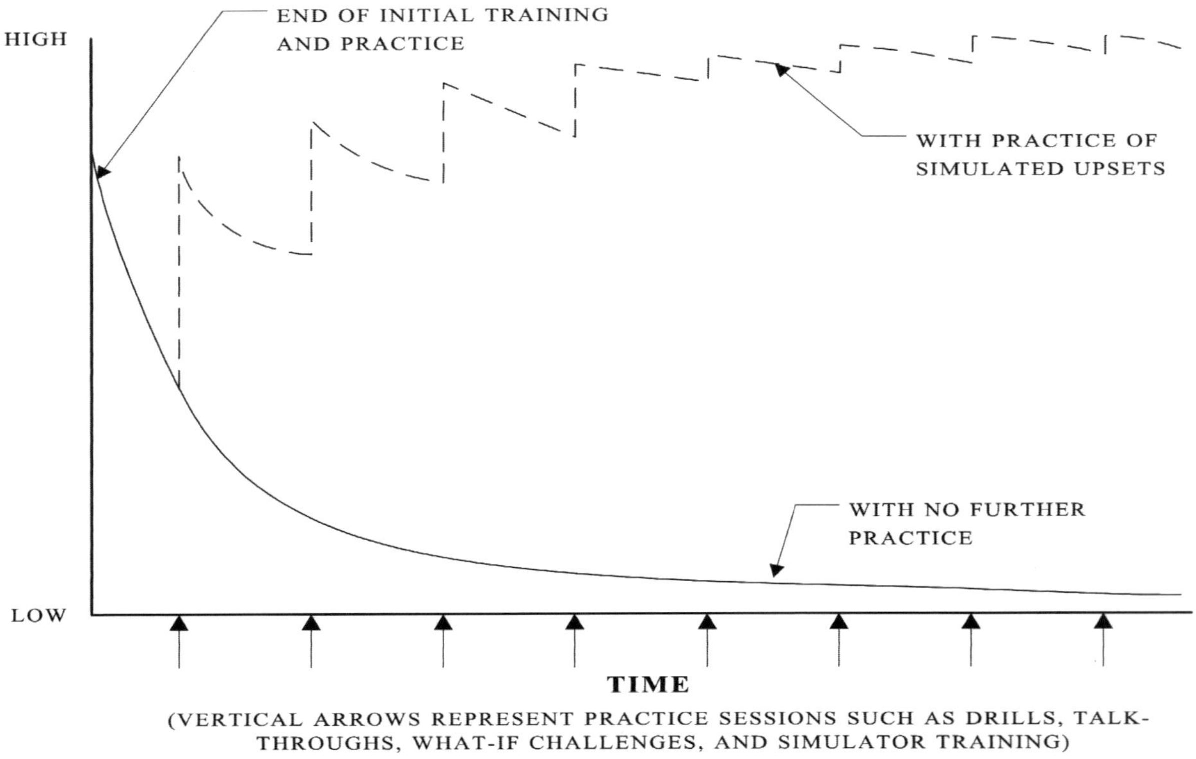

RESPONSE SKILLS

Figure 2

To round out a training program, management system training is required on a regular basis to ensure that workers can readily identify and follow relevant management systems. For example, on many safety-critical systems, a management system exists to help prevent systematic human variability. To prevent a common cause human error, many companies have policies for redundant checking of critical tasks by different personnel. For example, the equipment may be aligned by one worker, checked by another worker, and finally verified by the supervisor before beginning the procedure. Examples of other management systems include management of change, safe work practices, and pre-startup reviews. Many of these systems are intended to prevent human errors and should not be overlooked when developing a training program.

Element 4: Providing Ways to Detect and Correct Human Errors Before an Undesired Consequence Occurs

Many human errors can be prevented by implementing certain administrative controls and systems. For example, some companies have policies that require workers to work in pairs for certain activities. This "buddy system" can be effective in detecting a human error before an undesired consequence occurs. Many activities provide the opportunity to proof test the recently completed task. For example, after a maintenance technician finishes work on an instrument, it can be bench-tested to verify that the work was completed satisfactorily. Workers should identify opportunities to proof test their work to detect mistakes before batches are started and equipment is returned to service. Whenever possible, these proof tests should be incorporated into written procedures to help ensure that they are effectively performed and to provide training for new employees.

Another way workers can detect and correct human error is by using self-checking techniques. A self-checking technique is a practice in which a person consciously and deliberately reviews the intended action and expected response **before** performing a task. One self-checking technique used in the medical field is the "five-rights" used to administer medications to patients:

- ✓ Right patient
- ✓ Right medication
- ✓ Right dose
- ✓ Right route (oral, intravenous, dermal, etc.)
- ✓ Right time

A more general self-checking technique, developed by the Institute for Nuclear Power Operations, is the STAR (<u>S</u>top, <u>T</u>hink, <u>A</u>ct, <u>R</u>eview) technique, which includes the following four steps:

STOP	Organize your thoughts and try to eliminate distractions.
THINK	Locate the correct instruments, items, procedures, and people. Sense everything around you by listening, looking, smelling, and touching (as appropriate).
ACT	Confirm the correct time, amount, equipment, parts, chemicals, etc. Perform the task carefully and safely.
REVIEW	Observe what occurs when you have completed the task. Verify that you have performed the task correctly. Take appropriate follow-up action.

Obviously, a person could not (and should not) employ this technique for every task – making it through a normal day would become impossible! Therefore, knowing when to use this technique is important. Table 4 lists example questions or cues to facilitate using a self-checking technique. This list is a starting point from which to develop other cues that are specific to a particular work environment.

Table 4 When to Use a Self-checking Technique

1. Are you unsure about the intent of the steps or the task/performance standards?	6. Is this a routine (and maybe even boring) but critical task?
2. Are you confused or do you sense that something is "not right" (e.g., notice conflict between indications)?	7. Can you clearly see/identify the equipment you are working on (e.g., poor labeling, poor lighting, congestion)?
3. Has the task been interrupted, causing you to begin the task again, or is it a departure from a well-established routine?	8. Do you have insufficient indication of system/component status?
4. Have you received verbal instructions on performing a task?	9. Have you had an unexpected encounter with a system, system interlock, or system alarm?
5. Are you hurried or performing several tasks simultaneously?	10. Are you tired/fatigued (e.g., working the midnight shift)?

Element 5: Providing Avenues for Workers to Achieve Their Social and Psychological Needs

People seek satisfaction from life through what psychologist Dr. A. H. Maslow calls the "five basic needs,"[6] which are:

- *Existence needs*: food, drink, air
- *Security needs*: shelter, secure sources of the existence needs, freedom from fear, need for structure in life
- *Social needs*: affection, belonging to a group
- *Esteem needs*: need to be valued by self and others, competence, independence, recognition
- *Self-actualization needs*: self-fulfillment, achievement

Dr. Maslow was one of the first to recognize that various "system factors," such as equipment design, procedures, training, and organizational culture, are keys to motivating workers. Worker motivation will likely be high if (1) management applies accepted human factors principles to job tasks, (2) training provides the necessary skills to handle contingencies, and (3) workers are actively involved in their job through participation strategies. In addition, many first-line managers are given the opportunity daily to shape the work environment and to motivate workers to improve their level of performance. A motivated workforce with a positive attitude is less likely to commit errors such as those associated with internal human characteristics. Therefore, a team leader or supervisor who introduces various motivational factors into the work situation will help reduce the likelihood of human errors. For example:

Recognition for achievement – extending praise in the presence of his peers to the technician who reported a quality concern that led to the discovery of problems in the sampling process

Access to information – allowing and training a maintainer to obtain information on the computer system normally used only by the operators

Utilization of one's ability – allowing an operator who is familiar with computers to provide input into the selection of the hardware and software for the new computerized management system

Challenging assignments – giving the central stores staff the opportunity to assist in determining the optimum distribution of compressed gas bottles to the laboratories

Extended responsibilities – allowing a clerk, who has shown both the interest and the ability, to move into a position (a position normally filled by people with college degrees)

Freedom to act – empowering a technician to requisition new personal safety equipment without prior approval from his immediate supervisor

Involvement in planning, problem solving, or goal setting – inviting workers to participate in incident investigations, budget forecasting, or goal setting

However, beware of outcomes opposite of those intended. For example, lavishly praising "heroic" work in response to a compressor failure may devalue the "boring" preventive tasks that could have prevented the crisis from occurring. Praising individual initiative may encourage workers to bypass formal management of change systems. The supervisor's challenge is to positively reinforce all desired worker behaviors, not just a few special cases.

Conclusion

Most system performance problems and incidents are the result of human error. However, many times these human errors are the result of error-likely situations created in the way (1) systems are designed, operated, or maintained, (2) procedures are written, and (3) training is conducted. Today's management is challenged by a paradigm shift from blaming mistakes on carelessness and incompetence to implementing methods to identify the root cause of why a person made a mistake. Allocating time and resources to understanding human factors, and identifying and eliminating error-likely situations through methods such as the work-situation approach, will significantly help to improve overall system performance and process safety.

Biographies

Donald K. Lorenzo, P.E., is a senior risk and reliability engineer with ABS Consulting, Risk Consulting Division, Knoxville, Tennessee. He earned a master's degree in nuclear engineering from the Georgia Institute of Technology. He authored *A Manager's Guide to Reducing Human Errors: Improving Human Performance in the Chemical Industry* and is Director of the Process Safety Institute.

ABS Consulting, Risk Consulting Division
10301 Technology Drive
Knoxville, TN 37932-3392
Ph: 865-671-5842
Fax: 865-966-5287
E-mail: dlorenzo@absconsulting.com

Lee N. Vanden Heuvel is a senior risk and reliability engineer with ABS Consulting, Risk Consulting Division, Knoxville, Tennessee. He earned a master's degree in nuclear engineering from the University of Wisconsin. His current responsibilities are in the areas of root cause analyses, incident investigations, human factors, safety analyses, and economic/decision analyses. He coauthored the *Root Cause Analysis Handbook* and has provided instruction to thousands of individuals in root cause analysis, sentinel event investigations, and human error reduction techniques.

ABS Consulting, Risk Consulting Division
10301 Technology Drive
Knoxville, TN 37932-3392
Ph: 865-671-5848
Fax: 865-966-5287
E-mail: lnv@absconsulting.com

James J. Rooney, P.E., is a senior risk and reliability engineer with ABS Consulting, Risk Consulting Division, Knoxville, Tennessee. He earned a master's degree in nuclear engineering from the University of Tennessee, Knoxville. Mr. Rooney is a Fellow of ASQ and holds the following ASQ certifications: quality auditor, quality auditor-HACCP, quality engineer, quality improvement associate, quality manager, and reliability engineer.

ABS Consulting, Risk Consulting Division
10301 Technology Drive
Knoxville, TN 37932-3392
Ph: 865-671-5814
Fax: 865-966-5287
E-mail: jrooney@absconsulting.com

References

1. R. E. Butikofer, *Safety Digest of Lessons Learned*, API Publication 758, American Petroleum Institute, Washington, DC, 1986.

2. A. D. Swain and H. E. Guttmann, *Handbook of Human Reliability Analysis with Emphasis on Nuclear Power Plant Applications*, NUREG/CR-1278, U.S. Nuclear Regulatory Commission, Washington, DC, August 1983, with *Addendum #1 to NUREG/CR-1278, August 1983*, September 1, 1985, by A. D. Swain.

3. A. D. Swain, *Design Techniques for Improving Human Performance in Production*, A. D. Swain, Albuquerque, NM, January 1986 (Revised).

4. J. H. Burgess, *Designing for Humans: The Human Factor in Engineering*, Petrocelli Books, Inc. and McGraw-Hill, New York, NY, 1986.

5. D. K. Lorenzo, *A Manager's Guide to Reducing Human Errors: Improving Human Performance in the Chemical Industry*, Chemical Manufacturers Association, Inc. (now the American Chemistry Council), Washington, DC, 1990.

6. *Guidelines for Preventing Human Error in Process Safety*, American Institute of Chemical Engineers, New York, NY, 1994.

Human Reliability Analysis for Chlorine Bulk Unloading Procedures

Joseph J. Chaback
Tulsa County Local Emergency Planning Committee

Prepared for Presentation at
American Institute of Chemical Engineers
2005 Spring National Meeting
Process Plant Safety Symposium
Atlanta, Georgia
April 11-13, 2005

5 - February/2005

UNPUBLISHED

Introduction and Purpose

Provisions of the Superfund Amendment and Reauthorization Act (SARA) Title III of 1986 charge the various LEPC's with preparing an Emergency Response Plan for extremely hazardous substances (EHS) maintained at facilities and carried along transportation routes in their areas of responsibility. Hazard analysis is an important step in preparing the response plan. As part of that analysis, reported here are the results of a human reliability analysis (HRA) for the release of chlorine to the ambient atmosphere during the unloading operation to refill bulk chlorine storage vessels at a municipal water treatment plant. A delivery can lead to the transfer of up to nominally 16,000 kg of chlorine. Thus, potential for a substantial release during unloading exists.

Scenarios used in this study are based on Process Hazard Analysis (PHA) discussions (What If/Checklist Method) with plant staff. Certain procedural documents were also provided by them and include write-ups prepared by the plant staff and the chlorine vendor. These documents were used to develop a composite, simplified but realistic unloading procedure. Although based on a simplified procedure, the results indicate the types of human errors and equipment failures that could lead to a significant release of chlorine. The results also provide estimates of the frequency of such events. The study is limited to scenarios that could lead to releases large enough to have significant off-site consequences.

Principal Findings and Conclusions

1. An overall frequency for a hazardous release of chlorine during unloading is estimated at 2 E-03 events/yr. Such a frequency is equivalent to an event approximately once every 10,000 deliveries or once every 500 years.

2. A risk matrix representation of frequency versus chlorine release rates shows that the largest releases (associated with storage vessel or truck tank failure) have the lowest frequencies while the most likely events have the smallest releases.

3. The most likely failure paths leading to a chlorine release consist entirely of human errors which suggests additional second-party checking for completion of the identified activities as a simple approach for reducing the frequency of release events.

4. The failure path deemed of greatest risk (combined influence of frequency and size of release) also is heavily dependent on human error.

Methodology in Brief

Chlorine is delivered to the water treatment facilities in pressurized tank trucks. Chlorine is then transferred to the storage vessels through a manifold at the plant unloading station using compressed nitrogen to pad the truck tank to facilitate the delivery. Storage vessels are housed in an enclosed building. Figure 1 shows a schematic diagram of the chlorine bulk unloading station and the storage vessel. Two chlorine storage vessels are used; one off-line to receive the shipment and one to provide chlorine for water treatment. The dashed lines to processing are used to show operating mode; those lines are closed during off-line periods such as delivery. During delivery the driver is assumed to follow the composite procedure set out in Table 1 using check-off provisions.

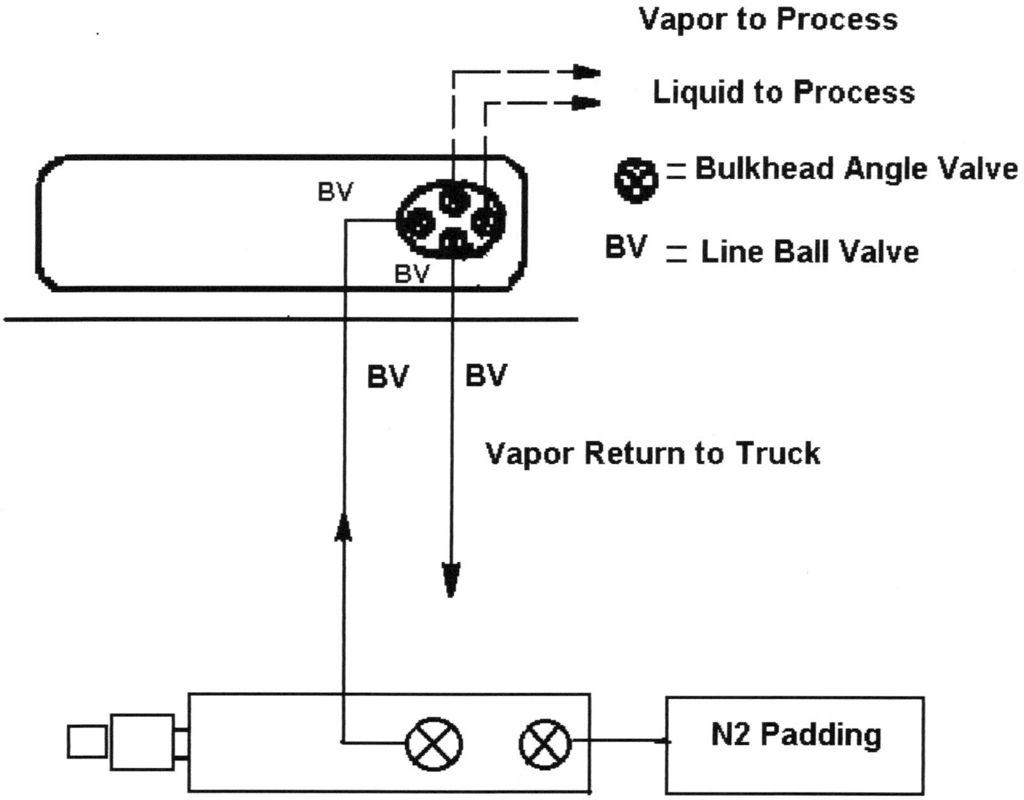

Figure 1. Chlorine Bulk Unloading Station – Schematic

Examination of that procedure suggests several possible scenarios leading to significant releases of chlorine including the six listed below.

1. The wrong storage vessel (that is the one nearly full and currently on-line) is selected to receive the delivery leading to overpressuring of either the storage vessel or truck tank. See Step 12.
2. Liquid chlorine flow is not established to the storage vessel leading to overpressuring of the truck tank. See Steps 12 and 13.
3. A faulty connection to the liquid phase valve on the truck tank or (Scenario 3A) on the unloading station manifold fails. See Step 9.
4. The truck rolls during unloading breaking liquid phase connections from the truck tank to the unloading station. See Step 2.
5. After delivery, the liquid phase valves on the storage vessel are left open or fail open creating a flow path to the atmosphere. See Steps 18 - 20.
6. After delivery, the truck tank liquid phase valve is left open or fails open creating a flow path to the atmosphere. See Steps 18 - 20.

For each scenario event trees were constructed based on the sequence of driver or plant operator errors and equipment failures needed to create the release; therefore, not all actions listed in Table 1 were incorporated into the event trees. Moreover, because of the large number of scenarios only primary failure paths were considered in developing frequencies for release events; the relatively small contributions of secondary paths were not considered.

Table 1. Chlorine Bulk Unloading - Simplified Procedure
(For HRA event tree development only)

Driver is to

1. Check-in and review paper work with operator.
2. Park truck/chock wheels/post warning cone.
3. Verify safety equipment is present and ready for use.
4. Verify proper storage vessel for connecting and unloading.
5. Inspect all hoses and connections to be used for unloading for serviceability.
6. All piping should be clean, dry, free of oil, and properly prepared.
7. Verify safety equipment and personal protective equipment (PPE) are present; an escape respirator is to be worn during all operations.
8. Turn on truck weigh scale.
9. Attach the flexible transfer hoses to the truck connections and to the unloading station connections: liquid to liquid and gas to gas.
10. Pressurize the liquid line up to the unloading station valves by briefly, but slowly opening the liquid phase valves so as not to activate the excess flow valve.
11. Check all connections for leaks.
12. At the bulk storage vessel to be filled open angle valves and ball valves for gas return and liquid inlet. As in Step 10, open valves slowly. Check for leaks.
13. Open the gas and liquid ball valves at the unloading station and check for leaks.
14. Connect nitrogen supply line to compressor, turn on nitrogen and start the compressor.
15. Using the compressor, pad trailer as needed to facilitate unloading.
16. Verify flow by noting a rise on the compressor pressure gauge; recheck all connections for leaks with flow now in progress.
17. Driver is not to leave the area; DOT requires monitoring during unloading.
18. When unloading is complete, turn off the compressor. Purge lines with nitrogen and close valves.
19. Disconnect hoses from the tanker truck taking care to bleed the pressure off slowly.
20. Disconnect hoses from the unloading station.

Those events contributing to a release of chlorine are listed in Table 2; nomenclature follows that given by Swain and Guttmann. The table also lists the probability assigned to each event. Figure 2 shows a portion of the HRA event tree describing Scenarios 3 and 4. Event trees for the other scenarios can be constructed from the event sequences given in Table 2. Table 3 collects the estimated event frequencies based on twenty (20) deliveries per year and also shows estimated release sizes. Deliveries typically take a full shift and are assumed to last ten hours for the purpose of estimating event frequencies.

Discussion

Description of Scenarios

Scenario 1 assumes that flow is initiated to the on-line storage vessel while Scenario 2 assumes that a flow path is not created from the truck tank to the storage vessel. In both cases detection of the no-flow condition or vessel failure is assumed before reaching safety relief valve settings. Therefore, for these overpressure scenarios loss of containment is attributed to

Table 2. HRA Event Tree Component Events

Event Code	Event Description	Probability	Data Source†
Scenario 1	Wrong Vessel Selected for Delivery		
a, A	Wrong vessel selected for delivery	0.003	Table20-13, item 2
r1, R1	Driver fails to detect selection of the wrong vessel	0.01	Table 20-7, item2, EF 3
r2, R2	Operator fails to detect selection of the wrong vessel	0.05	Table 20-22, item 3
φ, Φ	Storage vessel fails due to overpressure	4.E-06	See Text
φ1, Φ1	Truck tank fails due to overpressure	4.E-06	See Text
Scenario 2	Liquid Flow to Storage Vessel Not Established		
b, B	Driver fails to open liquid inlet valves to storage vessel	0.003	Table 20-7, item 2
σ1,Σ1	Liquid inlet angle valve sticks closed	0.001	Table 20-14, footnote *
σ2,Σ2	Liquid inlet ball valve sticks closed	0.001	Table 20-14, footnote *
σ3,Σ3	Liquid inlet unloading station valve sticks closed	0.001	Table 20-14, footnote *
r3, R3	Driver fails to detect no flow to storage vessel	0.003	Table 20-7, item 2
r4, R4	Operator fails to detect no flow to storage vessel	0.05	Table 20-22, item 3
c, C	Driver fails to detect stuck valve	0.005	Table 20-14, item 3
d, D	Driver fails to detect no flow to storage vessel	0.003	Table 20-7, item 2
e, E	Operator fails to detect no flow to storage vessel	0.05	Table 20-22, item 3
φ1, Φ1	Truck tank fails due to overpressure	4.E-06	See Text
Scenario 3	Liquid Phase Valves Improperly Connected		
g, G or g',G'	Liquid phase valve on truck (g,G) or (3A) loading station (g', G') improperly connected	0.01	Irwin et al.
r5, R5	Driver fails to detect bad liquid phase valve connection	0.003	Table 20-7, item 2
λ, Λ	Excess flow valve on truck fails on demand	0.00024	"Offshore"
Scenario 4	Truck Rolls During Unloading		
h, H	Driver fails to chock truck wheels	0.003	Table 20-7, item 2
i, I	Driver fails to set parking brake on truck	0.01	Table 20-7, item 5
γ, Γ	Parking brake on truck fails during unloading	0.0005	"Guidelines, 2000"
λ, Λ	Excess flow valve on truck fails on demand	0.00024	"Offshore"
Scenario 5	Liquid Inlet Valves Left/Fail Open on Storage Vessel		
σ4,Σ4	Liquid inlet angle valve sticks open	0.001	Table 20-14, footnote *
σ5,Σ5	Liquid inlet ball valve sticks open	0.001	Table 20-14, footnote *
σ6,Σ6	Liquid inlet unloading station valve sticks open	0.001	Table 20-14, footnote *
c, C	Driver fails to detect stuck valve	0.005	Table 20-14, item 3
j, J	Driver fails to close liquid phase angle valve	0.003	Table 20-7, item 2
k|J,K|J	Driver fails to close liquid phase ball valve	0.5	Table 20-21, item 4, HD
k,K	Driver fails to close liquid phase ball valve	0.003	Table 20-7, item 2
l|K, L|K l|J, L|J	Driver fails to close liquid phase unloading station valve	0.5	Table 20-21, item 4, HD
m, M	Driver fails to disconnect hoses from truck	0.003	Table 20-7, item 2
n, N	Driver continues disconnect with excessive bleed	0.001	Swain, p. 9-15
Scenario 6	Liquid Phase Valve on Truck Left/Fails Open		
δ, Δ	Liquid phase valve on truck sticks open	0.001	Table 20-14, footnote *
c, C	Driver fails to detect stuck valve	0.005	Table 20-14, item 3
p, P	Driver fails to close liquid phase valve on the truck	0.003	Table 20-7, item 2
m, M	Driver fails to disconnect hoses from truck	0.003	Table 20-7, item 2
n, N	Driver continues disconnect with excessive bleed	0.001	Swain, p. 9-15
λ, Λ	Excess flow valve on truck fails on demand	0.00024	"Offshore"

†All Table citations are from Swain and Guttmann

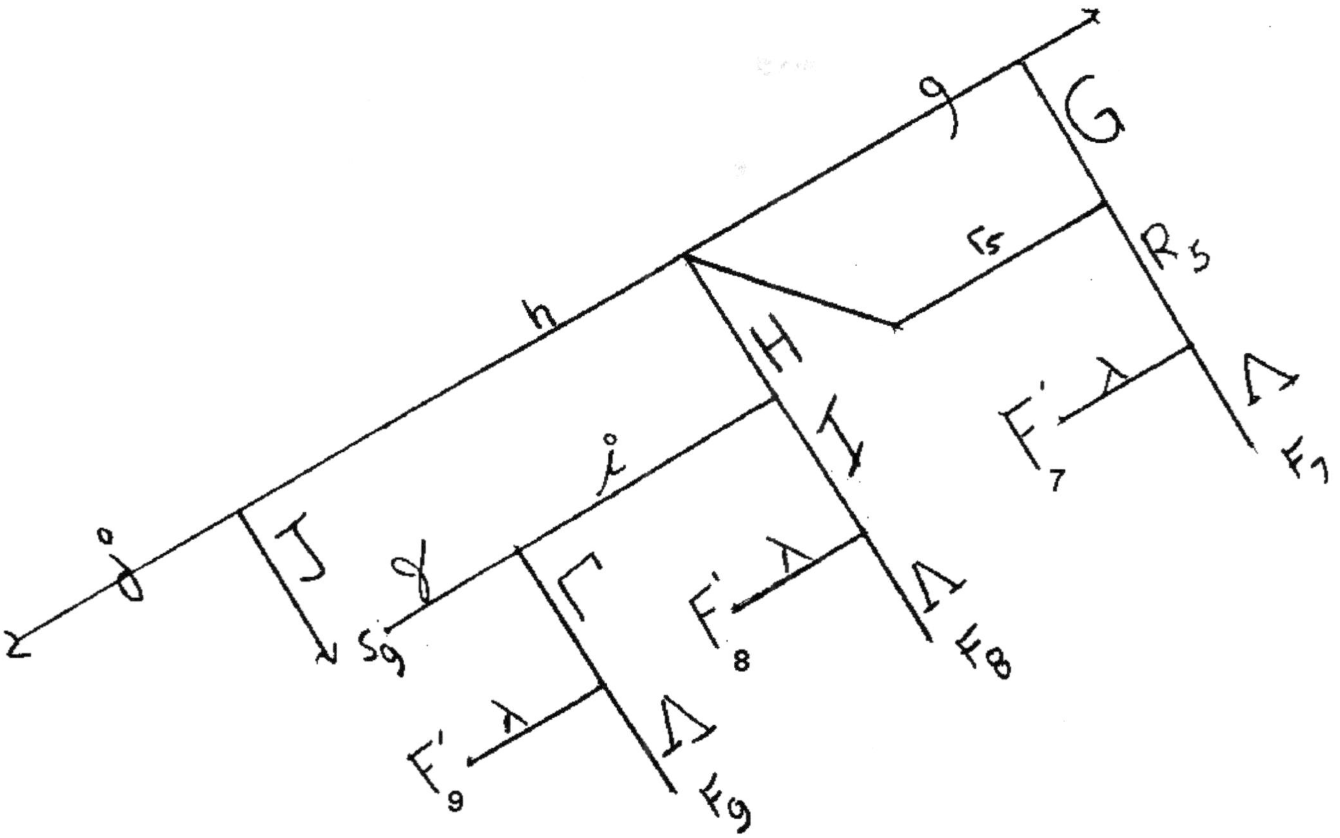

Figure 2. Event Tree Segment for Scenarios 3 and 4

vessel or tank failure at some lower pressure. Several hours of nitrogen padding would be needed to either (1) overfill the on-line storage vessel, maintained at no more than an 80% full condition, or (2) overpressure the truck tank. Moreover, while a local hazard, pressure relief valve events are not as likely to create significant off-site consequences and are, therefore, of lesser interest in this study.

In Scenario 3 either a faulty liquid connection at the truck or the unloading station fails. The error recovery factor for detecting the poor connection, item (r-5, R5) in Table 2, is based on omission of the leak check in Step 11; the connection subsequently fails because of the vibration of sustained flow out of the truck tank before the leak check in Step 16 is carried out. In Scenario 4, the study assumes that either the chocks or the parking brake are sufficient to keep the truck from rolling and that the truck will roll if neither brake nor chocks are deployed. Only event sequences leading to both the omission of the chocks (they are considered completely reliable once deployed) and failure of the parking brake either by error of omission or mechanical failure are considered in the event tree.

In both Scenarios 3 and 4 failure of the truck tank liquid phase valve connection is assumed to occur sufficiently close to the truck that the resulting increased flow is sufficient to activate the excess flow valve which then either seals flow or fails leading to a release from the truck tank. Moreover, because unloading is already in progress in both scenarios at the time of failure, it is assumed that flow out of the storage vessel through the unloading station will also ensue because insufficient flow is created to trigger its excess flow valve.

In Scenario 5 the driver closes the liquid phase valve on the truck tank but then either fails to close the liquid phase valves associated with the storage vessel or they stick open. Subsequently, the driver either leaves without attempting to disconnect the liquid phase line from the truck or, alternatively, during disconnection continues in the face of excessive bleed off of chlorine. Chlorine is released from the storage vessel at or below the excess flow valve setting for both drive-off and disconnect cases. Furthermore, given the failure to close the angle valve or the ball valve at the storage vessel, high dependence is assumed for the subsequent valve closures that are part of the storage vessel system. This dependence is reflected in the high probabilities assigned to failures "K|J", "L|J", and "L|K" in Table 2.

In Scenario 6 the truck tank is the source of the chlorine released; the valves on the storage vessel and unloading station are assumed to be closed according to the procedure, Step 18 of Table 1. Thus, only the residual liquid in the truck tank vents because the liquid phase valve on the truck is either left open or sticks open and the driver either continues the disconnection despite the excessive bleed-off or, alternatively, the driver fails to disconnect the hoses and drives off breaking the line connections.

For the case of improper disconnect, the connection nearest the truck is opened first; sustained release of chlorine occurs only if the excess flow valve fails leading to the high release rate given in Table 3. For the case of drive-off, a sufficient run of line is assumed attached to the truck such that venting chlorine from the truck tank occurs at or below the excess flow valve setting.

<u>Comments on Failure Data</u>

Table 2 lists the events contributing to chlorine releases during delivery and the estimated probability of those events. Data for human errors of omission and commission are, with two exceptions, taken from the compilation given by Swain and Guttmann; Table 2 cites the specific location of the data in their compilation.

Several of the entries in Table 2 require further comment. Note that an error factor of 3 is applied to the recovery factor R1 to account for the delayed appearance of alerting factors to selection of the wrong vessel. For that case flow would initially appear normal through the unloading lines. Only later would alerting factors such as pressure build up become evident. Error B treats opening the liquid phase valves from the unloading station to the storage vessel as a single step and failure to open all of them as a single error of omission. For Error I the probability in Table 2 is the adjusted value suggested by the footnote in Swain and Guttmann to item 5 of Table 20 - 7 for actions regarded as "second nature", such as setting a brake.

The two error probabilities not taken from Swain and Guttmann are used in Scenarios 3 and 3A for the faulty liquid phase valve connections, items (g, G) and (g', G'), and in Scenarios 5 and 6 for the continued disconnection of the hoses even though there is excessive chlorine bleed off, item (n, N). The datum for the faulty connection is taken from Irwin et al. Halting the disconnect operation in the face of excessive bleed off is considered a "skill of the craft" or "second nature" reaction and could be assigned a negligible error probability. However, consistent with an HRA convention suggested by Swain, a value of 0.001 is assigned to account for the increased chance of error the stress such an unexpected event could cause the driver.

Table 2 also shows entries for four mechanical failures: valves sticking open or closed, excess flow valve failure on demand, parking brake failure on demand, and storage vessel or truck tank failure when exposed to higher than normal pressures. The valve failure rate (denoted by σ, Σ or δ, Δ pairings) is taken from the first footnote of Table 20 - 14 in Swain and Guttmann. The excess flow valve datum (denoted by λ, Λ) is taken from "Offshore".

The parking brake failure rate (denoted by γ, Γ) is estimated from generic data given in "Guidelines, 2000" for "mechanical equipment". These data range from 0.1 to 1000 failures per 10,000 hours. A low-end value of 0.5 failures/10,000 hrs is selected for this simple braking system. To convert this failure rate datum to a probability per unloading, it is multiplied by 10 hours per delivery for a probability of failure during unloading of 0.0005.

Table 3. Estimated Frequencies and Release Rates for Loss of Containment Scenarios

Scenario	Failure No.	Path	Probability/Delivery	Events/yr	Release Rate, kg/min
1	F1	A*R1*R2*Φ	6.0E-12	1.2E-13†	816
	F2	A*R1*R2*φ*Φ1	6.0E-12	1.2E-10	1600
2	F3	B*R3*R4*Φ1	1.8E-12	3.6E-11	1600
	F4=F5=F6	Σ1*C*D*E*Φ1	3.0E-15	6.0E-14	1600
3	F7	G*R5*Λ	3.6E-07	1.4E-07	443
	F7'	G*R5*λ	3.0E-05	6.0E-04	53
3A	F7"	G'*R5	3.0E-05	6.0E-04	105
4	F8	H*I*Λ	7.2E-09	1.4E-07	443
	F8'	H*I* λ	3.0E-05	6.0E-04	53
	F9	H* i*Γ*Λ	3.6E-10	7.1E-09	443
	F9'	H* i*Γ* λ	1.5E-06	3.0E-05	53
5	F10	J*K|J*L|J*M	2.3E-06	4.5E-05	53
	F11	J*K|J*L|J*m*N	7.5E-07	1.5E-05	53
	F12	Σ4*C*K*L|K*M	2.3E-11	4.5E-10	53
	F12'	Σ4*C*K*L|K*m*N	7.5E-12	1.5E-10	53
6	F13	Δ*C*M	1.5E-08	3.0E-07	53
	F14	Δ*C*m*N*Λ	1.2E-12	2.4E-11	390
	F15	P*M	9.0E-06	1.8E-04	53
	F16	P*m*N*Λ	7.2E-10	1.4E-08	390
Total			1 E-04	2 E-03	

†This value reflects an additional factor of 0.001 to reflect the additional safety measures in-place to mitigate a chlorine release in the chemical storage facility housing the storage vessel.

Failure of either storage vessel or truck tank (denoted by φ, Φ pairings) when stressed at higher than normal pressure was estimated using the data in "Guidelines, 1989", Table 3.6.2.1. The upper bound value for failure of 0.0424 failures/1E06 hours was selected as a base value. This value is then multiplied by a factor of 100 to reflect increased risk of failure due to a rise in pressure above the normal operating condition. One hour of exposure to failure conditions per delivery was assumed. These considerations lead to an estimated probability of tank or vessel failure per delivery of 4E-06.

Comments on Failure Probabilities and Release Rates

The individual failure probabilities shown in Table 3 lead to an overall probability of a chlorine release of approximately once every 10,000 deliveries or for 20 deliveries per year once every 500 years. Note, too, that the largest contributors to the delivery failure rate occur during the connection and disconnection processes of Failure Paths 7', 7" of Scenario 3 and Failure 8' of Scenario 4, and Failure 15 of Scenario 6. Note further that these paths contain only human errors; no equipment failures are involved.

Thus, one simple event frequency reduction measure that could be taken without great effort is to add a recovery factor for those errors likely to cause the failures. One approach might be to have the plant operator check off each step as the driver carries it out. According to the data presented by Swain and Guttmann in Table 20-22 such redundancy would reduce the likelihood of a release by a factor of 20 just by having the operator verify that the driver has (1) inspected the liquid phase pipe and hose connections for wear or damage, (2) properly connected and disconnected the truck from the unloading station and (3) properly secured the truck. By also insuring the liquid phase valve on the truck is closed, the frequency of Failure Path15 could be reduced by an additional factor of 20 for a 400 fold reduction in frequency.

Table 3 also shows the estimated release rates arising from the failure paths developed in this analysis. The release rate for Failure 1 assumes a worst case event with the loss of the entire contents of a full storage vessel, approximately 22,500 kg, mitigated only by its occurrence in the enclosed chemical storage building; other mitigation measures such as activation of a scrubber system are also assumed to fail. Release rates for Failures 2-6 assume the entire contents of a nearly full truck tank are released over a ten minute period. Note that Failures 4-6 assume at least one of the valves needed to establish a flow path from the truck tank to the storage vessel is left closed; Table 3 shows the angle valve case.

For Failures 7, 8, 9, 14 and 16 the line break leading to the release is assumed to occur close enough to the truck that the excess flow valve, if it functions, would experience a flow rate sufficient to cause it to seal. Therefore, the ensuing flow rate for a failed valve would be that of two phase liquid flow out of a short run of pipe. "EPA", Exhibit 4-13, gives estimated flow rates for differing lengths of pipe. The maximum value of 390 kg/min is used and corresponds to a 25.4 cm run of 2.54 cm diameter pipe.

For Failures 7 – 9, 10 and 12 an additional 53 kg/min is assumed to vent from the storage vessel out the unloading station manifold because a sufficiently long run of pipe is in place to allow flow without triggering the excess flow valve. Flow is taken at the storage vessel setting of 3175 kg/hr (53 kg/min). Similarly, for Failures 13 and 16 the same excess flow setting is assumed for the truck tank excess flow valve.

The frequencies and release rates are shown in risk matrix format in Figure 7. Note that those events with the greatest release rates are associated with the smallest frequencies. As noted in Table 3 for Scenario 1, the smaller frequency and release rate for storage vessel failure reflects the subsequent failure of in-place safety equipment on frequency while the 816 kg/min rate includes the mitigating effect of an indoor release. The risk matrix also suggests that while Failures 7', 7", 8' and 15 have the greatest frequencies, Failure 7 may present the greatest overall risk because of its greater chlorine release rate of 443 kg/min. Its frequency would also be reduced by the additional operator checks suggested above.

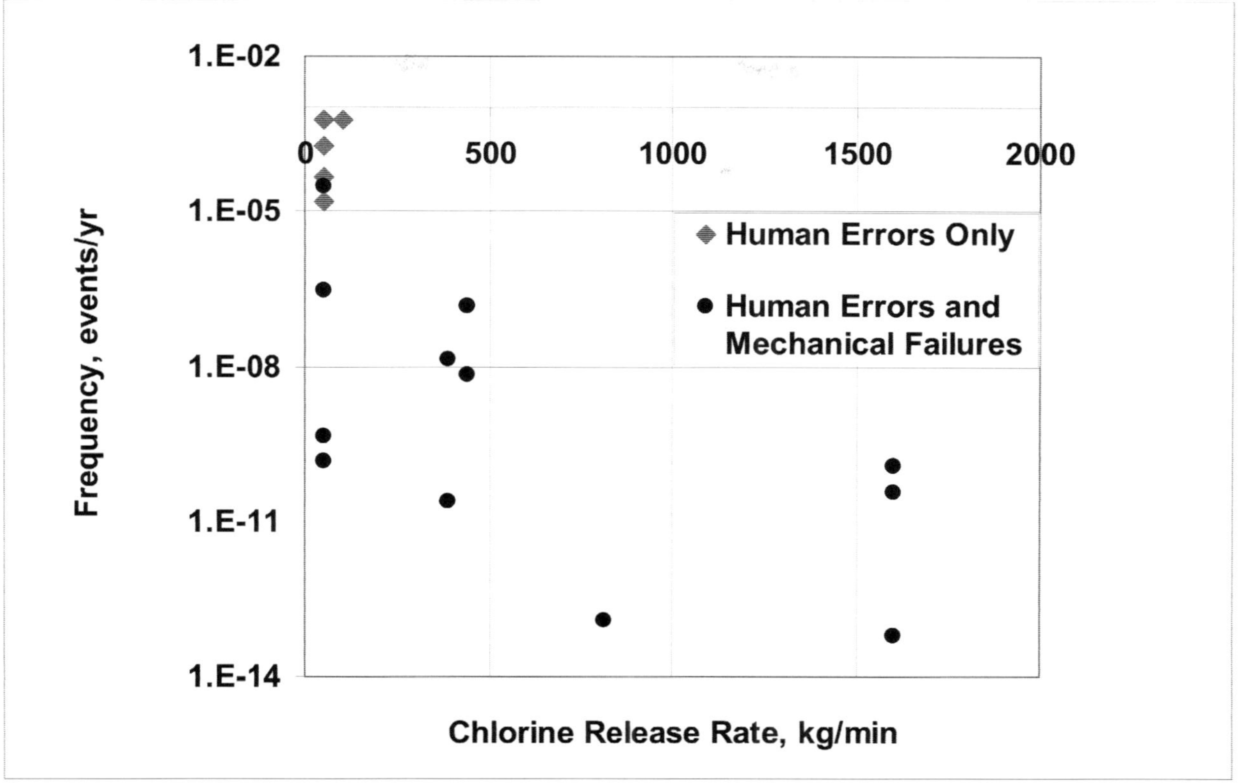

Figure 3. Risk Matrix for Chlorine Bulk Unloading Loss of Containment Scenarios

References

A. D. Swain and H. E. Guttmann, "Handbook of Human Reliability Analysis with Emphasis on Nuclear Power Plant Applications- Final Report", U. S. Nuclear Regulatory Commission, Washington D. C., August,1983, NUREG/CR-1278.

I. A. Irwin, et al., "Human Reliability in the Performance of Maintenance", Aerojet General Corp., Sacramento, CA, May 1964.

A. D. Swain, "Human Reliability Analysis", PSI Course 204 Notes, Process Safety Institute, Knoxville, TN, 2003.

"Offshore Reliability Data Handbook", Hovik, Norway: OREDA, 1984.

"Guidelines for Chemical Process Quantitative Risk Analysis", 2nd Edition, Center for Chemical Process Safety, AIChE, New York, NY, 2000, page 500.

"Guidelines for Process Equipment Reliability Data", Center for Chemical Process Safety, AIChE, New York, NY, 1989.

EPA 550-B-00-007 Risk Management Program Guidance for Wastewater Treatment Plants, May 2000, Chapter 4.

Safety Climate: It's NOT Just an Engineering Problem

Stephanie C. Payne, Mindy E. Bergman, and Jaime B. Henning
Department of Psychology, Texas A&M University

Prepared for Presentation at
American Institute of Chemical Engineers
2005 Spring National Meeting
Process Plant Safety Symposium
Atlanta, Georgia
April 11-13, 2005

(January 2005)

UNPUBLISHED

Abstract

The purpose of this paper is to provide a review and summary of the psychological literature and research regarding safety climate for non-psychologists. In psychology, important psychological construct distinctions are made between safety climate and safety culture (e.g., Rousseau, 1990; Schein, 1990). Safety climate is defined as shared employee perceptions of the policies, procedures, and practices concerning safety (Zohar, 2003). Policies describe organizational goals and means for goal attainment, whereas procedures provide tactical guidelines for actions related to these goals and means. Both of these exist at the organizational level. Practices are the implementation of policies and procedures by managers within each work group and therefore, describe a group level construct. It is these practices that seem especially important because not only do they guide everyday behavior of employees as supervisors structure the day-to-day work and work settings, but also because practices highlight any disconnect between what the company claims is important (i.e., policies and procedures) and what the organization's agents (i.e., supervisors) model as important. Safety climate can be understood by examining employees' perceptions of the types of safety behaviors that are rewarded, supported, and expected (Reichers & Schneider, 1990). Perceptions of safety climate can influence the way employees behave and perform their work, which can have a direct impact on safety outcomes such as accidents, near misses, and everyday safety behaviors.

Generally, definitions of safety culture refer to shared underlying values and assumptions that organizational members hold regarding safety issues. Culture is thought to be less observable and accessible than climate and therefore culture may be more difficult and slow to change in response to environmental events. We need to consider both the surface characteristics of the practices engaged in at an organization and the perceptions of these that employees hold (climate) as well as the underlying values of the organization (culture) when we want to change the safety situation at a company. Although it is possible to change safety climate without changing the underlying culture, enduring changes to a safety climate will likely require concomitant changes in the culture as well.

The level and the strength of safety climates are also differentiated and research examining each of these is reviewed. Whereas level of safety climate refers to the relative priority of safety signified by climate perceptions, climate strength refers to the amount of consensus surrounding climate perceptions. Additionally, there are a number of other variables that are likely to contribute to safety climate, safety behavior, and accidents which are not actually part of safety climate. Our review of some of these other factors focuses mainly on environmental characteristics and individual differences. These antecedents, correlates, and consequences of safety climate and safety behavior are discussed in an effort to demonstrate the systematized nature of safety issues. That is, it is essential to psychology's view of safety that safety is not a simple phenomenon that occurs in a vacuum, or that safety behavior can be easily affected by changing one feature of the workgroup or workplace. Rather, there are a variety of factors affecting safety in the workplace that dynamically influence safety climate, safety perceptions, and safety behavior.

Safety Climate: It's <u>NOT</u> Just an Engineering Problem

Industrial/Organizational (I/O) psychologists study human behavior and psychological processes in the workplace. We have been studying organizational climate and culture for over 30 years (Hellriegel & Slocum, 1974; James & Jones, 1974) and safety climate and culture for approximately 25 years (e.g., Zohar, 1980). The purpose of this paper is to review and summarize the psychological literature and research regarding safety climate. In doing so, we hope to demonstrate: (a) the "state of the science" of safety climate in the psychological literature and (b) that safety in organizations depends in large part on human behavior and psychological processes. This paper is intended as a review for non-psychologists, providing an overview of relevant psychological terminology as well as a look at the empirical research on safety climates.

Organizational Culture vs. Organizational Climate

I/O psychologists distinguish between organizational culture and organizational climate (e.g. Rousseau, 1990; Schein, 1990). *Organizational culture* refers to the shared underlying values and assumptions that organizational members hold. It is more concerned with the evolution of organizational systems over time, "thus it is rooted in history, collectively held, and sufficiently complex to resist many attempts at direct manipulation" (Denison, 1996; p. 644).

Organizational climate refers to shared employee perceptions of relevant policies, procedures, and practices, which are indicators of desired role behavior. *Policies* describe organizational goals and means for goal attainment (e.g., "We strive to produce the most innovative products in a safe manner."). *Procedures* provide tactical guidelines for actions related to these goals and means (e.g., "Protective gloves are to be worn by all employees in the warehouse."). *Practices* describe the implementation of policies and procedures by managers/supervisors within each work group (e.g., Supervisors reprimand employees who do not wear their protective gear in the warehouse). Thus organizational leadership has a strong influence on organizational climate. Because supervisors structure the day-to-day work and work settings, practices are especially important because they guide the everyday behavior of employees. Practices also highlight any disconnect between what the company claims is important (i.e., policies and procedures) and what the organization's agents (i.e., supervisors) model as important. Organizational climate is important in understanding employee behavior because organizational climate is positively related to role behavior, as described in a wide and varied theoretical and empirical literature (Bandura, 1986; Katz & Kahn, 1978; Vroom, 1964). Climate perceptions serve an adaptive function for employees by providing information about behavior-outcome expectancies. In other words, an organization's climate communicates to employees the probable consequences of working in a particular manner (e.g., safely vs. quickly).

Culture is thought to be less observable and accessible than climate and, therefore, can only be studied through qualitative methods (Denison, 1996) such as in-depth interviews with various organizational stakeholders. Therefore culture may be more difficult and slow to change in response to environmental events. Both the surface characteristics of the practices engaged in at an organization and the perceptions of these that employees hold (climate) as well as the underlying values of the organization (culture) need to be assessed and addressed when one wants to change the safety situation at a company. Although it is possible to change organizational climate without changing the underlying organizational culture, enduring changes to the climate will likely require concomitant changes in the culture as well.

In order to fully understand organizational climate and the research on safety climate, one must be familiar with the issue of *levels of analysis*. Levels of analysis include but are not limited to the individual, the work group, and the organization. It is important to note that the number of possible levels of analysis will differ across organizations, as different organizations have different structures and various divisions, units, businesses, and the like; terms like "group" do not necessarily reflect only the immediate workgroup to which an employee belongs but rather an identifiably related set of people. Psychological constructs examined in an organizational context exist at more than one level of analysis and these distinctions are conceptually and empirically meaningful. Referring back to the definition of organizational climate, policies, procedures, and practices do not exist at the same level of analysis. Policies and procedures exist at the organizational level as they are formulated by senior management and apply to everyone in the company. Practices (or the execution of policies and procedures) exist at the sub-unit level (e.g., work group). This suggests that organizational climate is a hierarchical construct such that policies and procedures operate at a higher level (organizational) than practices (group). Zohar and Luria (in press) propose that group level variables (e.g., organizational practices) are more proximal to employee behavior so they will have a stronger influence on employee behavior. This implies that changes in practices are likely to have a stronger impact on employee behavior than changes in policies or procedures. This also suggests that organizational practices mediate the relationship between organizational policies and procedures and subsequent employee behavior (Zohar & Luria, in press).

Distinct from this idea of levels of analysis is the *level* of organizational climate, that is, the mean level of importance held by the object of the climate (e.g., how important is the object of the climate?). Within an organization, different groups of people can have different levels of organizational climate. These different levels are likely to be influenced by a variety of factors, including sub-unit differences in practices implementing the overall organizational policies and procedures. Another property of climate is *strength*, or the extent to which a group shares the same climate perceptions. Climate strength reflects the amount of homogeneity there is in employees' perceptions' of the policies, procedures, and practices concerning the object of the climate. When employees report similar perceptions this suggests the organization has a strong organizational climate.

A given organization is likely to have many organizational climates. I/O psychologists have examined organizational climates for a wide variety of foci to include climate for achievement (Litwin & Stringer, 1968), safety (Zohar, 1980), service (Schneider, Bowen, Ehrhart, & Holcombe, 2000), innovation (González-Romá, Peiró, & Tordera, 2002), sexual harassment (Hulin, Fitzgerald, & Drasgow, 1996), and diversity (Kossek & Zonia, 1993). Thus, a referent is needed whenever organizational climate is discussed (Schneider & Reichers, 1983). Although there are many general lessons about organizational climate that can be taken from research in one area and applied to others, there are also specific antecedents, consequences, and interventions that are relevant to a particular domain. We turn next to a discussion of safety climate and then a review of the safety climate literature.

Safety Culture vs. Safety Climate

The concepts of organizational climate and culture can be applied to the safety sphere. *Safety culture* refers to the shared underlying values and assumptions that organizational members hold regarding safety issues. *Safety climate* refers to shared employee perceptions of the policies, practices, and procedures concerning safety. Consistent with this definition, Zohar (2000) recently offered a more parsimonious definition of safety climate as *the relative priority of safety in an organization*. Organizations often have competing operational goals and one goal that often competes with safety is productivity. Thus, the *level of safety climate* reflects the priority given to safety vs. productivity. *Strength of safety climate* reflects the extent to which relevant groups agree on these relative priorities. Perceptions of safety climate can influence the way employees behave and perform their work, which can have a direct impact on safety outcomes such as accidents, near misses, and everyday safety behaviors. Both level and strength of safety climate can influence employee behavior. Additionally, it is important to note that the inherent risk in an organization or a job will affect safety climate and culture.

The *level of safety climate* is typically operationalized as the mean of employee responses within a given workgroup to a series of safety climate items. This creates a metric that can be used to make comparisons between work groups, such that "Work Group A" with a score of 3.5 has a lower level of safety climate than "Work Group B" with a score of 4.5. This could be further interpreted as employees in "Work Group A" perceive that safety is less of a priority than employees in "Work Group B." Safety climate strength, as an indicator of consensus about the climate level, is typically operationalized as the level of interrater agreement between the respondents, or the standard deviation/variance among the items aggregated to form the safety climate score; although Zohar and Luria (in press) recently advocated for the use of the standard deviation among the items. The larger the variability in responding, the more heterogeneity there is in employees' perceptions, suggesting a weaker climate. Thus, a strong climate is characterized by very little variability in employee responses to the climate items, regardless of the level at which they rate those items.

Climate strength (variability) and climate level (mean) are somewhat independent, although the two may be related. An organization can have a strong safety climate that is low or high or a weak safety climate that is low or high. Research examining these two characteristics has shown that climate level tends to yield bigger effect sizes when predicting safety behavior than climate strength (Zohar & Luria, in press).

Theoretically, climate strength is predicted by the consistency of supervisory practices and decisions (Zohar & Luria, 2004). So, the strength of a work group's safety climate should largely be determined by the extent to which the work group's supervisor uniformly implements policies and procedures concerning safety. For example, if a company has a safety procedure of all employees wearing hard hats in the warehouse, the strength of the safety climate in that warehouse can be predicted by the extent to which supervisors enforces this procedure (i.e., put it into practice) every day. This includes praising employees who consistently abide by the rule and reprimanding employees who deviate from it. Thus, consensus about safety climate arises from stability and a lack of ambiguity when it comes to safety practices (Zohar & Luria, in press). It is also enhanced by procedural routinization and formalization (Zohar & Luria, in press). Researchers examining safety climate strength have found that it moderates the relationship between climate level and safety behavior (Zohar & Luria, 2004).

Safety Climate Measurement

Because safety climate reflects employee perceptions of the policies, practices, and procedures concerning safety, the best way to assess safety climate is to survey a representative sample of employees within a given work group and/or organization. In addition to asking employees about their perceptions of the policies, procedures, and practices concerning safety, employees should also be asked to report perceptions of the types of safety behaviors that are rewarded, supported, and expected (Reichers & Schneider, 1990). Other predictors of safe behavior (e.g., risk taking tendencies, etc.) should also be assessed, to examine the relative influence of safety climate on safety behavior. We elaborate on this later.

An examination of the literature on safety climate revealed numerous attempts to measure this construct. Most safety climate researchers have measured safety climate with a self-report questionnaire. Such inventories typically consist of a series of statements (e.g., "Top management in this plant reacts quickly to solve the problem when told about safety hazards." Zohar & Luria, in press) that employees respond to on a five-point agreement scale (1 = strongly disagree, 5 = strongly agree). Sometimes respondents are asked to indicate how descriptive each statement is of their work environment (e.g., "Workers who behave safely have a higher chance for promotion than those who don't." Zohar, 1980) on a five-point extent scale (1 = not at all descriptive, 5 = to a very great extent). The number of items utilized to capture safety climate have ranged from nine (Dedobbeleer & Beland, 1991) to forty (Zohar, 1980). Unfortunately, some of these items have been criticized for being unclear, double-barreled, and inappropriate to use as indicators of safety climate (Mueller, DaSilva, Townsend, & Tetrick, 1999). There have been so many different self-report measures of safety climate used in the literature that multiple articles have been written evaluating safety climate measures (Flin, Mearns, O'Connor, & Bryden, 2000; Guldenmond, 2000) and measurement models (Mueller et al., 1999). The items in these many inventories also tend to vary in terms of how industry-specific they are (Zohar, 2003), which sometimes limits their usability in other contexts. Thus, no one measure of safety climate has been identified as the most valid or useful.

Most safety climate researchers agree that safety climate is multidimensional, but the number of factors remains open to debate. As many as nine factors (Zohar, 1980) of safety climate and as few as one (Hoffman & Stetzer, 1996) have been proposed within one article, and empirical evidence has supported up to eight (Zohar, 1980). However, the robustness of these factors is suspect, as researchers do not always find empirical support for the same factors across samples. The numerous proposed dimensions of safety climate include: (a) perceived importance of safety training programs; (b) perceived management attitude toward safety; (c) perceived effects of safe conduct on promotion as well as (d) social status; (e) perceived level of risk at the workplace; (f) perceived effects of required work pace on promotion; perceived status of (g) the safety officer, (h) the safety committee, and (i) safety issues; worker (j) involvement, (k) competence level, (l) safety knowledge; (m) communication flow; and (n) rewards for working safely (Mueller et al., 1999; Zohar, 1980). It is important to reiterate at this juncture the meaning of safety climate: shared employee perceptions of the policies, practices, and procedures concerning safety. Thus, some of the factors previously proposed in the literature, such as worker characteristics, attitudes, and personal beliefs about safety, are not appropriately part of safety climate and have been confounded in the measurement of safety climate (Mueller et al., 1999; Neal & Griffin, 2004; Schneider & Reichers, 1983). This is not to say these factors are irrelevant; on the contrary, they likely are important determinants to safety behavior. But it is to say that they are not safety climate per se.

One relatively consistent empirical finding is the identification of a first order global factor described most frequently as management commitment to safety (Griffin & Neal, 2000; Zohar & Luria, in press). Some safety climate researchers have suggested that the factor structure of safety climate is not stable across organizations even when the organizations are from the same industry (Coyle, Sleeman, & Adams, 1995). Others have noted that a single factor is most useful for between-workgroup comparisons (Hoffman & Stetzer, 1996), whereas multiple factors are more useful for organizational safety diagnostics (Mueller et al., 1999). Recent research by Zohar and Luria (in press) on production workers in a manufacturing plant utilized a measure that differentiates organizational-level climate components (policies and procedures) from group-level climate components (practices) and includes three factors for each: active practices, proactive practices, and declarative practices. Additional research is needed to determine if this measurement approach is appropriate with other samples and in other industries.

In sum, the literature on safety climate measurement has produced a number of different assessments, none of which have emerged as the best practice. Various factors have been proposed, not all of which validly reflect the basic definition of safety climate, thus clouding the discussion of measurement issues. What is clear, as we review in the next section, is that although there is not currently a set of preferred methods to assess safety climate, the measures that do exist are doing an adequate job in capturing some of the safety climate construct and are related to safe behaviors, near misses, and accidents in organizations.

Safety Climate Research

Safety climate research has been conducted in a variety of industries, including manufacturing (Brown & Holmes, 1986; Zohar & Luria, in press), construction (Dedobbeleer & Beland, 1991), clerical and service (Coyle, Sleeman, & Adams, 1995), the military (Zohar & Luria, 2004), production (Brown & Holmes, 1986), off-shore environments (Cox & Cheyne, 2000), and students from a wide range of jobs (Mueller et al., 1999).

Employee ratings of safety climate have been found to correlate significantly with ratings of organizational safety program effectiveness made by safety inspectors (Zohar, 1980) and expert ratings of workplace hazards (Varonen & Mattila, 2000). Perceptions of safety climate have also been significantly associated with accidents at the individual (Brown & Holmes, 1986), group (Hofmann & Stetzer, 1996), and organizational (Varonen & Mattila, 2000) levels of analysis.

I/O Psychology's view of safety is that safety is not a simple phenomenon that occurs in a vacuum, or that safety behavior can be easily affected by changing one feature of the work, work group, or workplace. Rather, there are a variety of factors affecting safety in the workplace that dynamically influence safety climate, safety perceptions, and safety behavior.[1] Thus we briefly discuss a number of other variables that are likely to contribute to or be affected by safety climate and safety behavior or accidents but are *not* a part of safety climate itself. These antecedents, correlates, and consequences of safety climate and safety behavior are discussed in an effort to demonstrate the systematized nature of safety issues. There are a multitude of these factors, so it is not possible to review the entire range. However, there are two general categories that have been studied with some frequency: environmental characteristics and individual characteristics.

Environmental characteristics

Environmental characteristics related to safety behavior include characteristics of the job as well as the organization. Job characteristics include the inherent risk involved when completing the job. Certainly jobs that require employees to utilize sharp tools are more risky than jobs that do not require the use of such tools. Standardization of the work processes and work output are also of particular relevance. Standardization of work processes imply that the procedure for generating the output are specified in advance and there is no room for individual discretion as to how the work is performed. The work is designed in such a way that the same process is followed no matter who is performing the job. Standardization of work output means that the results of the work are specified in advance. The work is designed in such a way that the same output is achieved irrespective of differences in time, location, or the worker generating the output.

Organizational characteristics related to safety behavior include the nature of the supervision provided as well as the design of the work (Neal & Griffin, 2004). The nature of the communication about safety-related issues within the organizational hierarchy is also relevant. Communication can sometimes be limited to top-down or bottom-up, or be more bi-directional. The availability of safety training and equipment also play a role in safety behavior.

Individual characteristics

Individual characteristics related to safety behavior include employee attitudes toward safety (e.g., individual responsibility, personal immunity, fatalism, optimism), the job (e.g., job satisfaction), and the organization (e.g., organizational commitment). They also include dispositional variables like neuroticism, conscientiousness, locus of control, risk-taking tendencies, and Type A behavior patterns (Neal & Griffin, 2004). Employee knowledge and motivation about safety are also important individual characteristics.

Employee risk perceptions are evaluations of the environment, which may be influenced by exposure to hazards, work experience, and distraction from task activity (McLain, 1995). These risk perceptions are likely to be related to the risk inherent in their job or organization. It is important to recognize that while all organizations have some safety climate, whether strong

---

[1] Research suggests that safety behavior can be divided into two categories: (1) compliance (obeying safety regulations, following the correct procedures, and using appropriate equipment) and (2) participation (behavior that does not directly contribute to the individual's personal safety, such as volunteering to attend safety meetings) (Neal & Griffin, 2004). Most research has focused on compliance.)

or weak, high or low, all organizations also have differential needs for safety and different ways of protecting their workers. But these risk perceptions are also likely to be influenced by safety climate as well as organizational responses to unsafe incidents or accidents. Thus, it is important to note that employee risk perceptions, just like safety climate, are constantly in flux and affected by numerous events in one's work life.

As noted above, some of these environmental and individual characteristics have been included in safety climate measures. However, these are distinct from safety climate. These distinctions are important for a variety of reasons. First, some of these factors are not easily changed within environments, such as job characteristics, or within individuals, such as their dispositions. However, an organization can, for example, change the dispositional make-up of its workforce through selection procedures. Second, some of these factors—like many psychological phenomena, which are considered to be open systems (Katz & Kahn, 1978) — are multiply determined and affected by factors that are not (at least not directly) safety-related. For example, employees' attitudes about quitting an organization could be affected by their sense of belonging to the organization, but could also be affected by alternate job opportunities and the cost of quitting the job, such as losing stock options or benefits (Meyer & Allen, 1997). Finally, some of these factors are more useful for evaluating whether an organization's safety climate is operating as the management would want it to. Employee knowledge about and motivation toward safe behaviors is likely to be influenced by safety climate. A good safety climate at an organization will include policies, procedures, and practices of regular safety training and communication, as well as "check-ups" on safe behaviors. Thus, the level of employee knowledge and motivation may make suitable criteria for the evaluation of safety climate interventions.

Summary and Suggestions

This paper has been presented in an attempt to introduce I/O psychology's point of view regarding safety climate to the non-psychologist community. To that end, we have reviewed the general concepts of culture and climate, as well as how those concepts play out in safety research. We have touched on the literature regarding safety climate measurement, as well as some of the other factors that influence safety climate and safety behavior in a dynamic way. Our goal was to demonstrate that there is an important human side to safety in organizations and to review the "state of the science" in regards to safety climate. Based on this—albeit brief—review, it should be clear that psychology's interests in safety have paralleled interests in other fields, such as engineering.

Just as engineers recognize that the human element is important in creating safe organizations, psychologists realize that equipment and technical systems are also essential to creating safe organizations. However, we suspect that what might be occurring in organizations is a lack of alignment among various stakeholders when attempting to make the organization safer. The Human Resources Department of a company often employs, or has access through consultancy contracts, I/O psychologists who can bring their expertise to the table when safety questions arise. However, we contend that safety issues might arise less often if psychologists were included earlier when industrial processes are being created or changed. It may be the case that small changes in a planned design could reap great benefits in safety if the human elements are considered *before* the plan is implemented; it can be difficult to make changes later or to teach employees to act safely with systems that were not designed with human factors in mind. Thus, a final goal of this paper is to encourage scientists and practitioners working on the technical elements of safety to consider the human element earlier and more often in their work.

References

Bandura, A. (1986). *Social foundations of thought and action.* Englewood Cliffs, NJ: Prentice Hall.

Brown, R. L., & Holmes, H. (1986). The use of factor analytic procedure for assessing the validity of an employee safety climate model. *Accident Analysis & Prevention, 18,* 455-470.

Cox, S. J., & Cheyne, A. J. T. (2000). Assessing safety culture in offshore environments. *Safety Science, 34,* 1-3.

Coyle, I. R., Sleeman, S. D., & Adams, N. (1995). Safety climate. *Journal of Safety Research, 26*(4), 247-254.

Dedobbeleer, N., & Beland, F. (1991). A safety climate measure for construction sites. *Journal of Safety Research, 22,* 97-103.

Denison, D. R. (1996). What is the difference between organizational culture and organizational climate? A native's point of view on a decade of paradigm wars. *Academy of Management Review, 21,* 619-654.

Flin, R., Mearns, P., O'Connor, R., & Bryden, R. (2000). Measuring safety climate: Identifying the common features. *Safety Science, 34,* 177-192.

González-Romá, V., Peiró, J. M., & Tordera, N. (2002) An examination of the antecedents and moderator influences of climate strength. *Journal of Applied Psychology, 87,* 465-473.

Griffin, M.A. & Neal, A. (2000). Perceptions of safety at work: A framework for linking safety climate to safety performance, knowledge, and motivation. *Journal of Occupational Health Psychology, 5,* 347-358.

Guldenmond, F. W. (2000). The nature of safety culture: A review of theory and research. *Safety Science, 34,* 215-257.

Hellreigel, D., & Slocum, J. W., Jr. (1974). Organizational climate: Measures, research, and contingencies. *Academy of Management Journal, 19,* 255-280.

Hofmann, D. A. & Stetzer, A. (1996). A cross-level investigation of factors influencing unsafe behaviors and accidents. *Personnel Psychology, 49,* 307-339.

Hulin, C. L, Fitzgerald, L. F. & Drasgow, F. (1996). Organizational influences on sexual harassment. In M. S. Stockdale (Ed), *Sexual harassment in the workplace: Perspectives, frontiers, and response strategies. Women and work: A Research and Policy Series,* Vol. 5 (pp. 127-150). Thousand Oaks, CA: Sage Publications.

James, L., & Jones, A. (1974). Organizational climate: A review of theory and research. *Psychological Bulletin, 81,* 1096-1112.

Katz, D. & Kahn, R. L. (1978). *The social psychology of organizations* (2nd Ed.). New York: Wiley.

Kossek, E. E., & Zonia, S. C. (1993). Assessing diversity climate: A field study of reactions to employer efforts to promote diversity. *Journal of Organizational Behavior, 38,* 61-81.

Litwin, G. H. , & Stringer, R. A. (1968). *Motivation and organizational climate.* Boston: Harvard Business School, Division of Research.

McLain, D. (1995). Responses to health and safety risk in the work environment. *Academy of Management Journal, 38,* 1726-1743.

Meyer, J. P, & Allen, N. J. (1997). *Commitment in the workplace. Theory, research and application.* London: Sage Publications.

Mueller, L., DaSilva, N., Townsend, J., & Tetrick, L. (1999). *An empirical evaluation of competing safety climate measurement models.* Paper presented at the 14th annual meeting of the Society for Industrial and Organizational Psychology, Atlanta, GA.

Neal, A., & Griffin, M. A. (2004). Safety climate and safety at work. In J. Barling & M. R. Frone (Eds.), *The psychology of work place safety* (pp. 15-34). Washington, DC: American Psychological Association.

Reichers, A.E. & Schneider, B. (1990). Climate and culture: An evolution of constructs. In B. Schneider (Ed.), *Organizational climate and culture* (pp. 5-39). San Francisco: Jossey-Bass.

Rousseau, D.M. (1990). Assessing organizational culture: The case for multiple methods. In B. Schneider (Ed.), *Organizational climate and culture* (pp. 153- 192). San Francisco, CA: Jossey-Bass.

Schein, E.H. (1990). Organizational culture. *American Psychologist, 45*, 109-119.

Schneider, B., & Reichers, A. E. (1983). On the etiology of climates. *Personnel Psychology, 36,* 19-39.

Schneider, B., Bowen, D.E., Ehrhart, M.G., & Holcombe, K.M. (2000). The climate for service: Evolution of a construct. In N.M. Ashkanasy, C.P. Wilderom, M.F. Peterson (Eds), *Handbook of organizational culture and climate* (pp. 21-36). Thousand Oaks, CA: Sage.

Varonen, U., & Mattila, M. (2000). The safety climate and its relationship to safety practices, safety of the work environment and occupational accidents in eight wood-processing companies. *Accident Analysis and Prevention, 32,* 761-769.

Vroom, V. H. (1964). *Work and motivation.* New York: Wiley.

Zohar, D. (2003). Safety climate: Conceptual and measurement issues. In J. C. Quick, & L. E. Tetrick (Eds.), *Handbook of occupational health psychology; handbook of occupational health psychology.* (pp. 123-142). Washington, DC, US: American Psychological Association.

Zohar, D. (2000). A group-level model of safety climate: Testing the effect of group climate on microaccidents in manufacturing jobs. *Journal of Applied Psychology, 85*, 587-596.

Zohar, D. (1980). Safety climate in industrial organizations: Theoretical and applied implications. *Journal of Applied Psychology, 65*, 96-102.

Zohar, D., & Luria, G. (in press). A multilevel model of safety climate: Cross-level relationships between organization and group-level climates. *Journal of Applied Psychology.*

Zohar, D. & Luria, G. (2004). Climate as a social-cognitive construction of supervisory safety practices: Scripts as proxy of behavior patterns. *Journal of Applied Psychology, 89,* 322-333.

Paper TH005-2

OPERATIONAL DISCIPLINE AT DUPONT

James A. Klein
E. I. du Pont de Nemours and Company

Prepared for Presentation at
American Institute of Chemical Engineers
2005 Spring National Meeting
Process Plant Safety Symposium
Atlanta, Georgia
April 11-13, 2005

April, 2005

UNPUBLISHED

Abstract

DuPont defines operational discipline as the deeply rooted dedication and commitment by each member of the organization to carry out each task the right way, each time. As such, operational discipline reflects the strength of the safety culture in making safety systems effective and in providing tangible results for preventing injuries and incidents.

The purpose of this paper is to discuss the development, application, and continuous improvement of operational discipline principles at DuPont.

Introduction

Managing potentially hazardous materials and processes to prevent serious injuries and incidents is an essential part of safety management programs, requiring the continuous dedication and commitment of everyone in an organization. While founded over 200 years ago in 1802 for the manufacture of gunpowder on the banks of the Brandywine River in Delaware [1], this commitment – now often characterized as operational discipline – has always been an important piece of DuPont safety culture and programs. Reflecting on the hazards of the gunpowder business in 1808, the founder, E. I. Du Pont, wrote:

> The safety of our family, the safety of the farmers who live in our neighborhood, has imposed upon us the absolute duty of making choice of steady, sober men and of establishing the most rigid discipline among our workmen [2].

While the company has evolved in many ways since then, the need for operational discipline along with other core DuPont safety philosophies, including that management is responsible for safe operations, that it is necessary to train all people to work safely, and that all injuries and incidents can be prevented, have remained fundamental beliefs. For example, DuPont CEO Edward Jefferson stated in 1986:

> On one hand there is the day-to-day analysis required of each employee. A safe employee sees what he or she is looking at, instead of looking over or through it. This results in greater concentration, a habit of mind that will make a person work more safely as well as more intelligently. Everyone in every job must participate in this process of analysis. There are no exceptions [3].

Today, operational discipline is defined by DuPont as the deeply rooted dedication and commitment by every member of an organization to carry out each task the right way each time. No matter how comprehensive and well-designed safety programs may be, it is the day-to-day ability of everyone to practice operational discipline that successfully changes safety systems from concept to reality with tangible results in order to achieve program goals. Clearly, this ability not only helps prevent serious injuries and incidents, but also contributes to excellent business performance through increased operating excellence, including higher productivity, higher quality, reduced waste, and lower costs.

Process Safety Management

While operational discipline is fundamental to the success of all safety programs, it is especially important in process safety management (PSM) programs, where the risk associated with higher hazard processes could potentially result in catastrophic incidents with multiple injuries. Many of the current DuPont operational discipline practices were therefore developed by the PSM competency and then adopted and modified by other EHS competencies, as needed. PSM is the application of management systems and controls (programs, procedures, audits, evaluations) to a manufacturing or chemical process in a way that process hazards are identified, understood, and controlled, so that process-related injuries and incidents are prevented. Operational discipline is recognized as an essential part of the four steps used for implementation of PSM at DuPont [4]:

- Step 1 – Establishing the Safety Culture
- Step 2 – Providing Management Leadership and Commitment
- Step 3 – Implementing a Comprehensive PSRM Program
- Step 4 – Achieving Operational Excellence through Operational Discipline

The PSM program at DuPont today is defined by corporate safety standards that both reflect and extend OSHA 1910:119 PSM and other regulations. The principles and essential features of the PSM program are described by 14 elements grouped by Technology, Personnel, and Facilities, as shown in Figure 1. Figure 1 is typically called the PSM Wheel, with each of the 14 elements, such as Process Hazards Analysis, Operating Procedures and Safe Work Practices, and Personnel Training and Performance, arranged around the spokes of the wheel. Management leadership and commitment, necessary for implementing PSM and providing resources, is shown at the center of the PSM Wheel. Operational discipline is shown as the rim of the PSM Wheel, connecting all of the elements and translating the required management systems into real results for preventing injuries and incidents. The requirement for operational discipline in the corporate PSM standard is:

> Sites shall implement and maintain programs and systems to achieve and maintain a high degree of operational discipline for all PSM elements in a manner that supports business and operating objectives.

Development of Operational Discipline at DuPont

While operational discipline and operating excellence have long been important expectations at DuPont, formal programs for operational discipline were not developed until the late 1980's. PSM training in 1989 [5], for example, included a section on "The Importance of Operating Discipline," based on the principle that:

> The most technologically advanced plant in the world cannot be made completely safe unless the individuals managing and working there are dedicated and committed to keeping it free from accidents.

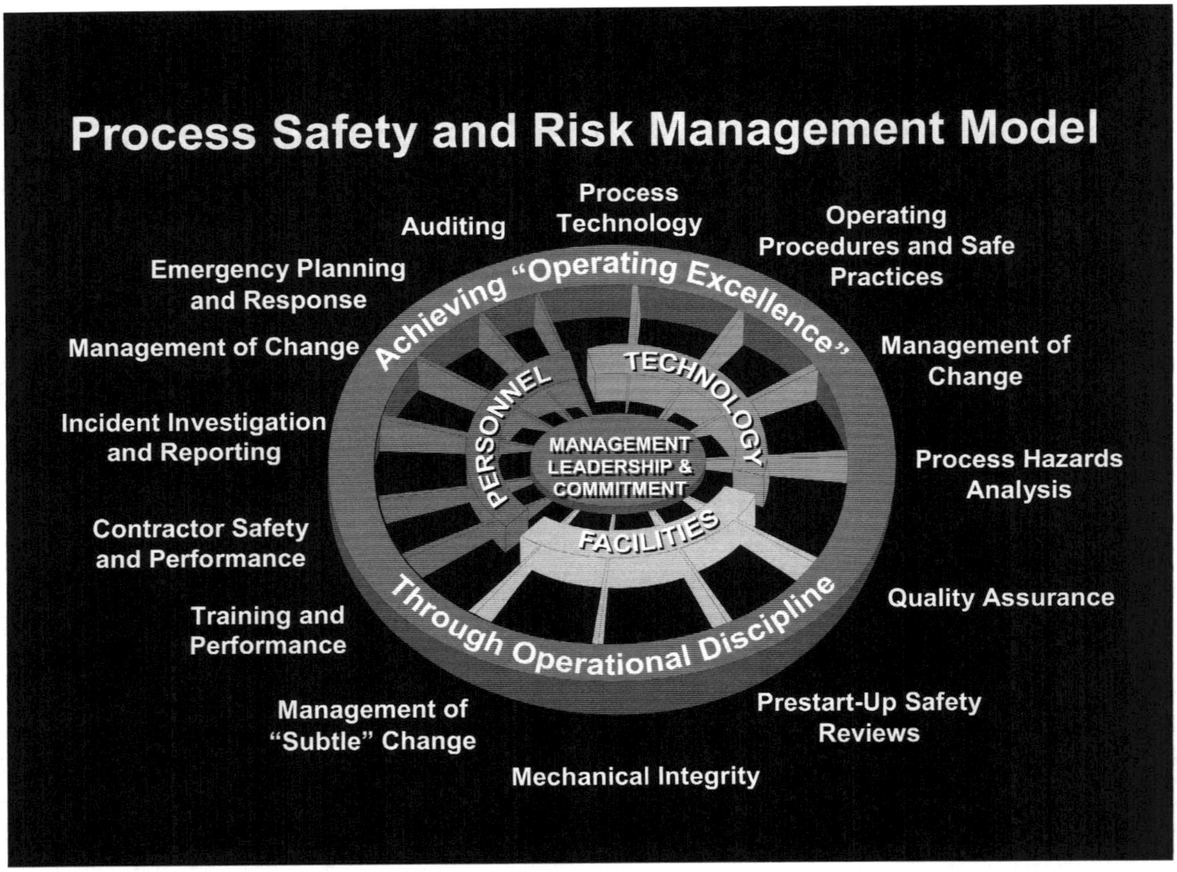

Figure 1 – DuPont PSM Wheel

Premises supporting the importance of operational discipline were based on the following:

- There is no pound worth an injury. (former CEO)
- Quality work can only be done when safety is step #1. (operator)
- Ye have the time to do the job right! (area manager)

Catastrophic incidents at Flixborough, Seveso, and Bhopal were cited as examples of where lack of operational discipline was a major contributing factor in the cause of the accident. The primary focus was:

> All of the principles and essential features [of PSM] mean nothing when it comes to operating a facility safely and preventing catastrophes – if the principles, procedures and practices established and set in place are not followed! Stated another way, the most technologically advanced plant in the world will not have a safe operating track record unless the individuals managing and working there are dedicated and committed to keeping it free from accident…

Essential practices for establishing high levels of operational discipline were:

- "Leadership by example" from the site's manager on down.
- Follow established principles, procedures, and practices as documented in the site's PSM Program.
- Production priorities should not be placed ahead of safety and health of personnel.
- If there is not sufficient time available to do the job the right way, then the job must be stopped until there is sufficient time.

By the mid-1990's, the four key steps for effective PSM programs had been developed [4, 6], with step #4 "Achieving Operating Excellence via Operational Discipline" receiving additional emphasis. The principle for operating excellence was:

> The manufacture, use, and handling of hazardous materials requires the dedication and commitment of those individuals managing and handling such material to complete each job the right way every time.

The characteristics of an organization that had achieved high levels of operational discipline were:

- Management exerts leadership by example.
- Management action consistent with words.
- Open lines of communication.
- Strong sense of teamwork.
- Employees dedicated and committed to doing the job the right way every time.

Operational discipline was also linked to operating excellence and overall business success, since "the benefits and rewards of achieving operating excellence are many and extend well beyond the areas of safety, health, and the environment." Operational discipline was viewed as consistent with major quality initiatives, trying to achieve the same end result of outstanding performance "in every aspect of every endeavor." Common points with other programs needed for business success were:

- Sound and up-to-date technology.
- Trained personnel.
- Equipment that is maintained and reliable.
- Standard procedures.
- Effective management of change.
- Audits, including control and feedback.
- Doing the job the "right way."
- Continuous improvement.

In late 1996, the corporate ZIP (Zero Incidents... Period!) Team was formed to focus on eliminating process-related injuries and incidents, while at the same time, building business value for PSM. The ZIP Team included members from all businesses and regions, ranging from senior corporate leadership, to plant managers, to operators and mechanics, and eventually included 70 members. The ZIP team made several recommendations, including that operational discipline must be strengthened as an essential element of sustainable improvement. Operational discipline was added to the PSM Wheel (see Figure 1), and business leadership was expected to lead implementation of operational discipline (and other) ZIP Team recommendations:

> The SBU (Strategic Business Unit) will develop a program to drive operational discipline across all sites.

The result was a renewed corporate effort to evaluate and improve operational discipline that led to additional training programs, assessment tools, and metrics.

A new 4-hour operational discipline workshop [7], for example, was introduced in 1998. Initially, 39 trainers were available to lead this workshop, and an informal operational discipline network was established to promote communication between instructors and provide a forum for discussing and updating instructional and resource materials. The workshop introduced the following characteristics of organizations that have achieved operating excellence, which are also discussed in more detail in the following section:

1. Leadership by example
2. Sufficient and capable resources
3. Employee involvement
4. Active lines of communication
5. Strong teamwork
6. Common shared values
7. Up-to-date documentation
8. Practice consistent with procedures
9. Absence of shortcuts
10. Excellent housekeeping
11. Pride in the organization

Videos and reports of serious incidents were introduced, and small groups were asked to consider what errors were made and where operational discipline had failed or could be improved. Additional exercises asked participants to consider how operational discipline applied to their job, what best practices were available at their sites, how well they were being used, and finally, what they could personally do to help improve operational discipline.

An operational discipline self-assessment tool was first developed in 2000, and was used by some businesses to evaluate the status of operational discipline and to identify potential improvements. The self-assessment tool was improved, and in 2003, the Operations Director of each business actively supported a global commitment to complete the self-assessments at all manufacturing and R&D sites, with the goal to identify key improvement opportunities at the site level. The self-assessments are discussed further in the next section.

In 2001, a Six Sigma team was formed to focus on operational discipline as a key to eliminating process-related injuries and incidents [8]. Using process incident data based on an incident scoring system required in DuPont, operational discipline was identified as a key factor contributing to over 50% of process incidents in 2000. Some recommendations from the Six Sigma team included:

- Consider modifying the corporate incident investigation standard to require identification of operational discipline characteristics that contributed to the incident and could be improved. A decision tree was also developed to support high quality incident investigations involving operational discipline.
- Consider requiring more detailed evaluation of operational discipline in second party PSM audits, where it had previously been optional.
- Raise operational discipline as an important issue with the DuPont Plant Managers' Sounding Board for awareness and support.

This last recommendation helped lead to the formation of a corporate Operational Discipline Leadership Team in late 2001. Team members included an Operations Director, several Plant Managers, regional representatives, operators, mechanics, and EHS professionals. The mission of this team was:

> To engage our global operations in a way that institutionalizes a sustainable process that significantly advances our capability and will to improve operational discipline.

The team worked to better define the current status of operational discipline within DuPont, develop a vision, identify gaps, benchmark best practices within and outside the company, and develop a plan for improvement. The current status of operational discipline was determined partly from the experience of the team, interviews, and a survey of plant managers to identify three top strengths, three top opportunities, and any other recommendations for improving operational discipline. Key recommendations of this team included:

- Develop a web site for operational discipline, including training materials, best practices, metrics, and reference materials.
- Add operational discipline requirements to the corporate PSM and incident investigation standards.
- Require evaluation of operational discipline in second party PSM audits for higher hazard processes.
- Revise the operational discipline self-assessment tool and require that all sites complete a self-assessment by the end of 2003.

This team also recommended that ownership of operational discipline activities be assigned to the corporate PSM Leadership Team, which has now also formed a team to pursue continuous improvement.

Current Operational Discipline Practice

Characteristics of Operational Discipline

Organizations that have a high level of operational discipline usually exhibit an identifiable set of positive behaviors or characteristics. While it's possible that these characteristics may vary from one organization to another, DuPont experience has shown that 11 key characteristics seem to be most important. Other organizations have identified different sets of characteristics [9], although there is usually some common ground. The key is their usefulness in helping to evaluate operational discipline through-out the organization and in helping to provide guidance for continuous improvement.

The 11 operational discipline characteristics currently used by DuPont are:

1. Leadership by Example – Management "walks the talk" in a visible and consistent manner. Leaders demonstrate an unrelenting passion for safety, clearly communicate standards and expectations, are visible in their commitment, are personally involved in safety, encourage high employee participation rates, evaluate and recognize good performance, listen to employee concerns, and follow-up on issues that have been identified.

2. Sufficient and Capable Resources – Enough resources are provided with the required capability and adequate time to complete safety, health, and environmental activities in a timely and effective manner consistent with a high priority for safety. Metrics are used to measure progress and periodic reviews are used to evaluate performance.

3. Employee Involvement – Employees at all levels of the organization are involved and enthusiastic about participating in safety-related activities. Managing processes encourage and facilitate high levels of participation by providing multiple opportunities for involvement, and employees routinely volunteer to help improve site safety.

4. Active Lines of Communication – Clear and effective communications occur at all levels of the organization. Everyone is open and actively listens to the concerns and input of others, multiple methods (e. g., e-mail, meetings, postings, etc.) are used to reinforce communications, and people feel accountable and valued.

5. Strong Teamwork – Teams are formed to work on common objectives with active participation from all levels of employees. Teams are chartered with clear goals and responsibilities, teams are accountable for results, team members with diverse experience build on individual strengths based on trust, and team contributions are recognized by the organization.

6. Common Shared Values – Everyone knows and understands the organization's commitment to safety with a goal of zero injuries. Leadership demonstrates safety is a core value, employees believe that all injuries can be prevented, and everyone works safely and helps their co-workers work safely.

7. Up-to-date Documentation – All operating documentation, such as procedures, checklists, process prints, etc., is current and reflects best practices. Systems exist to maintain and update documentation consistent with best practices, changes are properly reviewed and authorized, documentation is readily available, and use of current documentation is required.

8. Practice Consistent with Procedures – Work is completed as planned, following authorized procedures. Periodic reviews of procedures are made to ensure they are correct and complete, training on procedures is provided, field audits are made to confirm that procedures are being followed every time, and changes to procedures are reviewed and authorized by affected employees.

9. Absence of Shortcuts – Employees are encouraged to and always choose to follow established procedures, rules, and practices. Expectations for following procedures are clearly established, proposed changes and deviations from established procedures for tests are reviewed and authorized, and employees seek assistance in uncertain situations.

10. Excellent Housekeeping – Employees throughout the organization are proud of their workplace, maintaining consistently high levels of housekeeping in all areas. Housekeeping standards are established and clearly communicated, all areas of the site are routinely evaluated, and equipment is maintained in good operating condition.

11. Pride in Organization – Every employee displays pride in the organization, often developed through successful application of the other characteristics. Employees understand the organization's goals, know and share core values, and present a positive image of the organization in the workplace and community.

Clearly, many of these operational discipline characteristics are connected to one another, such as Up-to-date Documentation, Practice Consistent with Procedure, Absence of Shortcuts, and Sufficient and Capable Resources. If qualified engineering resources are not available for maintaining current process documentation, for example, then operations personnel would not have complete procedures and prints, training would be less effective, and possibly inadvertent deviations from procedures could result in increased risk of injury. Despite these connections, the 11 characteristics have proven useful in helping to evaluate operational discipline and provide guidance for continuous improvement.

An internal website has been set-up to support operational discipline efforts in DuPont. The website is accessible from the Global EHS website and includes the following main sections:

- Background – provides an introduction to operational discipline and background materials on activities supporting the development of operational discipline at DuPont.
- Self-assessment Tool – provides the operational discipline assessment tool and guidance for completing successful reviews and for evaluating results.

- Training and Education Tools – provides basic training materials on operational discipline for site use.
- Useful Tools and Best Practices – provides supporting information for the 11 characteristics of operational discipline.
- Resources / Reference Materials – provides recommended external sources for additional information, such as for employee involvement and successful teams.

The section on Useful Tools and Best Practices includes results from the Six Sigma project described in the previous section. This includes the components critical to quality (CTQ's) and the independent variables (critical X's) for each characteristic. A partial example for Leadership by Example includes:

- CTQ – Visible leadership presence and involvement across the organization.
- Critical X – Build value in leaders that this is a critical part of their role and hold them accountable with specific expectations.
- Critical X – Define and communicate what is "visible leadership" throughout the line organization.
- Critical X – Senior leadership spends time reinforcing EHS value at sites and communicating business value for EHS excellence.

Assumptions, metrics, leadership actions, and best practices are also listed for each characteristic. The result is a wealth of specific information available to support implementation, evaluation, and improvement of operational discipline at both site and business platform levels.

Operational Discipline Self-Assessments

Operational discipline self-assessments were developed in the late 1990's, recommended for broad corporate use in 2002, and required at all global manufacturing and R&D sites in 2003. The self-assessments are intended to provide a common methodology to assist sites in evaluating their level of operational discipline and to identify areas where improvement opportunities may exist to strengthen their EHS programs and performance. The self-assessment consists of approximately 75 questions, which are used to qualitatively score 10 characteristics of operational discipline (Pride in the Organization is considered to result from strength in the other 10 characteristics). Since operational discipline is not always easy to evaluate, pre-work to collect data and metrics on the current state of performance is recommended, including information such as progress on site EHS goals, open and overdue audit or incident recommendations, adherence to schedules for maintenance and training, and the extent of up-to-date documentation. In addition, multi-disciplinary teams are used representing a cross section of the site based on role and function, such as operations management, technical, EHS resources, operators, mechanics, and support staff. Depending on the size and complexity of the site, multiple self-assessments might be required, with the time commitment typically varying from about six to fifteen hours of meeting time. Several sessions are often scheduled to maintain team focus and energy.

A sample self-assessment question is shown in Figure 2. Each operational discipline characteristic is assigned a section (e.g., Excellent Housekeeping is section 10) and one of the Critical to Quality (CTQ) elements is identified, such as "Personal value and demonstration of personal housekeeping standards by each individual" in the figure. One of the independent variables (Critical X) is listed, such as "Giving status and recognition of good housekeeping. Reinforce link between good housekeeping and excellence in EHS results" in the figure. The Critical X is qualitatively scored from 1 to 5 (highest), with the ratings defined as:

1. Not addressed in current state.
2. Significant gaps or components missing.
3. Partially in place with multiple opportunities for improvement.
4. In place with minor opportunities for improvement.
5. Fully in place with strong results.

To assist with scoring, suggested questions are also provided to help teams better understand the intent of the Critical X and to evaluate performance. Active team discussion of site activities and performance is essential, and pre-work can also be used to help provide data on actual performance. Scoring is qualitative based on team consensus, and improvement opportunities are identified when scoring on individual questions is low. Typically, improvement opportunities are prioritized to identify the most significant system improvements that will have the greatest impact for improving site operational discipline. Results of the self-assessment are reported to site management for follow-up.

10-3 Personal value and demonstration of personal housekeeping standards by each individual.

- Giving status and recognition of good housekeeping. Reinforce link between good housekeeping and excellence in SHE results. 1 2 3 4 5

Consider :
Is everyone involved in housekeeping in their own areas or only a critical few?
How does the site or area recognize excellence in housekeeping?
Is there a periodic or annual focus on housekeeping?
Is housekeeping evaluated and included in incident investigations and reports, where applicable?
What percentage of time do individuals spend on personal housekeeping in their area?

Figure 2 – Sample Operational Discipline Self-Assessment Question

The results for a limited number of self-assessments are shown in Figure 3. This figure shows the range and average of results from about 25 self-assessments for 10 of the operational discipline characteristics, including data from the US/Canada, Europe, and Asia/Pacific Regions. Since the figure includes a wide range of global sites, including some relatively new to DuPont due to acquisitions, there is a fairly broad range for each characteristic. The average of each characteristic is between 3 and 4, suggesting that most sites have programs in place, but were able to identify several opportunities for improvement.

While comparison between sites and regions, etc., would be interesting, the qualitative nature of the scoring prevents generalizing too many conclusions. Results are primarily intended to help sites improve operational discipline locally, although corporate programs could be developed for characteristics that generally score lower.

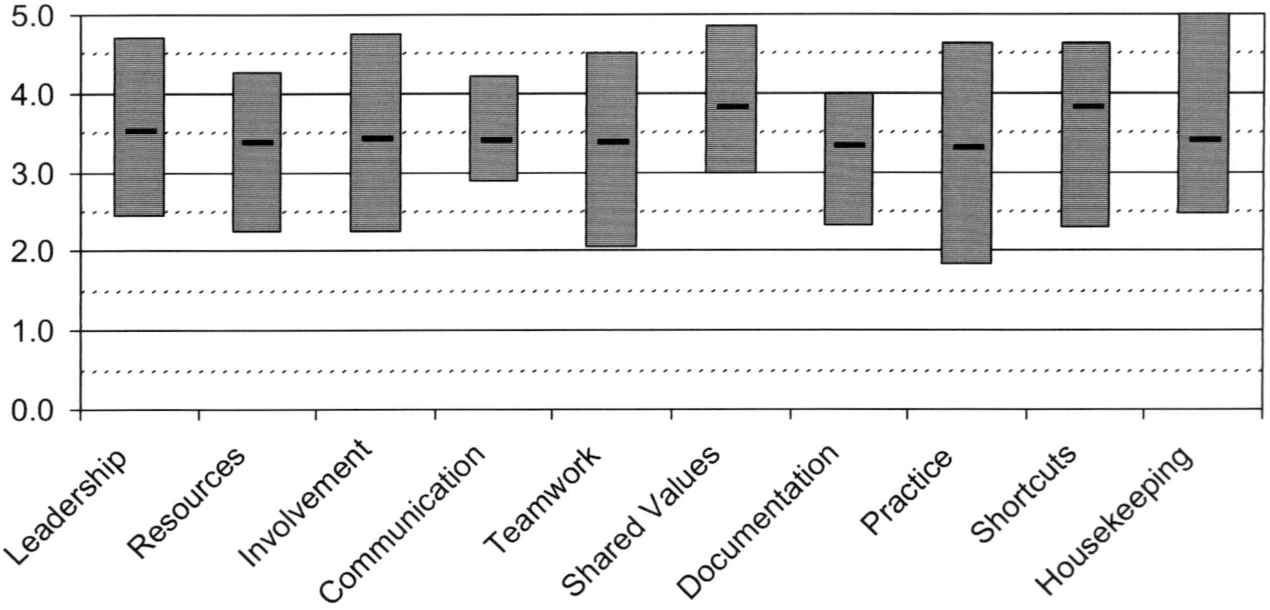

Figure 3– Operational Discipline Self-Assessment Scores

Incidents and Audits

A requirement to consider and document the contribution of operational discipline to incidents was added to the corporate incident investigation standard in 2003. Where lack of adequate operational discipline is noted as a key factor in the incident, incident investigation teams must identify which of the 11 characteristics needs to be strengthened. Typically, recommendations must be developed to address all key factors identified in the incident investigation, including operational discipline. Detailed analysis of incident data is on-going, but suggests that operational discipline continues to be identified as a key factor in approximately 50% of PSM incidents. The most commonly cited characteristic of operational discipline contributing to incidents is Practice Consistent with Procedures.

A requirement to include evaluation of operational discipline in second party PSM audits was established in 2004. PSM audits use a lengthy checklist, with questions scored from 0 to 100% to establish an audit score for each PSM element. Several specific questions have been added to the checklist in the Management section, which help provide a general assessment of operational discipline. A sample audit question is shown in Figure 4, where additional factors to consider when evaluating the question are also provided in the audit checklist. In addition, approximately 100 other specific questions under the various PSM elements relate directly to operational discipline, such as revision of procedures, communications, and overdue action items. The spreadsheet used for scoring questions for each PSM element has therefore been revised to also allow simultaneous scoring of the 11 characteristics of operational discipline. For example, there may be several questions for various PSM elements through-out the audit checklist that relate to Practice Consistent with Procedures. The score for these questions is used to score the separate PSM elements, and is then also combined in the spreadsheet to calculate a score for Practice Consistent with Procedures. The result for a limited number of audits is shown in Figure 5. This figure shows the range and average of results from about 10 PSM audits for 10 operational discipline characteristics. Since the figure again includes a wide range of sites, including some relatively new to DuPont due to acquisitions, there is a fairly broad range for some characteristics. This data is used to help audit teams evaluate operational discipline at the site and to provide additional recommendations, as needed.

1.18 Has the area/site established systems to achieve and maintain a high degree of operational discipline across the PSM program elements?

Consider:

⇒ Did the site/area conduct an operational discipline self-assessment in 2003? If so, what were the results/findings and improvement activities?
⇒ How do leaders communicate expectations and measure operational discipline across the site/area PSM program and systems?
⇒ Is operational discipline included as part of the training for new site employees and refresher training for existing employees who are involved with PSM?
⇒ Are the status of open/overdue metrics indicative of a high degree of operational discipline across the site line organization?

Figure 4 – Sample PSM Audit Question on Operational Discipline

A full operational discipline audit protocol involving employee interviews and requiring additional audit resources is also available as an option in PSM audits. The somewhat larger audit team interviews six people each for three days during the audit in addition to using and scoring the PSM audit checklist. The 18 people interviewed should include a mix of operators, mechanics, supervisors, technical, contractors, and site management, who are selected randomly, consistent with their rough proportion on the site. One hour is scheduled for each interview, focusing on both the strengths and opportunities for improving PSM and operational discipline. Sets of questions are provided for the auditors to assist with conducting the interviews. Sample questions for operational discipline include:

- Is site management providing opportunities for employee involvement in each and every one of the elements of PSM?
- Is site line organization paying as much (or more) attention to listening as to communicating?
- Do site teams which address PSM issues cut across both levels and functions, i.e., manager/employee and operations/maintenance?
- Does the site have an active communications system to document concerns/problems which interfere with following procedures as written?

- Do site policies on achieving excellence in housekeeping extend to all areas of the plant?

Some of the interviews are also supplemented with tours of the work area to talk to the employee in more informal conditions and also to allow a better assessment of area housekeeping and other practices. The interview discussions provide the basis for the audit team to develop strengths, opportunities for improvement, and recommendations for follow-up by the site for improving operational discipline. In all cases, specific interview discussions are kept confidential.

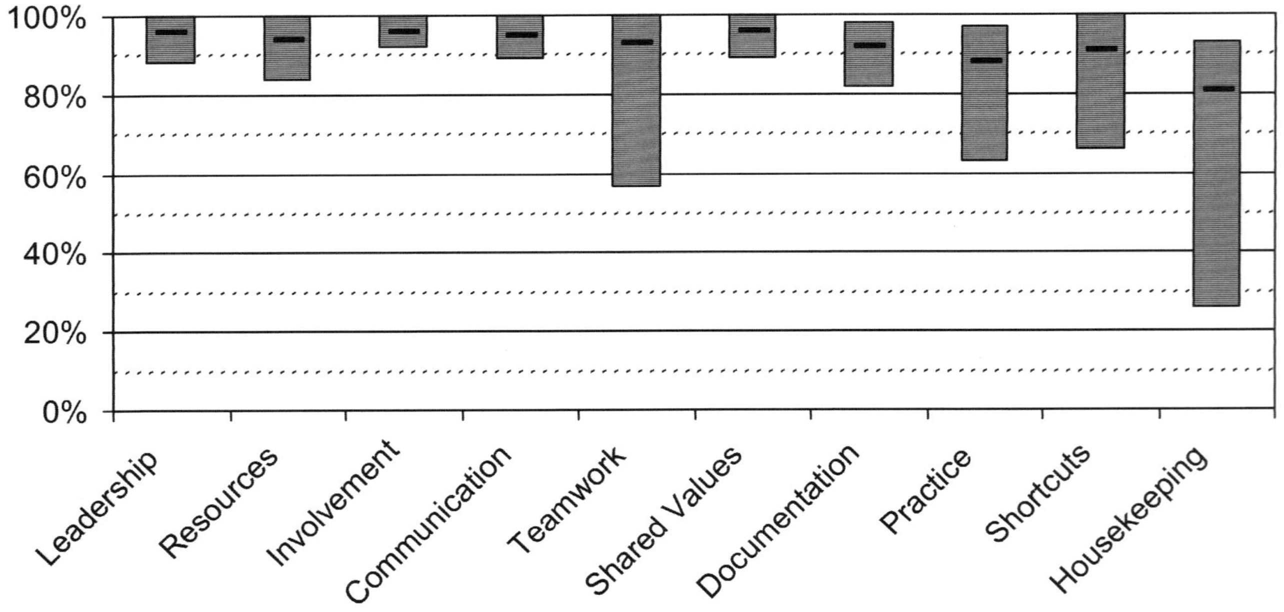

Figure 5 – Operational Discipline Audit Scores

Future Focus

Increased access to data on operational discipline from self-assessments, incident data, and audits provides detailed information for driving continuous improvement. Some specific areas for future work include:

- Improved analysis of data to identify continuous improvement opportunities.
- Improved training to assist sites in understanding the characteristics of operational discipline and evaluating their current level of performance.
- Further extension of operational discipline activities beyond PSM.

Ultimately, better guidance needs to be developed to help sites improve operational discipline performance when opportunities have been identified and when efficiency efforts and retirements may affect the availability of experienced resources. New acquisitions and business growth globally may also present challenges due to the different corporate safety cultures of acquired companies and due to different customs in other parts of the world. What is effective in the United States may not be effective elsewhere, and efforts to establish and continuously improve operational discipline must recognize these differences. This will likely require continued focus on safety culture, behavioral safety, and human error. The goal is to continue to motivate all employees to develop and maintain a deeply rooted dedication and commitment to carry out each task, the right way, each time. The result will be continued safe operations and the prevention of serious injuries and incidents.

References

1. Kinnane, Adrian, "DuPont: From the Banks of the Brandywine to Miracles of Science," DuPont, 2002
2. DuPont, E. I., letter to his father, 1808 (source: Hagley Museum, Wilmington, DE)
3. Jefferson, Edward G., "Attention to Safety Has Significant Benefits," Sunday News Journal, Nov. 2, 1986 (source: Hagley Museum, Wilmington, DE)
4. Burk, Arthur F., "Achieving Excellence in Process Safety & Risk Management," Managing Safety & Health in the Workplace, Chicago, IL, June 13-14, 1994
5. DuPont, "Process Safety Management," seminar manual, 1989
6. DuPont, "Process Safety and Risk Management," seminar manual, 1994
7. DuPont, "Achieving Operating Excellence Through Operational Discipline," seminar manual, 1997
8. Urbanik, Stanley A., "Operational Discipline Background and Program in DuPont," Paper #31b, Process Plant Safety Symposium, Mar. 31 - Apr. 2, 2003
9. Walter, Robert J., "Discovering Operational Discipline," HRD Press, 2002

Values and Behavior: Building a Culture that Promotes Safety

Terry McSween, Ph.D.
Quality Safety Edge
Montgomery, TX

Prepared for Presentation at
American Institute of Chemical Engineers
2005 Spring National Meeting
Process Plant Safety Symposium
Atlanta, Georgia
April 11-13, 2005

Safety professionals often talk about wanting to establish safety as a basic value within an organization. Typically, we talk about wanting to move safety from a priority to a value, using the rationale that priorities change while values remain unchanged over time. The logic is simple and straightforward. But what do we really mean when we talk about wanting to establish safety as a value? Usually, we want managers and employees never to compromise safety for convenience or to maintain production.

What are values?

To talk more precisely about values, we need to operationally define the term. Values are simply statements that describe practices, or sets of behavior, that are supported by our community. Typically, value statements describe behaviors that members of a community "feel good about." Value statements also describe or imply behaviors that are likely to be sanctioned or punished by other members of a group or culture. In talking about values, we might be talking about principles or ideals. Most members of our culture learn to respect the property of others at an early age. Often, part of this training involves the rule, "Thou shalt not steal" -- a practice that is typically punished by family members and others in the community. So when we talk about wanting safety to be a basic value in our organization, we mean that we want members of our organization to be uncompromising in following safety rules and in sanctioning safe practices.

We typically talk about two types of values: personal values and cultural values. When talking about personal values, we are talking about people's personal standards, or rules of conduct. These "rules" describe behavior that we feel good about, that is, behaviors that we find reinforcing. At other times, we are talking about cultural values. In this case, we are talking about social norms, or rules, that describe activities that are supported by the community. We are describing behavior that will be reinforced or punished by others in the culture where we live or work.

Both of these types of values are critical to our work in safety. Talking about which comes first, or which is more important makes little sense. We have an almost classic example of the chicken or the egg debate. One simply does not exist without the other.

What is the relationship between values and behavior?

Values often describe behavior that has become second nature, activities that we do almost without thinking about it. A common example in our culture is wearing clothes, an activity that is common to most members of our culture. Most of us would not consider leaving home without first putting on clothes. We would like all employees in our organizations to have the same consistency in their safety practices, and to have the same level of personal discomfort should they compromise their value for safety.

Our value for wearing clothes is in many ways a good analogy to our value for safety. Just as different organizations have different commitments to safety, different cultures value different forms of clothing. Some native cultures in Africa wear little or nothing. Just as an organization may have a unit or group that does not reflect the organization's commitment to safety, we have subcultures, such as nudist communities, where wearing clothes is not condoned. We also have individuals, who choose to streak, run naked through public places, in order to get attention, or be a "hero" within their subculture. In a similar way, some individuals will perform running maintenance in order to be a "hero" and save their organization money.

How do we establish safety as a personal value?

Values are a function of our beliefs, knowledge and skills. In safety, we address knowledge and skills through formal education and training. Unfortunately, beliefs are often more critical than knowledge and skills. Our beliefs will often have a greater impact on our behavior, than the rules we have learned through our formal educational experiences. Beliefs are acquired through a history of experience. One of the advantages of behavior-based approaches to safety is the experience they provide. Several studies suggest that conducting behavioral safety observations causes the observer to be more consistent in performing safety practices on the checklist, even when the observer gets no external feedback on those behaviors (cf. Alvero, Olson, & Austin, 1999). It appears that conducting behavioral safety observations provides and experience that helps people acquire more accurate beliefs about their safety practices and thereby strengthens their personal value for those practices.

The discussions that take place in the context of behavioral safety also help build stronger personal value for safety in other ways. When an employee provides verbal feedback to a peer promoting the use of fall protection, that employee is likely to remember times when he failed to use proper fall protection under similar circumstances. The recognition of this discrepancy between what the employee is publicly promoting and his personal practices contribute to the employee to being more consistent in the use of fall protection in the future. Scott Geller has referred to this phenomenon as the "hypocrisy effect" (Geller, 2000). This phenomenon that has been widely researched in the social psychology literature and is often referred to as cognitive dissonance. While some have hypothesized that cognitive mechanisms are responsible for this phenomenon, the behavior change that occurs is easily understood in terms of a learning history with a community that values consistency between words and action. Most of us value those who "walk the talk." Further, we are often critical of those who lack such consistency. For example, if a supervisor talks positively about safety in a safety meeting, but then compromises safety for to meet production targets, we are critical of this supervisor and criticize him for "talking out of both sides of his mouth" and try to improve the educational and accountability systems to encourage him be a better safety leader and more consistently "show the way."

This last example also illustrates the importance of having organizational values and formal systems that are aligned in support of practices consistent with our value for safety. Certainly our best strategy for encouraging stronger personal values for safety is through formal systems and cultural values that are aligned to support safety practices and decisions at every level of the organization

How do we establish safety as a cultural value?[1]

To establish safety as a value in organizations requires similar changes. It requires that we begin talking to one another more frequently about safety practices that are consistent with our value for the safety. Behavioral safety is one way of achieving that aim. When done well, behavioral safety provides an opportunity for employees at all levels to increase the frequency of talking with one another about safety on the job. Obviously, talking to one another about safety is a practice that already exists to some degree. We would always stop a coworker from doing something that would cause them harm. Behavioral safety processes reliably increase the frequency of such safety discussions, and typically expand the topics of discussion to include more positive feedback on safe practices and more problem-solving discussions on less significant risks. Whatever the means, frequent attention to safety is the primary characteristic of an organization that values safety. Behavioral safety is not a magic bullet, but it can be a significant element in promoting both the personal and organizational values that characterize a culture committed to minimizing the risks of injury.

To create a culture that truly values safety requires the alignment of the formal and informal systems to support safe work practices. The diagram in Figure 1 illustrates provides an illustration of the elements that must be considered. Thus if we wish to create an effective safety culture, we must first create a vision or mission that describes our ideal, then establish a process to achieve our results, and then define values that clarify how we will work together. Behavior is a particularly important element of our model because it is key to both our processes and living our values. Good managers and healthy organizations achieve a balanced emphasis on each of these.

Unfortunately, many managers tend to emphasize either process or results. People who report to these managers then reflect the manager's style. In addition, new managers tend to learn their management skills from those in the existing hierarchy. The organization may not focus its attention solely on process or results, but, because managers tend to hire and promote people like themselves, the emphasis becomes and remains unbalanced. The result is that many managers spend too much time on one of these elements.

Overemphasis on results. Many American managers manage their employees according to the results achieved by the employees. In many cases, we train managers to manage by results, and then use antiquated systems of annual objectives and appraisals that maintain an unbalanced emphasis on results. Managers who enter the private sector from the military often have a strong results orientation and industries that hire them, such as aerospace companies, defense contractors, and nuclear utilities, have frequently adopted such an orientation. Companies that overemphasize results can be very successful, but they often achieve their success only through Herculean efforts by top managers and key employees who often frequently work extra-long hours. The orientation also frequently has a decided personal cost. These companies typically have high levels of stress, often characterized by a high rate of employee burnout, high turnover, and higher than average rates of participation in Employee Assistance Programs. Furthermore, this approach often creates a perception among employees that "All management really cares about is the numbers," the reference being to budget figures, production records, efficiency targets, or safety statistics.

---

[1] This section based on McSween (2003).

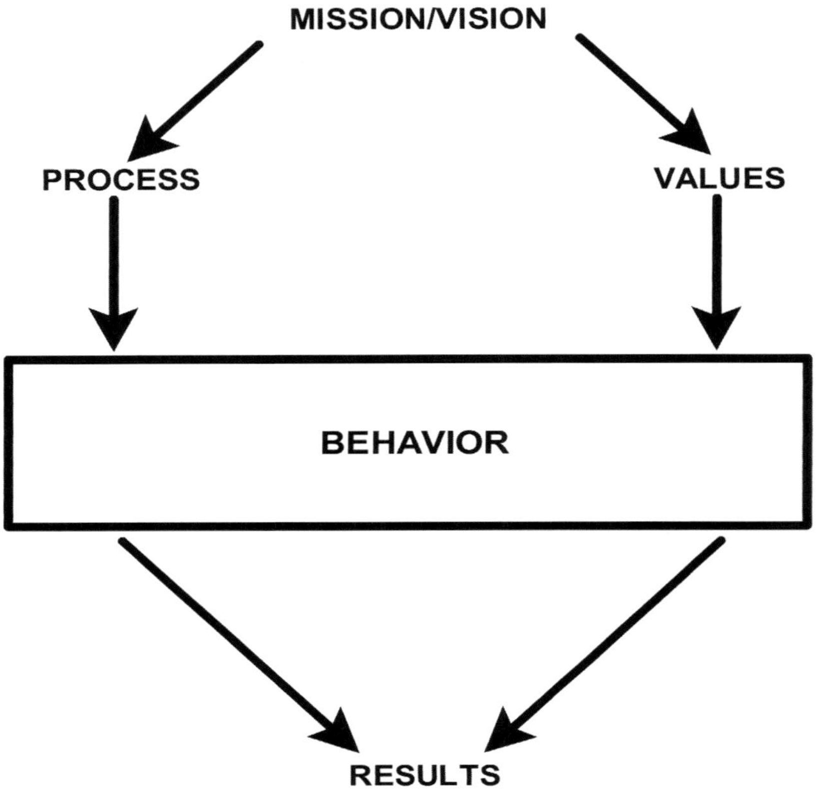

Figure 1 - Elements that must be aligned to create an effective culture.

In results-oriented companies, when the results are not achieved, the tendency is to find who to blame. The resulting fear and distrust often make improvement efforts difficult. Employees resist efforts to establish effective performance measures because they are skeptical about how managers will use those measures. Employees feel frustrated and believe that the organization does not care about them. In fact, results-oriented organizations seldom systematically address how managers and employees should treat one another. The result is that the quality of personal relationships becomes a further barrier to significant culture change.

In safety, "condition of employment" policies and incentive systems that unintentionally encourage employees to hide or overlook incidents often indicate an organization that overemphasizes results. Such policies, when not supported by effective management practices, implicitly threaten employees with the potential loss of their jobs, income, or promotional opportunities if they report an injury that results from a failure to comply with a company safety requirement. So, they may claim that the injury occurred at home. Or they may inaccurately report that the incident occurred in spite of full compliance with company procedure, saying something like "I was wearing my safety glasses, but this metal filing somehow got around them." They may make similar responses if an injury jeopardizes a group's chances of getting a tangible safety award, particularly if the award has significant value to group members.

Overemphasis on process. The other focus for many American managers is on process. As quality improvement efforts have grown during the last decade, American managers learned to look more closely at the process by which their organizations achieve results. Using such methods as clarifying requirements, standardizing procedures, and establishing measures for key steps in the work process, this approach does bring more of the consistency in work process required for continuous improvement.

While a process emphasis can work very well, many of the problems in today's organizations are not in process but in the behavior of leadership. Most of today's problem-solving tools are useful for addressing process issues, but they are often not adequate for addressing behavioral issues. Many of the tools assume that all problems are a function of the system while the efforts to resolve them often add complexity, bureaucracy and paperwork. Too often the tendency is to simply create a new system, rather than conduct an in depth analysis to identify and deal with the root cause. For example, we have seen organizations experience a significant increase in injuries resulting from unsafe work practices. The site management responded by creating a new safety incentive system, without considering how much attention they paid to safety on the plant floor or the frequency of safety discussions and feedback. Solutions such as this well-intentioned safety award program increase costs and compete with other activities that have more value to an organization.

The seemly ever-increasing bureaucracy in process-oriented organizations can create the perception that "All management cares about is the paperwork." Employees who work in a process-oriented culture tend to attack problems and change processes without identifying or even considering that the problems may be rooted in behavior and personal relationships -- behavior and relationships that are the primary responsibility of leadership.

Many companies have adopted an observation-based safety process to improve compliance with safety procedures. Alas, simply implementing the basic elements of a behavioral observation process does not always work either. Companies can encounter problems in getting employees to conduct needed observations, in getting employees to effectively provide feedback to other employees, and with employees filling out observation checklists informally without conducting observations. These problems may indicate an overemphasis on the reporting process and a lack of adequate attention to the values required to successfully implement safety improvements.

Other general management problems. Individual managers may conversely place too much emphasis on values, interpersonal relationships, or behavior without adequate focus on process and results. Such an emphasis does not usually characterize an entire organization because the organization simply would not survive. It can occur, however, in the leadership of such smaller groups as teams or committees within the organization. Normally such groups have little conflict and everyone likes one another -- but they are not very productive. They often seem to take on a life of their own and go on forever, with frequent changes in direction but little actual progress toward problem resolution.

The special role of leaders

Leaders have two special roles in creating a culture that values safety. First, in their personal actions, they must demonstrate value for safety through what they do and say. They must consistently make statements supportive of safety and exhibit behavior consistent with those stated values for safety. This aspect of the leaders' responsibility is represented in the left half of the diagram in Figure 1. Second, leadership is responsible for the formal systems that support safety (the left half of the diagram in Figure 1). By fulfilling these two roles, leadership can typically achieved a balanced emphasis on each of the elements in Figure 1.

Leaders communicate what they value through what they say and do. This is often referred to as "Say/Do" correspondence. The matrix presented in Table 1 graphically represents this correspondence. While important for all employees, this correspondence is even more critical for those in a leadership role. A manager in an organization can talk about the importance of safety or not, and the manager can act in a safe manner or not. If the organization has a strong value safety, it will respond as indicated in each of the cells represented in Table 1. Effective leaders that both verbalize the importance of safety and act in ways that support safety should be recognized, rewarded, and ultimately promoted (assuming that they are performing well in other areas). If a leader is not talking with employees about safety, promoting safety practices, providing feedback, discussing its importance in safety meetings, and verbally supporting safety in other ways, an organization that values safety will ensure that the leader gets training and coaching that encourages more active verbal support of safety. Similarly, if a leader is not acting in ways that demonstrate the importance of safety through personal actions such as paying attention to safe and unsafe work practices, consistently following safety procedures, following up on safety related work orders, ensuring employees have time to participate in safety meetings, and other actions, again, the organization that values safety will ensure that the leader receives additional training and coaching to help them "show the way" through their personal actions. Hopefully, an employee who did not verbalize support for safety, nor demonstrate its importance through personal actions would be screened out during the hiring or selection process. If a leader falls into this cell of the matrix, an organization with an effective safety culture would address them quickly through training and coaching and, in extreme cases, even termination.

Leadership's second role is to take responsibility for the formal systems illustrated on the left side of the diagram in Figure 1 above. Management is responsible to ensure that formal systems establish responsibility and accountability for safety. This responsibility requires that they ensure that the formal management systems and processes are well designed, effectively implemented, and that they continue to be supported until formally dropped, modified, or replaced.

		Act Safely	
		Yes	No
Talk Safety	Yes	Recognize Reward Promote	Train & coach
	No	Train & coach	(Pre-select) Train Coach Terminate

Table 1 – The "Say/Do" matrix shows the possible correspondence between safety talk and safety actions.

One of the biggest dangers is systems drift, changes that occurs gradually, over time. Often effective systems deteriorate so gradually that no one recognizes the change until the system routinely allows shortcuts that ultimately result in serious injury. A few years ago, a major utility had an effective observation process that required the lead to observe anytime a lineman was working on a utility pole. The company's safety process was very effective in minimizing incidents and the organization became one of the best in the utility industry. Over a period of several years, the organization went through a series of changes that increased the emphasis on productivity, this combined with the lack of serious injuries, created an environment in which the lead began to get involved in other tasks that took him away from the routine observations, allowing the jobs to be completed more quickly. As a result, other safety practices begin to drift, until the organization experienced seven very serious injuries in a single year. Predictably, the initial response to the first incident was to consider discipline for the lead who had failed to conduct observations. On closer examination, the organization recognized that the entire system had failed. Over a period of years, managers had gotten caught fighting brush first and not regularly monitored the leads, leads had stopped observing linemen, and linemen had grown complacent in their attention to safety. These changes did not occur all at once; rather they were gradual, occurring so slowly that no one noticed the degradation of the process until serous injuries began to occur on a regular basis.

Notice that this story illustrates drift in both the formal monitoring systems and behavior of managers, supervisors and working employees. The stated values did not change, but personnel working in this system became complacent. The complacency was in part a result of the organization's success, that is, it came about because employees did not see anyone getting injured. No one recognized that behavior was no longer aligned with the formal procedures. Nor did anyone recognize employees had grown too comfortable with shortcuts that placed them at risk of serious injury. Behavior was no longer aligned with the formal process nor the stated values.

Conclusions

Creating an accountability process that prevents this kind of gradual systems drift is perhaps one of our greatest challenges. Stated differently, one of our greatest challenges is creating systems that ensure consistency of purpose. Had the organization in the example above had a more effective process for collecting and analyzing how well their process was functioning at each level of the organization, they would have been better able to recognize the growing complacency. Such data might have helped the organization respond proactively to prevent the unfortunate incidents that costs several millions of dollars. Having the right information to assess complacency, and a process that ensures its use, is our most effective solution.

Bibliography:

Alvero, A., Olson, R., & Austin, J. (1999, Oct). Behavioral Safety Research: Discoveries To Guide Your Decisions. Behavioral Safety NOW Conference, Las Vegas, NV.

Cook, S. and McSween T. E. (2000) The Role of Supervisors in Behavioral Safety Observations *Professional Safety*, Oct.

Geller, E. S. (2000, Oct) Behavior-Based is Not Enough: Non-behavioral Principles to Strengthen Your Safety Process. Behavioral Safety NOW Conference, Reno, NV.

McSween, T. E. (2003) The Values-Based Safety Process: Improving Your Safety Culture with Behavior Based Safety. John-Wiley & Sons, New York, NY.

Paper TH005 - 4

Integrated Management Systems and Safety Culture

Operational Discipline as the Foundation of Manufacturing Excellence

Author: Robert Walter

Company: AntiEntropics, Inc.

Prepared for Presentation at
American Institute of Chemical Engineers
2005 Spring National Meeting
Process Plant Safety Symposium
Atlanta, Georgia
April 11-13, 2005

December 2004

Integrated Management Systems and Safety Culture

Operational Discipline is the Key.

How important is safety culture to the success of a true integrated management system? For several years organizations have been realizing the similarity in basic requirements of performance based, management system oriented regulations and industry standards. Compare the elements of the following:

- OSHA's Process safety management regulation,
- USEPA's Environmental risk management program rule,
- ISO 9000:2000 quality management system requirements
- ISO 14000 environmental management system requirements
- Office of Pipeline Safety's pipeline integrity management programs regulations

You will see that these mostly common sense rules and standards are all asking for the same thing. A business management system with a philosophy of:

- know what you have,
- analyze it as necessary to understand it, its risks, and its needs,
- operate it properly,
- maintain it properly,
- know how you change it over time,
- communicate the changes appropriately,
- train your employees and key contractors on it as needed, and
- step back to look at yourself once in a while to see how you are doing.

Throw in profitability as a concern that needs management and could be served by these common elements and guess what? You end up with a total business management system – an integrated management system.

Many organizations were duplicating efforts with personnel safety, process safety, environmental responsibility, quality, and profitability improvement systems. But now they are seeing the benefits of building a total business management system to take advantage of these similarities by integrating their management systems. We talk about a good safety culture, but in reality when we take the stance of an integrated management system, we really mean simply *a good organizational culture*. That is, a culture founded on a high level of operational discipline.

One company in the midst of building an integrated management system asked us to assist their management team to create a common focus by using our operational discipline workshop. They were faced with a challenging task, but after the session, the team understood how the concept of a good organizational culture, one with a high level of operational discipline, could not only help them develop the integrated management system, but that a high level of operational discipline in each employee would help it succeed. A high level of operational discipline supports a good safety culture, environmental culture, quality culture and profitability culture.

What is operational discipline?

The 15 characteristics of operational discipline themselves are something people unconsciously seek in individuals, teams, and organizations during any interaction. It is most like something sensed or perceived, as being present or not, of being satisfactory or not, but not something you can truly quantify. A person's, teams', or organization's level of operational discipline is in constant flux as perceived by an outsider based upon observing their actions.

People, teams, and organizations make many decisions to act in a day, week, or year - their actions display their true value for a high level or low level of operational discipline.

The positive impact of a high level of operational discipline is seen and measured in the next level - what we call Operating Excellence. This consists of the everyday things we can measure or quantify – this is where we begin measuring our *Safety Culture* – this is also where we begin building an integrated management system. Safety data, yields, productivity, operating tasks, maintenance tasks, product quality specs, uptime, profit, compliance with internal or external rules or systems (like OSHA, ISO or EPA) can all be tied together. These are the things that can be easily quantified (if a company chooses to do so) and actions can be taken to correct deficiencies, in many cases, through statistical process control methods. A high level of operational discipline will produce the best results possible in terms of operating excellence and, even in the event of a catastrophic process safety or business event; help ensure an appropriate response through an integrated management system.

Looking back on experience working with some high-performing client teams in the chemical processing and refining industries, we realized those teams and their members had certain characteristics in common. They had some sense of common personal values and attitudes that inspired those around them to want to give the best effort possible to anything they were faced with. Whether the project involved safety management, quality management, environmental management, or was profit-oriented did not matter.

We began to create a list of these underlying characteristics. To validate the concept, we presented the initial list to hourly employees assigned to our procedure development project. They innately understood it. After only a brief overview, they were discussing operational discipline easily and giving real-life examples of how it reinforces a positive safety culture and overall work environment.

We realized we were identifying specific characteristics in manufacturing workers and managers that fit the highest level of the *affective* domain of Bloom's taxonomy. In the 1950s, educational psychologist Benjamin Bloom led a team to develop the taxonomy (or classification) for performance objectives that became the basis of all performance-based training. To summarize his team's work, they identified three distinct domains within which learning and level of performance move from simple to more complex. Bloom's taxonomy is briefly summarized below:

- **Cognitive Domain**
    - This domain addresses pure thought, from simply gaining knowledge to synthesis and evaluation. (For example, from knowing the hazards of a task to evaluating specific risks and troubleshooting.)

- **Psychomotor Domain**
    - This domain addresses skill performance; it ranges from simple imitation to naturalization, or mastery, of the skill. (For example, from operating a lathe to producing an object on the lathe to specification with minimal waste and no injuries.)

- **Affective Domain**
    - This domain is often overlooked in workplace training. The concept of operational discipline seems to fill that gap. The affective domain addresses the demonstration and internalization of attitudes and values. It ranges from the level of receiving (or awareness of an attitude) to the highest level, characterizing, that is, total ownership of a value. (For example, regularly demonstrating one's respect for another's world view.)

Business Excellence: The Peak

How does the concept apply in the workplace? All organizations seek to attain levels of productivity, safety, and quality that cause employees, outside investors, and other companies in their industry to respect the overall business excellence they have achieved. When a company is recognized for achieving this goal, its practices become a model for others to imitate.

When we ask workshop participants for examples of organizations that, in their opinion, have attained business excellence, they respond with names such as IBM, NASA, Ford, Microsoft, and DuPont. Participants in the operational discipline workshops often mention DuPont as an example of a company that has attained business excellence particularly for its strong safety culture.

Operating Excellence: The Path

However, business excellence can only be attained by first achieving another type of excellence--operating excellence. Operating excellence is dependent upon measurable activities such as:

- maintaining well-designed facilities and equipment
- hiring qualified people at all levels
- providing training required for workers to do their jobs
- providing accurate company information to all employees
- adhering to personnel and process safety requirements
- adhering to risk management program requirements
- adhering to environmental standards, and
- adhering to quality standards.

The defining feature for elements of operating excellence is that they are measurable and observable. This is the level at which behavior-based safety starts measuring behaviors. An audit of any of the above items provides feedback on areas that may need improvement to reach operational excellence or show where it has been achieved.

Operational Discipline: The Power

We came to the conclusion that common principles, values, and attitudes held by an organization's employees are key factors in achieving operating excellence. While these characteristics are not easily quantifiable, they have a discernible impact on the ability of an organization to achieve operating and business excellence. Where operational discipline is "high," success is more likely. Where operational discipline is "low," success is less certain. The principles, values, and attitudes can be seen as what drives operating excellence--that is, *operational discipline*.

The Definition of Operational Discipline

The concept is simple. It applies to you directly, both in the workplace and in life in general. Operational discipline is: *a consistent pattern of desirable behavioral choices that supports successful human activity.*

Achieving a pattern of behavior that represents operational discipline relies upon developing and displaying values and attitudes in one's performance that enhance individual, team, and organizational success. Understanding the concept of operational discipline encourages a holistic view of the concept of quality, an integrated business management system (see Figure 1).

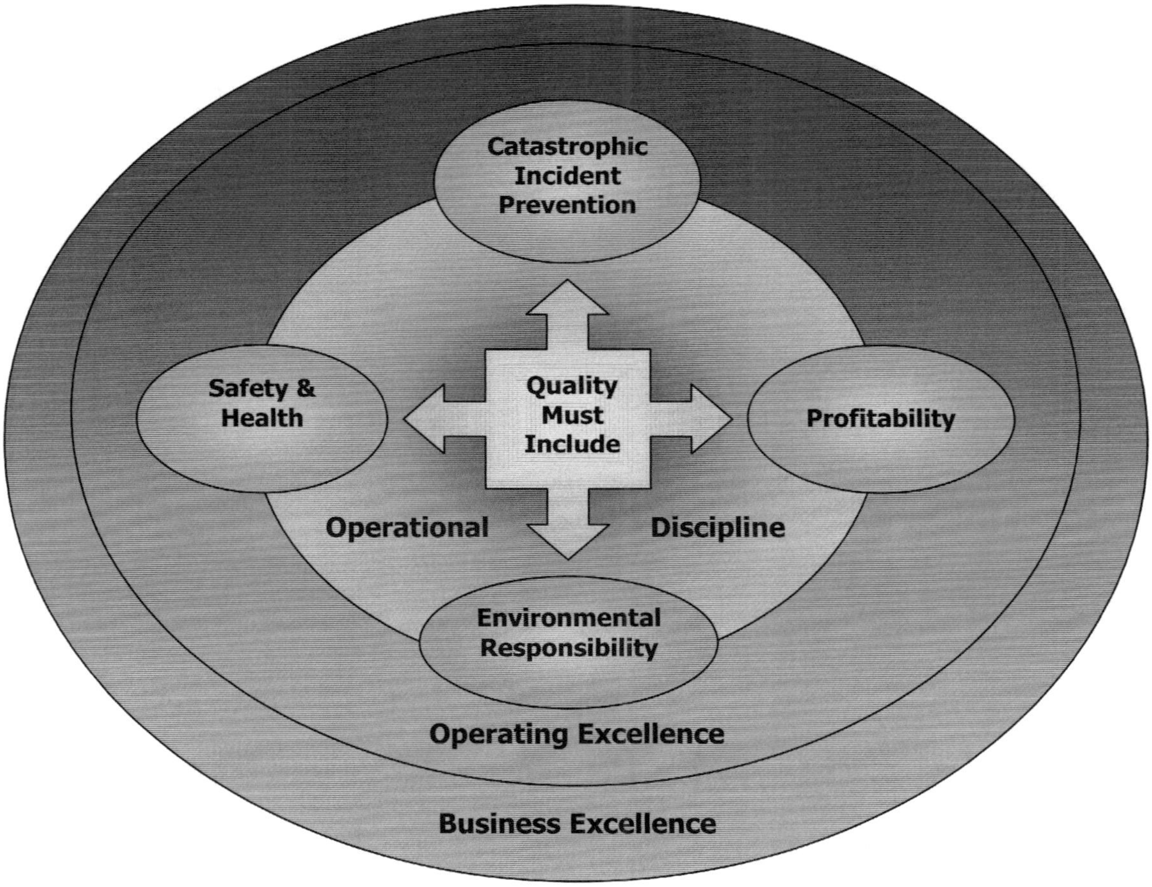

Figure 1.

Enhancing Performance through Shared Values

The specific principles, values, and attitudes are called ***characteristics of operational discipline***. These characteristics are what each of us unconsciously seeks in others during every interaction. These characteristics help to achieve operating and business excellence in the areas of process safety, personnel safety, environmental responsibility, quality, and profitability. When displayed consistently, they support an organization's goal of attaining a state of sustainable growth.

The 15 characteristics of operational discipline were drawn from experience working with many organizations and observing how people interacted within their job description, as well as how they interact when teamed with co-workers. Whether a company embraced the formal "team" approach or simply assigned a group of people to a common goal, we observed that some teams worked like a dream while others resembled a nightmare. Can we know which behaviors naturally support a healthy team, a team effective at pursuing operating excellence and possibly resulting in business excellence?

An individual, work team, or company displays operational discipline and promotes it in others when he demonstrates 15 specific characteristics in his or her actions. We call the groupings of characteristics Interactive Dependency Clusters. Every individual, team, or organization must strive to foster value for displaying all of the characteristics when appropriate. However, they can be grouped based upon the increasing level of interaction implied as follows:

Interactive Dependency Clusters

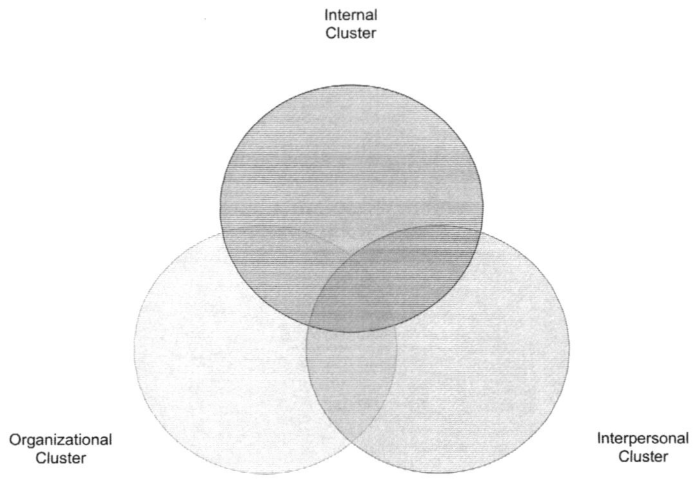

Figure 2.

Internal Cluster

These characteristics come from within an individual and are not necessarily reliant upon interaction to display.

- Take personal responsibility for actions
- Honor commitments
- Seek outcome-based results
- View problems as opportunities
- Use effective time management techniques

Interpersonal Cluster

Opportunities to display these characteristics are fueled by interaction with one or more people.

- Respect and understand others' worldviews
- Seek fairness in all exchanges
- Share recognition
- Value the life and health of all
- Use active, two-way communication

Organizational Cluster

Opportunities to display these characteristics typically arise within groups of people working together—whether families, teams, or companies.

- Seek to perform required tasks for your position
- Assume leader and follower roles as appropriate
- Use systems to reach goals and fix a system when needed
- Desire to use resources efficiently
- Trust that others have a high degree of operational discipline

People teams and organizations exhibit a high level of operational discipline when they consistently display behaviors that demonstrate their value for these characteristics.

The Characteristics Apply to Everyone

The 15 characteristics are desirable when found in any employee, from the president of the company to the organization's newest member in an entry-level position.

Perfection is not the goal. The best analogy is that the characteristics are like tools you have in your tool kit but use only when needed. The test for whether you value any one characteristic (and to what degree) is taken when opportunities arise to display it. Did it come easily? Personal use of the characteristics can be evaluated this way.

The workshop invites participants to use the concept to analyze the following:

- characteristics displayed by a personal role model
- personal strengths in regard to operational discipline
- tools to help practice operational discipline
- how it enhances safety, the environment, quality, and profit
- how it enhances team performance in everyday tasks
- operational discipline in company values, and
- operational discipline in company management systems.

Results of Implementing the Workshop

Since the concept was developed, several different types of refining and chemical processing facilities have piloted the one-day workshop with their hourly and management teams. Although created for the chemical processing industry to help address the four ever present business drivers of safety, environmental responsibility, quality and profitability, the concept has also been used with telecommunication companies, software development firms, the U.S. army, non-profit organizations, universities, other government agencies, and unions throughout the US, Europe, and Asia.

We recently facilitated four one-day pilot workshops for 50 refinery management team members. The group had a very positive response. This experience was not only enlightening for them but also encouraging to our team, as evidenced by the following excerpts from the course evaluations:

- I feel I can use operational discipline as a valuable tool in my job.
- Thought provoking--I like the level playing field that the 15 characteristics provide for all.
- This session brought to light and gave terminology to strengths we sense in individuals we like to work with.
- The role model exercise made me realize why my role model succeeded.

- I think it will help me to lead my own team and work with management.
- The self-assessment was a real eye opener.
- The 15 characteristics emphasize better decision making by focusing on these elements before acting.

Even more significant were unsolicited comments from senior managers at the facility. They saw this approach as a welcome departure from the typical "pep talk" being widely used today to influence plant culture. "This down-to-earth approach to improving human performance is refreshing and easily understood at all levels," one manager commented. "Everyone needs some measurable criteria with which to gauge his or her own performance," was another interesting comment from a senior manager who attended one of the sessions. The collective observation of the refinery training staff was, "The program is easily presented and cuts across all levels of employees."

The Connection to Behavior Based Safety

Using the concept of operational discipline by engaging in deductive learning experiences – that is, learning by way of discovery through guided discussion and team exercises – has had special appeal to facilities attempting to implement behavior based safety programs.

It is well-known that the behavior based safety approach just doesn't automatically "take" at every site. One client reported rapid acceptance at 50% of their facilities but instant rejection at the others. What was the difference? As one client stated, it may be that the characteristics of operational discipline need to be reinforced at those sites. Awareness must exist from the highest levels of corporate management, through the business units, to the teams, and the individual workers that these characteristics are valued and will be sought internally. A high level of operational discipline seems to be the underlying values-system that enables behavior based safety to succeed.

The Seeds of a Total Values-Based Integrated Management System

Futurist authors Wacker and Taylor in their book *The 500 Year Delta: What happens after what comes next;* predict a management style that we will eventually see all organizations move toward. You may detect that your company is already planting seeds to achieve it.

They predict that the rapid pace of change we face in technology, social structure, government leadership, and many other aspects of life will drive businesses to realize that they must base their decisions on something more permanent, something that does not need to change with the seasons. The basis they identify is a shared set of core values. This acceptance and realization will eventually result in a management style based not on objectives, but on a finite number of incontrovertible beliefs never subject to a proof test. These must be sensed as important and shared by everyone from the CEO to the newest employee in the company. In a chaos-based world in which objectives are constantly overwhelmed by variables, a values-based management style assures that decisions ultimately arrange themselves to serve the good of the whole. The concept of operational discipline helps people evaluate their decisions and actions to align with those fundamental values we each seek in one another. It supports the move toward a values-based integrated management style.

Summary

To answer our opening question: How important is safety culture to the success of a true integrated management system? The answer is: extremely important. The challenge is for each individual to find a personalized way to encourage a positive impact on organizational culture through their actions, their team's actions, and the organization's collective behavior.

Literature Cited:

Wacker, Watts and Taylor, Jim, 1997. *The 500 Year Delta: What happens after what comes next?* HarperCollins Publishers, Inc.

Portions of this paper were excerpted with permission from:

Walter, Robert, 2002. *Discovering Operational Discipline*, © AntiEntropics, Inc. HRD Press, Amherst, Massachusetts

Walter, Robert, Foottit, Richard, and Dannar, Jerry, *Discovering Operational Discipline, Occupational Health and Safety Magazine, Vol. 71, 4:86-89,* © Stevens Publishing, April 2002

About the Author:

Robert Walter (*rjwalter@antientropics.com*) is president of AntiEntropics, Inc. (*www.antientropics.com*) a performance enhancement consulting firm specializing in OSHA, EPA, and ISO management system development. Voice: 301-865-0015

The book *Discovering Operational Discipline* is available online at Amazon.com and at HRD Press – 1-800-822-2801

Paper TH005 - 5

Essential Elements of a Sound Safety Culture

W. L. Frank, P.E.
ABS Consulting
Wilmington, Delaware
wfrank@absconsulting.com

Prepared for Presentation at
American Institute of Chemical Engineers
2005 Spring National Meeting
Process Plant Safety Symposium
Atlanta, Georgia
April 11-13, 2005

February, 2005

UNPUBLISHED

Essential Elements of a Sound Safety Culture

W. L. Frank, P.E.
ABS Consulting
Wilmington, Delaware
wfrank@absconsulting.com

ABSTRACT

This paper addresses management's responsibility for facilitating and nurturing a sound safety culture within their organization. Essential activities that promote a safety culture are discussed, thematically grouped under five management responsibilities: lead, listen, help, inspire, and improve. Probing questions for self-evaluation are proposed.

The paper cites some of the leading writers in the field on the role of culture in attaining organizational goals, and draws from personal experiences in auditing the effectiveness of safety systems.

1. INTRODUCTION

The OSHA process safety management (PSM) standard (29 CFR 1910.119) went into effect on May 26, 1992. The standard mandated the implementation of chemical accident prevention program requirements that were patterned after existing practices followed by industry leaders and endorsed by numerous industry technical organizations and trade associations. The EPA risk management program rule (40 CFR Part 68), which was issued on May 24, 1996, required a similar chemical accident prevention program and stipulated full compliance by June 21, 1999.

More than 12 years have passed since the promulgation of the PSM standard, and many (presumably, most) covered facilities have devoted considerable time, effort, and expense to developing and implementing chemical accident prevention programs intended to comply with the regulatory requirements. Nevertheless, serious near misses, incidents, and program audits all too frequently reveal that some regulated facilities are not realizing the anticipated benefits of PSM.

A reasonable question to ask as we approach the 13th anniversary of OSHA PSM is "What is limiting PSM performance for certain organizations?" While the answers can be myriad, it is the author's perception, based upon numerous PSM compliance audits, safety program reviews, and organizational culture evaluations, that the cause is not uncommonly rooted in the beliefs and attitudes of the organization. PSM programs can be highly sophisticated, well documented, and proudly showcased, yet not produce the intended results because the organizational culture does not fully value, support, and establish an imperative for PSM excellence.

This paper addresses the role of organizational culture as a critical determinant of safety performance.

2. DEFINING SAFETY CULTURE

A significant, and growing, body of literature exists on the topic of safety culture. In fact, individuals with time on their hands may wish to sample the nearly 15,000,000 "hits" they would find through a Google™ search on the term *safety culture*.

Not surprisingly, the literature contains a variety of definitions for *safety culture*; consequently, we should begin by defining what we mean by the term.

The UK Health and Safety Executive defines safety culture as " . . . the product of the individual and group values, attitudes, competencies and patterns of behavior that determine the commitment to, and the style and proficiency of, an organization's health and safety programs." (1)

The International Atomic Energy Agency defines safety culture as ". . . that assembly of characteristics and attitudes in organizations and individuals which establishes, as an overriding priority, that safety issues receive the attention warranted by their significance." (2)

Speaking specifically of process safety, Jones describes safety culture as ". . . the combination of group values and behaviors that determine the manner in which process safety is managed." (3) For our purposes, Jones' rather more succinct definition will be used.

These three definitions, and likely many of the variants that we could find through our Google search, have a common, basic theme that implies a causal relationship: the values of the group (e.g., corporation, plant, shift) help shape the attitudes of the individual which, in turn, play a significant role in determining individual actions.

3. WHY SHOULD WE BE CONCERNED WITH SAFETY CULTURE?

Successful implementation of a process safety program requires a multitude of individual actions; hopefully, most in compliance with the policies and procedures established to ensure the success of the program.

The basic premise of this paper is that engineered controls, safety systems, management systems (such as PSM), and procedures can only take us so far with respect to safety performance, in that these systems and controls all can be made less effective (perhaps rendered ineffective) by human action or inaction. Consequently, a sound safety culture, as a positive determinant of individual actions, is essential to maximizing the effectiveness of such systems, controls, and procedures.

Safety literature is full of graphic examples of the catastrophic consequences of safety culture deficiencies. Investigation reports for events such as the disaster at the Chernobyl nuclear reactor or the explosion and fire on the Piper Alpha oil platform commonly point to key decisions or actions made by individuals or groups that either caused or exacerbated the event. Readers can likely point to incidents within their own organizations that were caused by the ill-considered decisions and actions of an individual or small group. Such actions or decisions, often lumped under the term "lack of operating discipline," may be the primary focus of a less than rigorous incident investigation. However, closer analysis (i.e., an attempt to identify the true root causes of the incident) might reveal that the actions or decisions, while contrary to established policies or procedures, were entirely consistent with the cultural norms of the organization. In fact, the actions or decisions might even be predictable in light of the cultural norms.

A nonindustrial example that most people can relate to would be vehicle speeds on major expressways and interstates. Barring traffic congestion, we can predict that a significant percentage of drivers will operate their vehicles appreciably above the posted speed limit. This occurs even though most of these drivers have some degree of awareness that the risk of accident and injury increases at higher speeds. At the same time, most of these drivers have likely never experienced, either personally or vicariously through the experience of a friend or loved one, the consequences of a high speed crash. Furthermore, many of these drivers may see a positive benefit-to-cost ratio, considering the time they feel they are saving compared to their perception of a low potential for receiving a speeding ticket (". . . just as long as there are a few cars faster than me . . ."). Regrettably, some degree of speeding has become a cultural norm in our society. We may grumble when a particularly aggressive driver blows past us but, let's face the facts, we are not likely to dial 9-1-1 on our cell phone and make a report, as we ourselves are cruising along just above the speed limit ("I'd be a hazard if I weren't keeping up with the traffic flow . . .").

To bring our focus back to process safety, how is the above example different from the operator who takes a shortcut starting up the process knowing full well that, while unlikely, there are serious potential consequences to doing so (". . . everybody does it . . . nothing bad has happened yet . . . it's midnight shift and I'm not likely to get caught . . .")? Weak safety cultures permit (perhaps, facilitate) this sort of thinking. [‡]

Examples of safety culture failure need not be catastrophic in their scope. Some organizations, having what they believe to be "complete" systems (e.g., "fully implemented" PSM), may be frustrated by plateaued performance (or dismayed by deteriorating performance). Attention to, and enhancement of, the safety culture may be the most effective solution to such performance problems.

For example, a plant may have a well-defined piping inspection program with a sophisticated database for tracking inspection schedules, recording thickness measurements, and trending the results. However, the system as provided is of little value in supporting the plant mechanical integrity program if the plant culture does not reinforce the responsibilities for completing inspections on schedule, promptly analyzing the data, and implementing proper responses to anomalous conditions.

[‡] The freeway driving example becomes more relevant to industrial safety culture when it is noted that some organizations known for their particularly sound safety cultures include a strong emphasis on off-the-job safety in their programs.

Civil engineering professor Dr. Henry Petroski, in his book *To Engineer Is Human*, notes that systems designed to address high consequence events, such as bridges designed with the intention of avoiding collapse, often include thousands of elements.[4] Each element is important, but failure of a key element can shift an unanticipated load to other elements, exceeding their design capacity and causing successive failures that result in system collapse. An example is the link pin that failed on the I-95 bridge over the Mianus River in Connecticut in June 1983. The failure of this single pin, which held up one corner of a span of the roadway decking, led to the collapse of a 100-foot section of the bridge and caused three fatalities.

Analogously, we should seek to ensure that safety culture does not become the weak element whose failure leads to the progressive collapse of the safety program.

4. THE CHALLENGE AND THE GOAL

The challenge facing managers is to foster and sustain a safety culture that establishes and supports high standards of individual and organizational safety performance. This challenge resides with management since they, as the business leaders, have the responsibility to:

- Set the safety tone for the organization,
- Establish standards of performance,
- Provide the resources necessary to achieve goals, and
- Provide the reinforcement necessary to ensure performance.

As Dr. Edgar Schein, professor of management at MIT, has noted ". . . one could argue that the only thing of real importance that leaders do is to create and manage culture . . ." [5]

What would success look like when we get there? Lagging indicators of safety performance are easy to identify: reduced frequencies of near misses and incidents; fewer injuries; and increased productivity, to suggest a few. But what can we look to as leading indicators of a sound safety culture?

Some authors have written about the *safety conscious work environment*. Carroll and Hatakenaka have described such a work environment as having the following characteristics: [6]

- Safety and quality are valued above cost and schedule;
- Employees believe they can bring concerns to management and feel safe from harassment, intimidation, retaliation and discrimination;
- Employees believe that management will act appropriately on their concerns;
- An employee-concerns program offers a legitimate alternative to bringing concerns to management; and
- There are mechanisms to monitor and maintain these characteristics.

Note the significant element of worker trust in management that is embodied in these characteristics.

While much of this paper deals with the roles and responsibilities of management with respect to the safety culture issue, it must be noted that management can only enable and nurture a sound safety culture; they cannot mandate it. Individual behavior is largely determined by individual values and beliefs. Managers cannot effectively control by dictates the values and beliefs that determine an individual's behavior on the Sunday midnight shift.

Consequently, the path to a sound safety culture involves a process of maturation. The lowest level of maturity is assigned to rule-driven, discipline-based systems where individual actions are motivated by fear of reprisal for detected offenses ("detected" perhaps being the operative concept). A degree of safety culture maturation brings us to the stage where workers recognize the self-preservation benefits associated with following safety rules. Ultimately, the level of safety culture maturity that organizations should aspire to is one where safety has become a moral imperative (i.e., it has become part of the value structure of the organization and individual employees).

To draw a conventional worker safety analogy, these three levels of increasing safety culture maturity could be typified, respectively, by the following attitudes:

- "I don't need to wear my safety glasses – the foreman isn't around."
- "I need to put on my safety glasses before I start up that grinder."
- "Hey, Joe . . . Where are your safety glasses?"

The highest degree of safety culture maturity is evidenced when we, the workers, assume not just the responsibility for our own safety, but also recognize a responsibility to our coworkers, to the organization, and to society.

Michael Deak, formerly corporate director of safety and health for DuPont, has observed that organizations that excel in a certain activity invariably have integrated value for that activity into their corporate culture, whether it be safety, quality, customer service, etc. [(7)] Managers may draw favorable conclusions about the increasing maturity of the safety culture based upon reduced frequencies of loss events and near misses. However, greater confidence will be warranted when observed behaviors and discussions with staff reveal that the process safety program is being actively supported *because people believe that it is the right thing to do.*

5. FIVE KEY MANAGEMENT RESPONSIBILITIES FOR NURTURING SAFETY CULTURE

As noted previously, the amount of material written on the topic of organizational culture in general, and safety culture in particular, borders on overwhelming. Reviews of a small fraction of this material, observations from audits of safety programs, and the fortunate opportunity to "grow up" in an organization with a strong safety culture allows the distillation of five key responsibilities for managers seeking to nurture a strong safety culture. These are:

- Lead
- Listen
- Help
- Inspire
- Improve

As illustrated in Figure 1, these responsibilities are overlapping and mutually supportive. Management behaviors intended to support one responsibility often support several.

The balance of this paper focuses on defining the roles required to satisfy these responsibilities. Probing questions for use in self-evaluation are also provided.

Figure 1: Five Key Management Responsibilities

5.1 Leading

True safety leaders establish (or sustain) the organization's vision and values with respect to safety performance. Leaders must clearly define for the organization issues such as: What does safety mean to us? What does success look like? What do we expect of the organization and each individual within it?

True safety leaders communicate the organization's vision and values in a variety of ways, including, but not limited to, the following:

- In writing – via policy documents, procedures, and written communications;
- Orally – through presentations and day-to-day communications; and
- By their actions – through the priorities they set, by providing resources to support the safety effort, by reinforcing safety responsibilities within performance reviews, and through active involvement in safety-related activities.

Figure 2: NASA Screen Saver

It is important that management communications on safety be credible, consistent, and persistent. We must remember that even if we profess "Safety is No. 1," safety initiatives can remain in competition for attention and resources. The panel that investigated the Columbia shuttle disaster noted that NASA management, which on one hand stressed the agency's values for mission safety, chose to distribute to all managers the screen saver shown in Figure 2. The screen saver continuously counted down to the scheduled completion date for the NASA contribution to the International Space Station. This was a matter of significance to the NASA space shuttle group, since the shuttles were required to haul materials to the space station, and delays in shuttle missions jeopardized the space station schedule. [8] Is it unreasonable to suspect that this screen saver potentially sent a conflicting message to shuttle managers about the relative importance of schedule and safety?

Management's persistent support of safety in the face of external drivers is important. Management priorities for safety must be sustained through good times and bad, and in periods of calm and crisis. This can be a particular issue in organizations seeking to make substantive changes in safety culture, as workers may monitor management performance to gauge management's sincerity in support of the safety program. There can be a tendency for the workforce to skeptically hold back, awaiting evidence that management really is serious about what they profess to accomplish. Any management wavering (e.g., delay of programs because of poor financial performance in a quarter) may be taken as evidence of lack of sincerity in the safety message. This can create the perception that this is just another "program du jour" that can be waited out until it is replaced by the next.

Leadership includes setting, maintaining, and enforcing standards of performance. Key among the causal factors of the two space shuttle disasters was the concept of *normalization of deviance*. Briefly, this term refers to the gradual slippage in standards that can occur when a system failure (engineered or human) repeatedly occurs, without the realization of the potential negative consequences projected for such a failure. Repeated system failures, without severe consequences, can engender an organizational myopia that ultimately prompts the organization as a whole to discount the potential significance of the system failure.

With Challenger, the issue was thermal degradation of the O-rings in the solid rocket boosters. Repeated discoveries of O-ring damage without serious consequences eventually led NASA staff to discount the potential for loss of mission spacecraft and crew . . . until it happened. With Columbia, the issue was insulation from the external fuel tank breaking free and striking the shuttle. Repeated known occurrences of insulation strikes without severe damage to the shuttle eventually led NASA staff to discount the potential for loss of mission spacecraft and crew . . . until it happened.

Repeated nonconsequential deviations from design intent (O-rings were not supposed to burn through and insulation was not supposed to strike the shuttle) can lead to the perception that such events are now "normal" and do not warrant concern. The adoption of such an attitude requires a lack of awareness (or an unwillingness to admit) that most catastrophic events require a coincident combination of undesired factors to occur and that loosening the controls on any factor makes the undesired catastrophic consequences, however remote, less remote.

The parallels in our industry are myriad. Process hazard analysis (PHA) recommendations are not "designed" to go unaddressed for years, yet the failure to implement a given recommendation may not immediately result in a process incident. Failure to implement a safe work practice (e.g., hot work permit) as "designed" may not cause an injury or injury . . . at least not this time. It may be decades before the failure to rigorously comply with a relief valve testing schedule results in an in-service valve failing to provide its intended protective function and causing the catastrophic failure of a pressure vessel. Management awareness of and failure to address such situations constitute a normalization of deviance that weakens both the safety culture and the safety program.

Managers must carefully monitor safety performance and initiate corrective actions when warranted in order to prevent normalization of deviance and the performance deterioration it can bring.

Professor Schein suggests five main ways in which leaders can influence culture: [9]

What You Pay Attention To. "Leaders communicate their priorities, values, and concerns [by what they] ask about, measure, comment on, praise, and criticize . . . By not paying attention to something, a leader sends the message that it is not important." Is it surprising, in an environment of conflicting priorities and with reduced resources and time, that "smart" employees only put their full efforts into what they perceive their manager values?

Your Reactions to Crises. "A leader who faithfully supports espoused values even when under pressure for expedient action communicates clearly that the values are really important." If we are not going to tolerate employee shortcuts, how can we accept management shortcuts?

How You Behave (Role Modeling). "Leaders can communicate values and expectations by their own actions . . . A leader who institutes a policy or procedure but fails to act in accordance with it is communicating the message that it is not really important or necessary." Consider the example of the manager of a remote petroleum production facility whom employees regularly observed speeding to work in the morning. "Don't do as I do, do as I tell you" is not an effective management style.

How You Allocate Awards. What you reward signals what you value. "Formal recognition in ceremonies and informal praise communicate a leader's concerns and priorities . . . Failure to recognize contributions and achievements sends a message that they are not important."

Criteria for Selection and Dismissal. "Leaders can influence culture by their choice of criteria for recruiting, selecting, promoting, and dismissing people." Clearly communicate your safety values and concerns by reflecting them in criteria and requirements for success in the organization.

Table 1 suggests some critical evaluation questions addressing the topic of management leadership in safety culture development and maintenance.

5.2 Listening

Leadership, as described above, clearly requires communication of values, priorities, and concerns to the organization. However, to be effective, communications must flow both ways. Management usually does not have all of the information, understanding, and experience to make all of the proper decisions in a vacuum, and should seek and accept input from those closest to the hazards. In addressing high performance work systems (i.e., management systems designed to address the unique issues of very high risk endeavors) the Institute of Nuclear Power Operations (INPO) noted that "The worker is the best source of information about the weakness of the organization." [10]

Table 1: Evaluation Questions – Leading

- Has the corporate safety vision been clearly and unambiguously stated?
  - How has/is this communicated? How, and how frequently, is the message reinforced?
- Does every employee understand his or her responsibilities?
  - How do you know? How do you confirm this?
- Do job descriptions include a substantive element addressing safety responsibilities?
  - Are these roles and responsibilities reinforced in performance reviews?
  - Does good or poor safety performance actually have a tangible impact on employee compensation?
- How often do employees take safety shortcuts?
  - Do they do so because they think they can "get away with it" or don't appreciate the potential harm to others?
  - What is management's response to such transgressions?
- Are risk takers rewarded for their successes?
  - How likely is a manager to look the other way if an employee successfully takes a risk that produces a short-term gain for the organization?
- Is the safety message subject to "reinterpretation" as it moves down the organization?
  - Is safety subject to "situational ethics"?
  - How would you know? What do you have in place to prevent this?
- Is there sufficient active involvement in the safety program at all levels of management/supervision?
  - Does line management perceive safety to be someone else's job (e.g., the Safety Department)?
- Are conflicting priorities allowed to confuse the safety message?
  - Does safety truly equal or exceed production in importance?
  - Do management actions consistently demonstrate this?
- Does the organization perceive the safety program as an initiative that is here to stay or a "program du jour" that can be waited out?
  - How do you know?
- Do managers understand the concept of "normalization of deviance"?
  - What controls exist to detect and respond to patterns of abnormal conditions or practices before they become the norm?
- At what level of maturity do you perceive the safety culture to be?

Management should provide and monitor redundant channels for routine communications and, if circumstances warrant, provide alternate communication paths for those who feel their message is not getting through. Routine communications channels could include employee meetings, near-miss reporting and safety suggestion programs, and the trite-sounding but important "management by walking around." With respect to alternate channels, some organizations have found it necessary and effective to establish an anonymous telephone "hot line" to provide a safe communications path for those feeling the need for such.

Some organizations have "bad news filters" within the communications chain between the shop floor and the management offices. This may be a feature of the organizational culture as a whole, or it may be caused by a few individuals who feel they can best protect their career by handling all issues locally. Management should be alert to the potential for such filters. A prudent manager who seldom (or never) hears bad news should have the curiosity to find out why this is the case.

An organization open to effective communications will accept communications that do not go through channels. A culture that stimulates effective communications accepts, indeed encourages, the sharing of bad news. However, policy statements notwithstanding, some employees may initially not feel comfortable sharing bad news or contrary opinions with managers. Managers must make it clear, in their statements and through their actions, that messengers will not be shot. To do so, they must be willing to accept negative feedback and give people license to speak openly.

Listening without responding is only the appearance of listening. Management should respond to employee concerns in a substantive and timely fashion. Lack of response or consistently tardy responses will discourage future communications. Consider how ineffective a safety suggestion program would be for stimulating communications if the safety suggestion forms were allowed to accumulate in someone's in-box and were not responded to.

A nuclear power plant was shut down by order of the Nuclear Regulatory Commission for nearly 2 years to remedy safety culture issues. The total cost to the company due to lost revenue and remediation programs was over $1 billion. The following statement, made by a safety culture consultant regarding the former management of the nuclear power utility, is suggested as a case study in communications failure:

> "Never forget that previous management failed so miserably, not because they were not intelligent and not because they did not understand clearly what successful economics looked like in a competitive environment. They failed because they were arrogant, dismissive and refused to listen to the issues and concerns of the people who make this place run." [11]

Finally, consistent with the guidance in the prior section of this paper, management should consider identifying and publicizing communications successes. It was previously noted that the characteristics of a safety-conscious work environment included a significant theme of worker trust of management. Effective, two-way communications on safety issues should help engender such employee confidence.

Table 2 suggests some critical evaluation questions addressing the responsibility of management to listen to employees as part of an effective communications program in support of safety culture development and maintenance.

5.3 Helping

Management should take care to provide employees the resources needed to achieve safety program goals. Such resources include, but are not limited to, staffing, funding, tools, education, and time. Clearly defining roles and responsibilities and providing frank, timely performance feedback are also important.

Table 2: Evaluation Questions – Listening

- What means do employees have to share information and concerns with management?
  - How effective are they? How do you know?
  - Are communications that bypass the chain of command tolerated?
- How often do you hear bad news?
  - How would you detect a bad news filter in the communications chain?
- Is there a safe, secure alternative conduit for those who feel their message is not getting through normal channels?
  - If so, is this publicized sufficiently to be effective?
  - How frequently and how well is it monitored?
- How are those with bad news or a contrary opinion on safety issues treated?
  - Discouraged? Ignored? Tolerated? Rewarded? Punished?
- Are there systems in place to ensure timely response to issues raised?
  - Is response time monitored?
  - Are slow responders counseled to do better?
- How do you prevent organizational rank from being an intimidating factor in direct communications?
- Are significant communications successes shared, celebrated, and/or rewarded?

Management must also provide employees the authority and support commensurate with their safety responsibilities. Patrick O'Hara suggests:

> "Management sets the tone, and needs to focus more on removing barriers to safety excellence than controlling behavior. Thus, management has to treat employees fairly, be consistent in their actions, and be a good role model for the safety culture it desires." [12]

Perhaps there is no behavior more unfair than to assign a responsibility to a person, then deny them the authority necessary to successfully fulfill that responsibility. Some common examples of failure to support employees in their safety responsibilities include:

- Making a person responsible for ensuring the timely inspection of pressure vessels or the testing of relief valves, but failing to provide the authority to require that equipment be made available for testing and inspection
- Shifting the burden of proof for safety concerns to "Prove to me that this is unsafe"

We tell operating staff that they have the authority to shut down the process if they believe it is necessary to ensure the safety of personnel and the plant. Management can most effectively help employees fulfill this responsibility by giving them the basis for a high degree of confidence that they will be supported, and not second guessed after the fact, should they implement the decision to shut down.

To be perceived as being important, safety responsibilities must be treated as important. Adequately staffing to satisfy safety roles communicates the importance of the tasks, while assigning new responsibilities to an individual known to already be overworked says "This really isn't important." Note that adequate staffing is not just a matter of headcount. Requisite knowledge and skill must also be considered and, in some instances, organizational level may also be an issue. For example, a first-line supervisor, acting as PHA coordinator, might be disadvantaged when negotiating with department heads regarding provision of team members for a PHA.

Managers help employees by facilitating their success. Such facilitation includes providing employees the training necessary to understand the significance of their safety responsibilities and to develop the skills necessary to perform those responsibilities.

One of the most effective means of helping employees fulfill their responsibilities is to set rational priorities and to ensure that required tasks are truly important and still needed. Ray Witter, formerly safety director for Monsanto and a CCPS staff member, relates the story of a young engineer at a remote site who was told that one of his most critical responsibilities was to change daily the chart on a circular chart recorder and mail it to a senior staff member at the downtown offices. One day, needing to visit the downtown offices for other reasons, he decided to hand deliver the chart. In doing so, he had the opportunity to ask the senior staff member why the information on the chart was so important. This, in turn, gave the senior staff member the opportunity to ask why the plant was sending him these charts. No one could determine the original reason for the requirement, only that it was no longer necessary. [13] Management should periodically review program requirements and discontinue tasks/initiatives that are no longer adding value.

As noted before, managers communicate through their actions, perhaps more eloquently than by their words. Consider the case of the newly hired engineer who was given the responsibility for coordinating a PHA because, as he was bluntly told, he did not yet know enough about the process to be helpful elsewhere. How was this individual helped to see the importance of process safety?

Table 3 suggests some critical evaluation questions addressing the topic of how management can help employees and thus foster safety culture development and maintenance.

Table 3: Evaluation Questions – Helping

- Are there any "unfunded mandates" in your safety program?
  - Are the time, resources, and costs associated with safety tasks explicitly budgeted?
  - Or, are the responsible parties expected to be sufficiently creative to handle such tasks within existing resource constraints?
- When safety program deliverables are late, is a root cause analysis performed to determine why?
  - Are system problems addressed to improve performance?
- Are safety responsibilities an explicit part of job descriptions?
  - Is feedback on the performance of safety responsibilities given regularly (more frequently than the annual performance review)?
- Are staff members provided the time, funding, and opportunities necessary to receive training on their safety responsibilities?
- When a safety concern is raised, where is the burden of proof placed?
  - Does the concerned party have to prove something is unsafe for the issue to get full consideration?
  - Do employees have the authority to shut down an activity that they feel is unsafe?
  - Is there a system in place to ensure independent review of major safety-related decisions?

5.4 Inspiring

Management has the responsibility to inspire the organization to the desired level of safety performance. As shown in Figure 1, there is considerable overlap between the five essential activities. Inspiring performance draws, in particular, on *leading* (by example), *listening* (and responding), and *helping*.

This paper addresses three particular aspects of inspiring performance:

- Keeping the organization focused on what is at stake
- Celebrating successes
- Challenging the organization to stretch for higher performance

Perhaps one of the greatest obstacles to maintaining an exemplary safety performance is . . . an exemplary safety performance. Professor Petroski suggests, with respect to civil engineering and the design of structures such as bridges, that "failures appear to be inevitable in the wake of prolonged success, which encourages lower margins of safety." (14) His point is that, absent failures, there are market and aesthetic drivers for building structures ever more cost-effectively and more elegantly (both often meaning, less massively).

While the term "inevitable" might be overly pessimistic, are there not parallels in the process safety field to the point Professor Petroski is trying to make for civil engineers? Managers need to be sensitive to the potential for "removing concrete and rebar" from our safety programs as a consequence of (1) an inflated sense of security prompted by past performance, (2) increased emphasis on the control of safety program costs, and (3) distraction by other high priority issues.

Managers are faced with the challenge of maintaining a sense of vigilance to hazards within the organization. In attempting to do so, managers should keep in mind the distinction between just knowing of a hazard and appreciating its significance. Knowing that a dust is combustible is not the same as appreciating that a sufficient amount of that dust, if suspended in air and ignited, could cause an explosion that might devastate the facility. Management needs to create and maintain within the organization an appropriate sense of vulnerability to ensure that safety responsibilities are perceived to have an appropriate degree of import in light of the hazards they are intended to mitigate.

Many companies with particularly strong safety cultures have one or more significant events in their past that catalyzed the development of that culture, and work to keep the organizational and individual memory of such events fresh. Other companies draw lessons from industry and share them within their organization.

Success is good and should be recognized and celebrated as positive reinforcement of good performance. However, success should not be allowed to engender complacency. It has been suggested that organizations should maintain ". . . a 'humble' attitude towards safety, meaning that good safety performance is never taken for granted." (15) The underlying theme here is that staff must understand that, in process safety as in investing, "past performance is no guarantee of future success."

One possible antidote to complacency is to challenge the organization to stretch for higher performance, which provides the linkage to the next section of the paper. First, however, Table 4 suggests some critical evaluation questions addressing the topic of how management can inspire performance in support of safety culture development and maintenance.

Table 4: Evaluation Questions – Inspiring

- If your safety performance has been good, is the organization becoming complacent; i.e., trending towards the assumption that such performance should naturally be anticipated to continue?
  - How would you know?
  - Are you confident that the organization is still kicking the safety program flywheel often enough to keep it up to speed?
- Have you trimmed back on any safety programs, initiatives, or staffing?
  - If so, how was this justified?
  - Was a management of change process used to evaluate the prudence of the cutbacks?
- How often are relevant incident descriptions or statistics shared with staff?
  - In what manner?
  - Is relevant information from other company locations, or elsewhere in industry, shared?
- What percentage of the organization has directly seen the consequences of a major process incident?
  - How do you bring home the reality of process hazards to the rest?
- Do risk analyses (e.g., PHAs) include the consideration of credible major events?
  - Is the consideration substantive, or are such events dismissed as "unlikely"?
  - Do reviews of PHAs support an assertion that teams appreciate the worst-case hazard potential of the process?
- Do investigations of incidents and near misses reveal a lack of appreciation of the potential consequences of not adequately controlling process hazards?

5.5 Improving

It is hard to run in place, and safety programs that do not emphasize improvement risk sliding backwards. Building on the theme from the prior section, organizations should not declare safety success and significantly curtail their level of effort based upon past performance or based upon the satisfaction of a milestone such as "PSM has been implemented."

Similarly, organizations should avoid the deterioration of performance that can come from a "no more than compliance" mindset. A prior paper expressed concern for the potential that some companies may be forgoing their traditional safety consciousness in favor of such a compliance mentality. [16] Figure 3, for conceptual purposes only, suggests that some companies may be moving towards a common performance asymptote – defined by strict compliance with regulatory PSM requirements. . . no more, no less. Whereas previous company programs for

PSM leaders reflected a cultural heritage and were based upon a strong desire to prevent accidents, a compliance mentality becomes focused on merely conforming to the letter of law.

To stimulate improvement, managers should:

- Establish reasonable, attainable goals that challenge the organization to higher standards of performance;
- Study and learn from mistakes (being especially alert to recurrent themes that may indicate systemic problems); and
- Monitor performance and remove obstacles (*lead*, *listen*, *help*, and *inspire*).

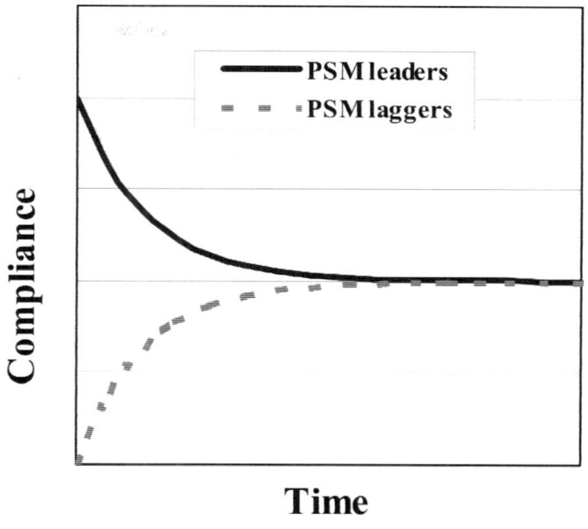

Figure 3: The Compliance Convergence

Lord Kelvin is frequently quoted as having stated: "To measure is to know . . . If you can not measure it, you can not improve it." Managers require tangible performance data to (1) monitor and substantiate progress, (2) justify expenditures (fight the "safety as a cost" myth), and (3) reinforce their interest in safety to the organization.

There are many common errors in measuring safety performance, and one of the more common involves confusing PSM and traditional safety metrics. Many organizations focus solely or predominantly on "slips, trips, and falls" statistics; however, LTI rates and other injury statistics are not adequate discriminators of catastrophic event potential. Catastrophic consequences are too infrequent (thankfully) and costly to monitor as a sole indicator of safety system health. We need to consider alternate metrics such as the number of unplanned shutdowns or the frequency of safety system activations to monitor the effectiveness of PSM programs.

Another common error is measuring only safety system outputs, such as the number of process incidents. The concern is that such metrics are reactive, in that often the damage has already been done. Organizations should monitor a mix of leading and lagging indicators. [17, 18, 19, 20] In addition, it is prudent to select these lagging indicators from as low as possible in the traditional event hierarchy, or pyramid, shown in Figure 4. Ultimately, attitudes can be the most leading of indicators, and some organizations have found carefully designed and implemented attitude surveys to be effective.

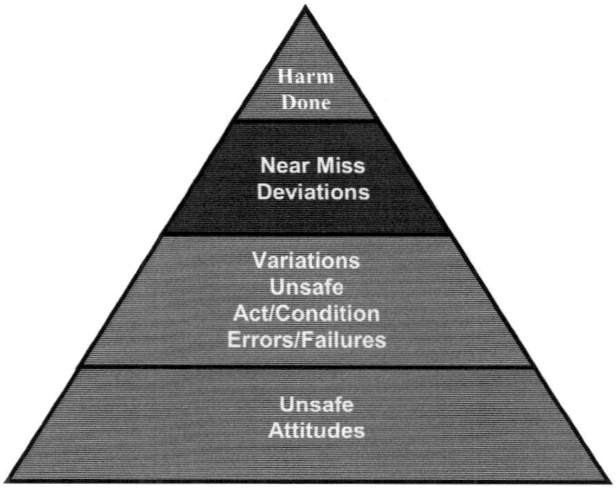

Figure 4: Event Hierarchy

All too frequently, organizations use weak process safety measurement tools such as audits that appear to focus on putting a paper in the files to demonstrate that an audit was done, or incident investigation systems that don't (1) stimulate reporting of all incidents and near misses, (2) identify and correct root causes, or (3) trend results.

Other errors include not measuring/monitoring frequently enough and not paying attention to the results; i.e., inadequate accountability or lack of follow-up to resolve identified issues.

Commonly encountered indicators that PSM performance is not being sufficiently tracked include:

- Significant numbers of overdue PHA, audit, or incident investigation recommendations and/or significant delinquencies;
- Systemic problems with overdue equipment inspections (e.g., pressure vessels, relief valves, piping);
- Equipment inspection records showing deficient conditions, which are filed away but not resolved; and
- Tardy responses (or no responses) to employee safety concerns.

Just like process incidents, failures to meet safety system objectives should be investigated to identify root causes. For example, perhaps the issue is that relief valves are not consistently tested according to the established schedule. Why is this? Is there a logistical issue with the inspection system, or is there a cultural issue that fails to place enough emphasis on meeting the schedule? In either case, management should identify and address the cause in order to improve safety performance.

The causal parallels in the Challenger and Columbia shuttle disasters, detailed in the Columbia investigation report (notably, the recurrent themes of normalization of deviance and dysfunctional communications and decision making), graphically illustrate what can happen when identified cultural deficiencies are not addressed and corrected.

Sound safety cultures place a value on accurately monitoring and improving performance and avoid the types of measurement and follow-up mistakes described above.

Table 5 suggests some critical evaluation questions addressing the topic of how management can promote improved performance in safety culture development and maintenance.

6. SUMMARY

Industry has devoted considerable time and expense to the development and implementation of PSM systems. While there are sufficient examples of successful PSM programs to inspire us all to continued efforts, some organizations are still struggling. Some organizations with ostensibly complete PSM programs still do not achieve the benefits anticipated from PSM. Experience has shown that one possible explanation could be an organizational culture that does not support the implementation of the PSM system.

Table 5: Evaluation Questions – Improving

- Does your safety program clearly contain a continuous improvement theme?
  - Are goals kept attainable, while being ambitious enough to challenge the organization?
- Are relevant metrics prepared and regularly shared with leadership?
  - How frequently?
- Do PSM metrics receive attention comparable to that given to more traditional safety measures?
- Do PSM metrics receive attention comparable to that given to production and quality goals?
- Is there a properly balanced mix of system inputs (leading indicators) and system results (lagging indicators) being monitored?
- Is there real accountability associated with avoidable failures to meet safety goals?
- Do you display curiosity?
  - If a key safety-related report (e.g., relief valve testing status) is not delivered, do you inquire about it?
- Do you hear questions like "Where in the regulation does it say we have to do this?"
- Do incident investigation and audit protocols provide for sufficient expertise and objectivity?
  - How is this ensured?
- Is there a conscious effort to trend results?
  - To determine if improvement is occurring?
  - To identify recurrent, systemic problems?

This paper suggests that, while each member of the organization has a role to play, management must take the lead in fostering a sound safety culture in support of PSM objectives. It is further suggested that management's responsibilities for doing so can be grouped under five overlapping initiatives:

- Leading
- Listening
- Helping
- Inspiring
- Improving

These initiatives have been discussed in detail, and sets of self-evaluation questions have been proposed to aid in determining the status of each within a particular culture.

7. ACKNOWLEDGEMENTS

The author wishes to acknowledge Mr. Steve Arendt for his thoughtful review of, and comments on, this manuscript; Ms. Leslie Adair for her careful editorial attention; and Ms. Angie Nicely for her skill in preparing this manuscript.

8. REFERENCES

1. UK Health and Safety Executive, *Safety Culture: A Review of the Literature*, HSL/2002/25, 2002.
2. International Nuclear Safety Advisory Group, *Safety Culture*, Safety Series No. 75, NSAG-4, International Atomic Energy Agency, Vienna, 1991.
3. Jones, David, *Turning the Titanic – Three Case Histories in Cultural Change*, CCPS International Conference and Workshop, Toronto, 2001.
4. Petroski, Henry, *To Engineer Is Human: The Role of Failure in Successful Design*, Vintage, 1992.
5. Quoted by Collins, David, *Managing Safety Culture*, presentation made to the 503rd meeting of the Advisory Committee on Reactor Safeguards, June 12-13, 2003, http://nuclearsafetyculture.freeyellow.com/ManagingSafetyCulture.ppt.
6. Carroll, John S. and Hatakenaka, Sachi, "Driving Organizational Change in the Midst of Crisis," *MIT Sloan Management Review*, Spring 2001.
7. Deak, Michael S., personal communication.
8. *Columbia Accident Investigation Board, Report Volume 1*, Government Printing Office, 2003.
9. Based upon Schein, as quoted by Gary Yukl, *Leadership in Organizations*, Fifth Edition, Prentice Hall, 2002.
10. Collins, *ibid.* The specific INPO reference that Collins quotes is *Excellence in Human Performance*, 1997. INPO publications are only available to INPO members.
11. Collins, *ibid.*
12. O'Hara, Patrick W., *Nuclear Safety Culture: Competitiveness Through Safety*, January 10, 2000, http://nuclearsafetyculture.freeyellow.com/Nuclear.htm.
13. Witter, Ray E., personal communication.
14. Petroski, *ibid.*
15. Barraclough, Ian and Carnino, Annick, "Safety Culture – Keys for Sustaining Progress," *IEAE Bulletin*, 40/2/1998.
16. Jones, D., Frank, W., and Hobbs, D., "What Are Process Safety Management Audits Telling Operators?," *Hydrocarbon Processing*, October 2000.
17. Arendt, J. Steven, Barrett, Kari, and Lodal, Peter N., *Measuring Process Safety Progress: First Steps*, CCPS International Conference and Workshop, Toronto, 2001.
18. Perry, Bob G., Arendt, J. Steven, Campbell, David J., and Schreiber, Sanford, *If You Can't Measure It, You Can't Control It: ProSmart Process Safety Management*, CCPS International Conference and Workshop, Toronto, 2001.
19. *OECD Guidance on Safety Performance Indicators*, Organisation for Economic Co-operation and Development (OECD) Environment, Health and Safety Publications, Series on Chemical Accidents No. 11, Paris, France, 2003.
20. *Safety Performance Measurement*, Edited by Jacques van Steen, TNO, for the European Process Safety Centre, Institution of Chemical Engineers, Rugby, Warwickshire, United Kingdom, 1996.

Paper

Enhanced Layer of Protection Analysis (LOPA) Method

Al Waller
FMC Corporation

Prepared for Presentation at
American Institute of Chemical Engineers
2005 Spring National Meeting
Process Plant Safety Symposium
Atlanta, Georgia
April 11-13, 2005

December 2004

UNPUBLISHED

ABSTRACT

Layer of Protection Analysis (LOPA) is a valuable risk analysis methodology for the chemical process industries that bridges the gap between qualitative methods and fully quantitative risk assessment (QRA) methods. FMC Corporation has developed an enhanced application of LOPA for performing Safety Instrumented Systems (SIS) evaluations that further reduces the remaining gap with QRA. This paper describes our enhanced LOPA method that remains easy to use but provides more quantitative rigor to address moderately complex dependencies that are often encountered between control systems and protection layers. Initiating event failures are divided into four categories — sensor, logic solver, final element, and others — to allow identification of each independent layer of protection that applies specifically to each initiating event failure category. This negates the tendency for SIS evaluation teams to either under-credit or over-credit layers of protection when using “coarse” standard LOPA methods. An example is provided to compare standard LOPA versus enhanced LOPA methods. An overview of implementation in Microsoft Access is discussed including its benefits in providing a structured yet flexible tool for SIS evaluation and documentation.

SUMMARY

FMC Corporation has developed a methodology based on Layer of Protection Analysis (LOPA) that combines some elements of simplified LOPA techniques with some elements of more rigorous Quantitative Risk Assessment (QRA) techniques. This enhanced LOPA method has a wide range of applications including:

- Safety Instrumented Systems (SIS) evaluations
- Process Hazard Analysis (PHA) recommendations prioritization
- Facility siting evaluations

Strengths of this enhanced LOPA method include:

- Larger number of leaders can be trained (and retained) compared to QRA
- Minimum training required to get active participation from plant operations, maintenance, and engineering personnel compared to QRA
- More realistic results compared to standard LOPA

BACKGROUND

Process Hazard Analysis (PHA) methods cover a wide range of application rigor. What if, Checklist, and HAZOP (Hazard and Operability) methods are generally qualitative while FMEA (Failure Modes and Effective Analysis) and Event Tree Analysis/Fault Tree Analysis methods are generally quantitative.

The qualitative methods are particularly useful for evaluating processes that have a large number of equipment items or process steps. They can be used to screen a large number of scenarios. However, most of these qualitative methods also incorporate semi-quantitative estimates of risk to enable prioritization and selective recommendations (Fryman, 1996). PHA teams often estimate a consequence and frequency for each scenario based solely on judgment.

The quantitative methods FMEA and Event Tree Analysis/Fault Tree Analysis are more rigorous, and therefore more tedious, so they are impractical to apply to a broad range of processes. Typically, the screening results from a qualitative analysis identify relatively few processes for additional quantitative evaluation. Fault Tree Analysis is useful for analyzing complex interactions but requires specialized training for the lead analyst. Fault Tree is also the most difficult for team members to grasp.

Layer of Protection Analysis (LOPA) was developed as a methodology to bridge the gap between the qualitative methods and the quantitative methods. LOPA is a semi-quantitative method that retains some of the advantages of qualitative methods and some of the advantages of quantitative methods. Any company using LOPA or any other risk analysis methodology must adopt a risk criteria standard to allow decision making based on comparison of results.

LOPA also may be used as a consistent method for conducting Safety Instrumented Systems (SIS) evaluations. It can be applied to identify the need for Safety Instrumented Functions (SIF) and for designating a Safety Integrity Level (SIL) for each SIF. It can be applied at the design stage for new installations or as a supplement to a PHA of existing systems.

STANDARD LOPA INTRODUCTION

LOPA is a method of evaluating the effectiveness of independent protection layer(s) in reducing the likelihood or severity of an undesired event. First, a consequence is estimated for an undesired scenario of interest. Next, events are identified that can initiate the scenario and safeguards are identified that may break the chain of events. LOPA uses simple calculations to estimate the frequency of occurrence from the initiating event frequency and the probability of failure on demand (PFD) of qualified safeguards. Only safeguards that meet several tests for independence are included in the calculations. The standard LOPA methodology requires complete independence of each independent protection layer (IPL) from initiating events and from other protection layers.

$$f_{\text{Event}} = f_{\text{Initiating event}} \times PFD_1 \times PFD_2 \times PFD_3 \ldots\ldots \times PFD_i$$

where

f_{Event} = frequency estimate for the undesired event (events/year)

$f_{\text{Initiating event}}$ = frequency estimate for an initiating event (events per year)

PFD_i = probability of failure on demand of each applicable IPL (a fraction from 0 to 1)

The strictest use of standard LOPA uses order of magnitude estimates for initiating event frequencies and for PFD. This strict use results in frequency estimates in factors of 10. However, this is not a requirement of standard LOPA and is not considered as standard LOPA in this paper.

For convenience, PFD can be replaced by RRF (risk reduction factor) to allow use of integers. Also, IPLs can be separated into non-SIS protection layers and SIS protection layers. Separating non-SIS from SIS helps enforce stricter rules for estimating RRF. A Basic Process Control System (BPCS) might be claimed as a non-SIS IPL but should be limited to a maximum RRF of 10. A SIL1 SIF will have an RRF from 10 to 100, a SIL2 will have an RRF from 100 to 1000, and a SIL3 will have an RRF from 1000 to 10,000 (ISA, 1996).

$$f_{\text{Event}} = f_{\text{Initiating event}} \div RRF_{\text{non-SIS 1}} \div RRF_{\text{non-SIS 2}} \div RRF_{\text{SIF 1}} \ldots\ldots$$

where

RRF_i = risk reduction factor of each applicable layer = $1 \div PFD_i$

ENHANCED LOPA INTRODUCTION

In 1997, FMC Corporation began developing a methodology to implement ISA S84.01 (ISA, 1996). A working group of representatives from corporate safety and engineering functions plus representatives from selected site instrumentation and process engineering functions was assembled. Over a period of time, a draft methodology was prepared and pilot tested at several different times as revisions were made. The draft methodology was based on LOPA.

Initially the standard LOPA methodology was used in the early pilot test. The standard LOPA applies each independent protection layer as a complete layer. If there is any dependence by any part of protection layer with either an initiating event or part of another protection layer, then standard LOPA prohibits taking credit. For the FMC working group, this simplification was a significant barrier, especially for modern BPCS systems where a BPCS control loop failure is often an initiating event but the BPCS also provides safeguards. The LOPA book (CCPS, 2001) introduces two alternatives when interdependency is [occasionally] encountered but these were not practical for widespread use at FMC. A hybrid of LOPA and Fault Tree Analysis addresses the lack of precision of standard LOPA (Rothschild, 2004) but requires a Fault Tree Analysis expert to perform the calculations, so this is not practical either for widespread use.

ENHANCED LOPA

Calculations for enhanced LOPA are only slightly more complicated than for standard LOPA. As was shown above:

$$f_{\text{Event}} = f_{\text{Initiating event}} \div \text{RRF}_{\text{non-SIS 1}} \div \text{RRF}_{\text{non-SIS 2}} \div \text{RRF}_{\text{SIF 1}} \;\ldots\ldots$$

By separating initiating events into categories, additional IPLs may be claimed that are only effective for certain initiating event categories.

$$\begin{aligned} f_{\text{Event}} = \; & (f_{\text{IE sensor}} \div \text{RRF}_{\text{non-SIS 1}} \div \text{RRF}_{\text{non-SIS 2}} \div \text{RRF}_{\text{SIF 1}} \;\ldots\ldots) \\ & + (f_{\text{IE logic solver}} \div \text{RRF}_{\text{non-SIS 1}} \div \text{RRF}_{\text{non-SIS 2}} \div \text{RRF}_{\text{SIF 1}} \;\ldots\ldots) \\ & + (f_{\text{IE final element}} \div \text{RRF}_{\text{non-SIS 1}} \div \text{RRF}_{\text{non-SIS 2}} \div \text{RRF}_{\text{SIF 1}} \;\ldots\ldots) \\ & + (f_{\text{IE others}} \div \text{RRF}_{\text{non-SIS 1}} \div \text{RRF}_{\text{non-SIS 2}} \div \text{RRF}_{\text{SIF 1}} \;\ldots\ldots) \end{aligned}$$

where

$f_{\text{IE sensor}}$ = frequency for initiating event of BPCS sensor failure (/year)

$f_{\text{IE logic solver}}$ = frequency for initiating event of BPCS logic solver failure including input/output card failures (/year)

$f_{\text{IE final element}}$ = frequency for initiating event of BPCS final element failure (/year)

$f_{\text{IE others}}$ = frequency for initiating event of other failures (/year)

EXAMPLE

The following simplified example demonstrates some of the differences between standard LOPA and enhanced LOPA. Figures 1A through 2B illustrate a tank (T-1) that is fed liquid from a supply header. Similar to the example in CCPS (2001, see Continuing Example 1), an upset in level control could result in overfill of the tank and liquid release through the pressure relief valve. For the purposes of this example, the liquid is flammable and almost completely rains out during an overfill release from the pressure relief valve; no flammable cloud is produced at the relief valve. A dike is present that will contain the overfill release for up to 30 minutes. The scenario of interest is an overfill of less than 30 minutes that will result in a pool fire with damage to the tank and instrumentation estimated to cost $20,000. Ignition is assumed to occur and a firewater deluge system is excluded from consideration to provide conservatism. Other scenarios that could result in the overfilled liquid flowing beyond the dike and causing greater consequences should also be evaluated but are beyond the scope of this example.

Figures 1A through 2B are explained in paragraphs following the figures. Each figure includes a matrix below the diagram that indicates each layer of protection's applicability to various initiating events.

All values used for initiating event frequencies and IPL risk reduction factors are intended for this example only. Users should evaluate values from a variety of sources and modify appropriately for local conditions and maintenance practices.

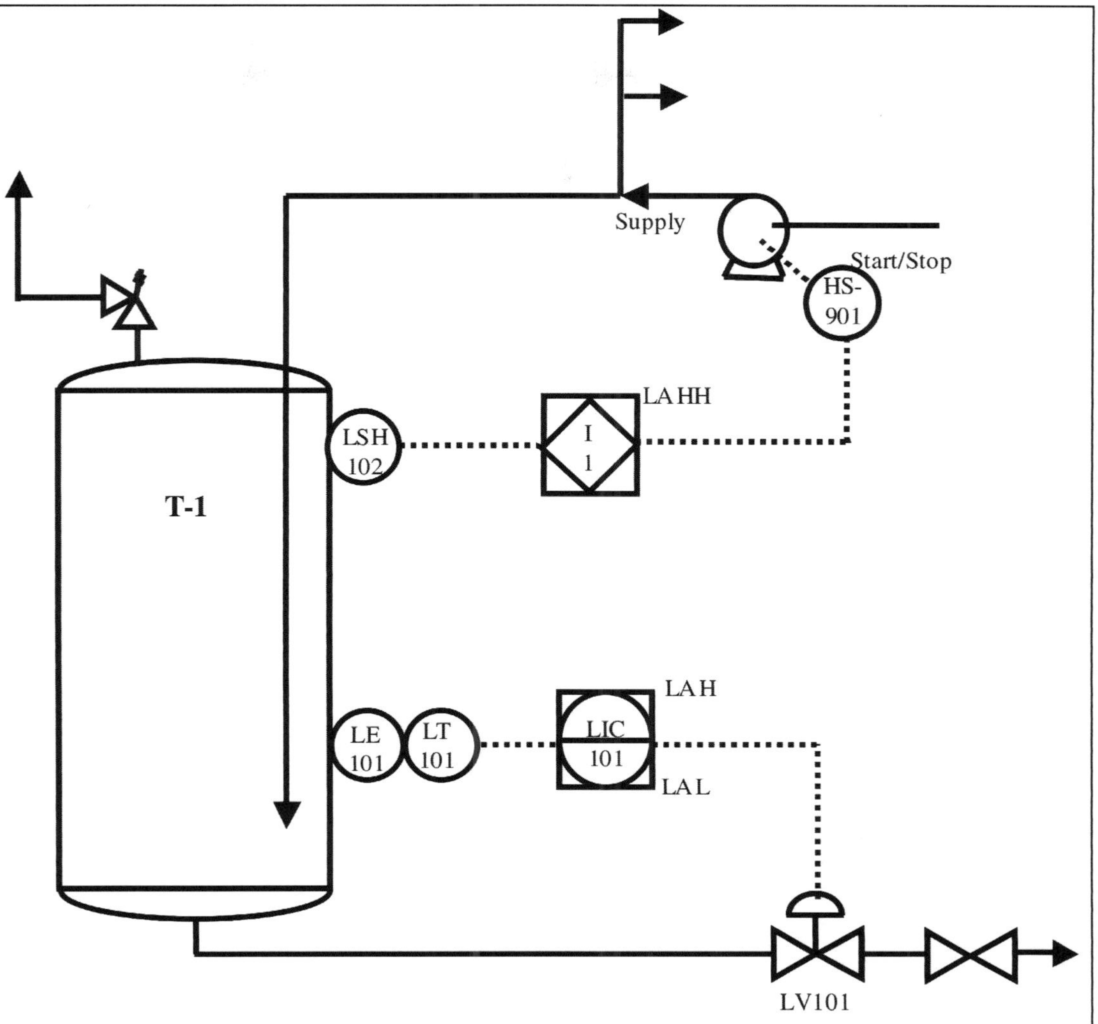

Layer Category	Description	BPCS	Other
Initiating Events	BPCS loop LIC101 fails closed	✓	
	Human error, closes valve		✓
Non-SIS IPLs	BPCS I-1 Overfill Protection		✓
SIS			

FIGURE 1A – Existing Example Using Standard LOPA

Figure 1A illustrates the analysis for the existing system using standard LOPA. Either a failure of BPCS control loop LIC101 (consisting of a level element/transmitter as the sensor, a BPCS controller, and a control valve) or a human error closing a manual valve could initiate the overfill scenario. An interlock is available in the BPCS that uses an independent level sensor and an independent starter switch to trip the supply pump on demand. Standard LOPA could credit the BPCS interlock as an IPL for the human error initiating event but not for the BPCS level control loop failure because the logic solver is common.

Table 1A quantifies the frequency and risk of the scenario based on the matrix included as part of Figure 1A.

TABLE 1A – Estimated Risk Using Standard LOPA for Existing System

Layer Category	Description	Initiating Events		
		BPCS	Other	
Initiating Events	BPCS Loop LIC101 fails closed	1×10^{-1} /yr		
	Human error, closes valve		1×10^{-1} /yr	
RRF of Non-SIS IPLs	BPCS I-1 Overfill Protection		10	
RRF of SIS				
	Frequency subtotals	1×10^{-1} /yr	1×10^{-2} /yr	1.1×10^{-1} /yr
	Total Risk (@ $20,000 per event)			$2200 /yr

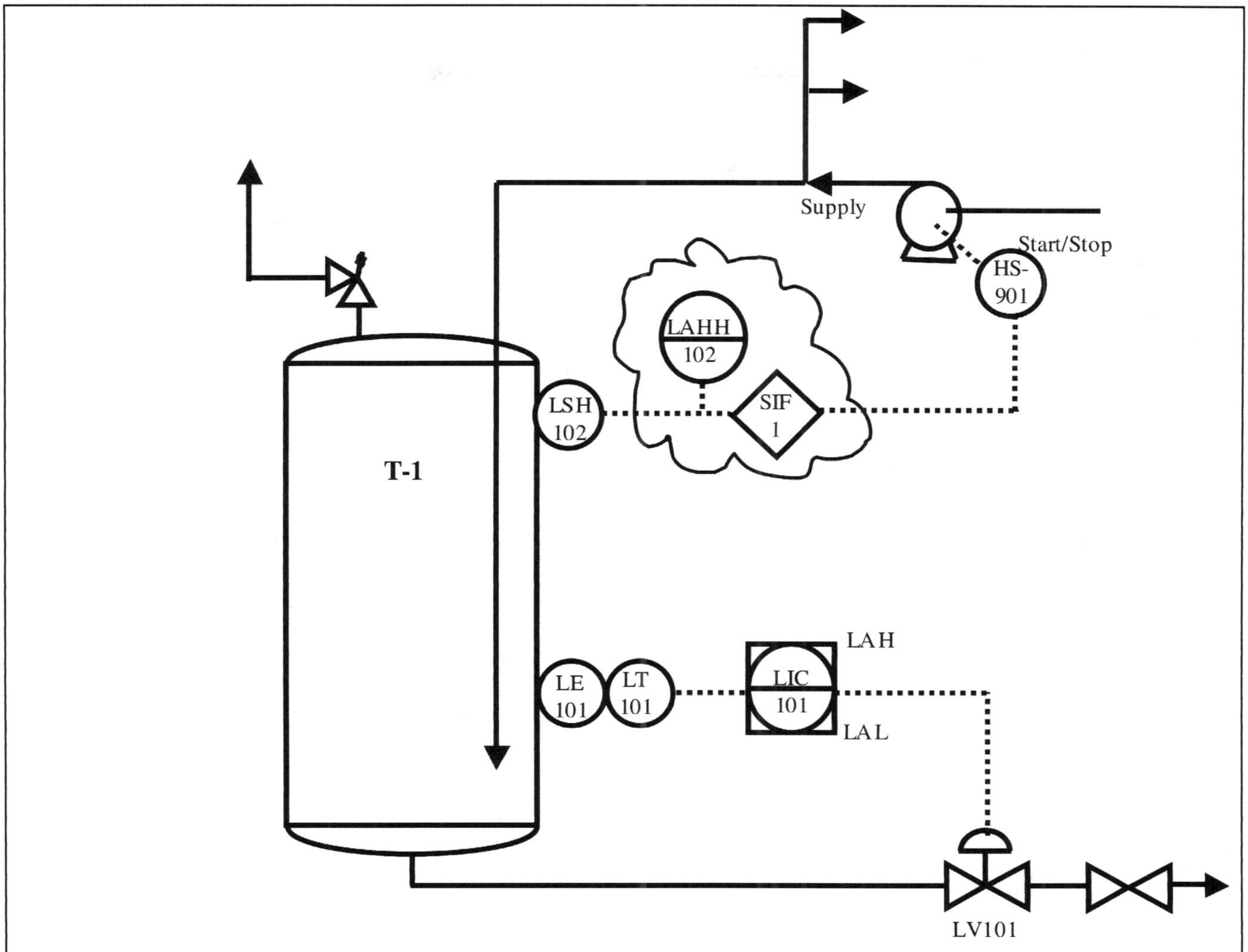

Layer Category	Description	BPCS	Other
Initiating Events	BPCS loop LIC101 fails closed	✓	
	Human error, closes valve		✓
Non-SIS IPLs			
SIS	SIF-1 Overfill Protection	✓	✓

FIGURE 1B – Mitigated Example Using Standard LOPA

Figure 1B illustrates one proposal to mitigate the existing system and the corresponding analysis using standard LOPA. Converting the BPCS interlock to an SIF by adding a relay or Programmable Electronic System provides complete independence for the Overfill Protection IPL. Standard LOPA can credit the SIF as risk reduction for the human error initiating event and for the BPCS level control loop failure. A panel alarm (LAHH102) is included in the proposal to alert the operator that the supply pump shutdown has been triggered.

Table 1B quantifies the frequency and risk of the mitigated scenario based on the matrix included with Figure 1B. For this example, a risk reduction factor of 32 is used for the proposed SIF to represent the geometric mean of the range allowed for SIL 1 (10 to 100). A better estimate for the proposed SIF-1 can be estimated depending on the type of logic solver selected and on the test interval chosen.

TABLE 1B – Estimated Risk Using Standard LOPA for Mitigated System

Layer Category	Description	Initiating Events		
		BPCS	Other	
Initiating Events	BPCS Loop LIC101 fails closed	1×10^{-1} /yr		
	Human error, closes valve		1×10^{-1} /yr	
RRF of Non-SIS IPLs				
RRF of SIS	SIF-1 Overfill Protection (generic SIL 1)	32	32	
	Frequency subtotals	3.1×10^{-3} /yr	3.1×10^{-3} /yr	6.3×10^{-3} /yr
	Total Risk (@ $20,000 per event)			$125 /yr

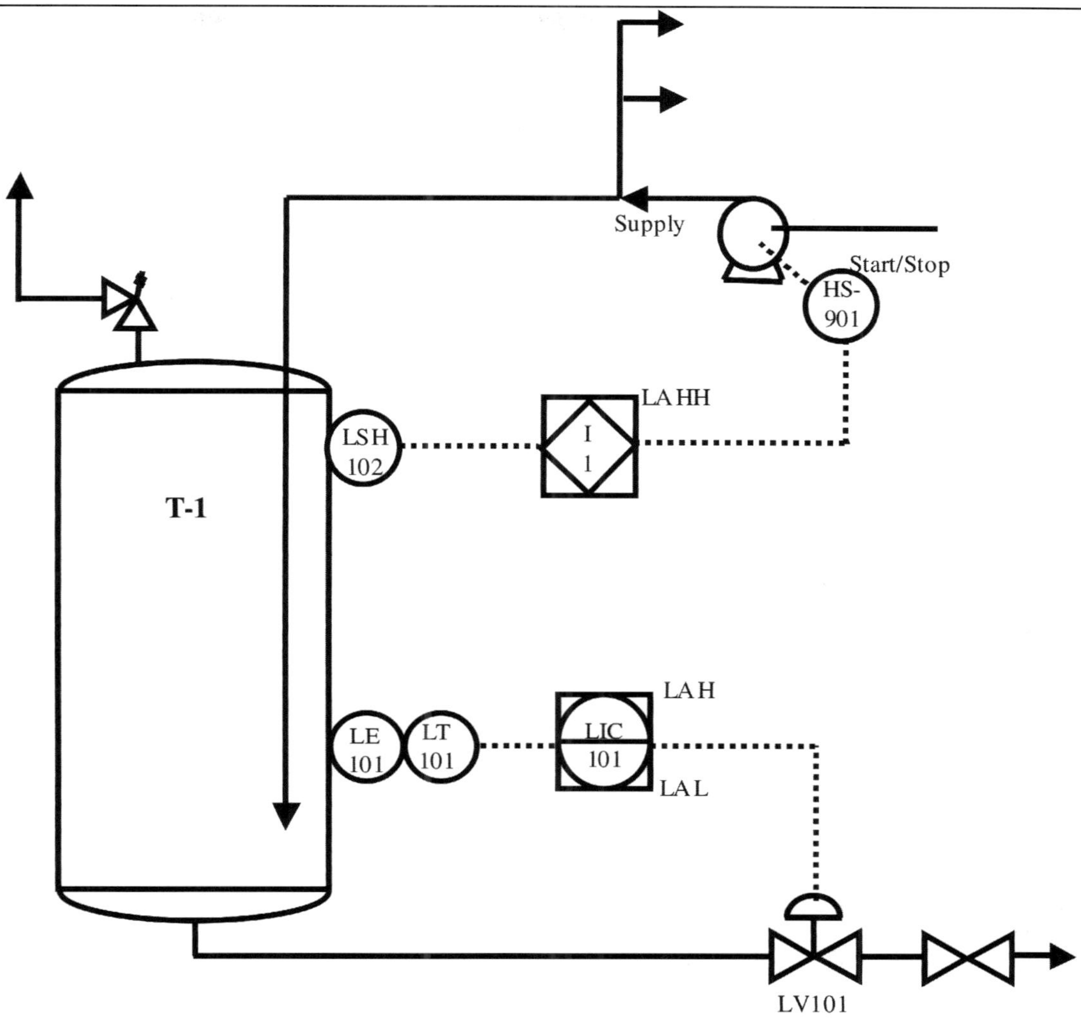

Layer Category	Description	Sensor	Logic Solver	Final Element	Other
Initiating Events	BPCS loop LIC101 fails closed	✓	✓	✓	
	Human error, closes valve				✓
Non-SIS IPLs	BPCS I-1 Overfill Protection	✓		✓	✓
SIS					

FIGURE 2A – Existing Example Using Enhanced LOPA

Figure 2A illustrates the analysis for the existing system using enhanced LOPA. Either a failure in the BPCS control loop LIC101 (consisting of a level element/transmitter as the sensor, a BPCS controller, and a control valve) or a human error closing a manual valve could initiate the overfill scenario. An interlock is available in the BPCS that uses an independent level sensor and an independent starter switch to trip the supply pump on demand. Enhanced LOPA credits the BPCS interlock for all initiating events except the BPCS logic solver failure because the logic solver is common.

Table 2A quantifies the frequency and risk of the scenario based on the matrix included with Figure 2A. This risk estimate from enhanced LOPA is significantly less than Table 1A results from standard LOPA due to the more realistic modeling.

TABLE 2A – Estimated Risk Using Enhanced LOPA for Existing System

Layer Category	Description	Initiating Event Categories				
		BPCS Sensor	BPCS Logic Solver	BPCS Final Element	Other	
Initiating Events	BPCS Loop LIC101 fails closed	5×10^{-2} /yr	2×10^{-2} /yr	3×10^{-2} /yr		
	Human error, closes valve				1×10^{-1} /yr	
RRF of Non-SIS IPLs	BPCS I-1 Overfill Protection	10		10	10	
RRF of SIS						
	Frequency subtotals	5×10^{-3} /yr	2×10^{-2} /yr	3×10^{-3} /yr	1×10^{-2} /yr	3.8×10^{-2} /yr
	Total Risk (@ $20,000 per event)					$760 /yr

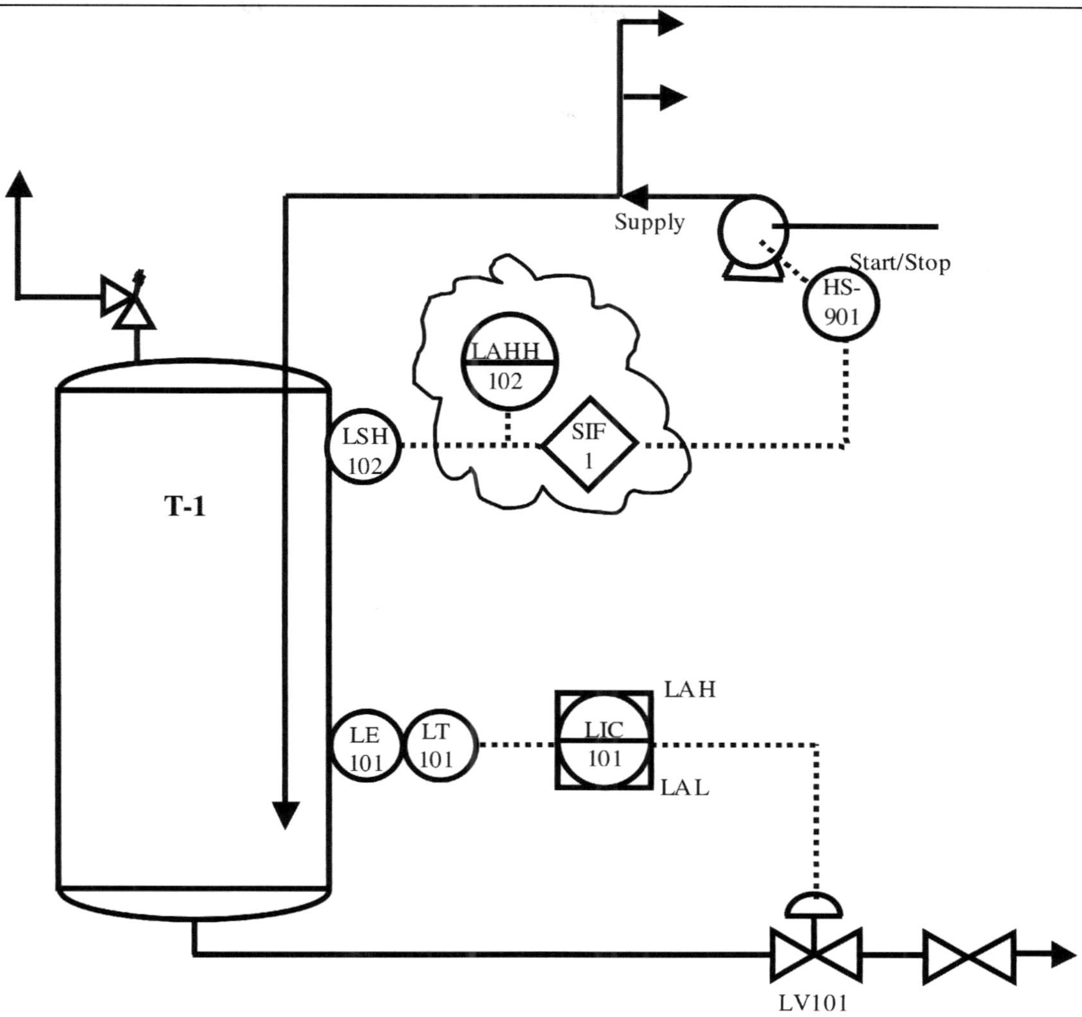

Layer Category	Description	Sensor	Logic Solver	Final Element	Other
Initiating Events	BPCS loop LIC101 fails closed	✓	✓	✓	
	Human error, closes valve				✓
Non-SIS IPLs					
SIS	SIF-1 Overfill Protection	✓	✓	✓	✓

FIGURE 2B – Mitigated Example Using Enhanced LOPA

Figure 2B illustrates the same approach as Figure 1B to mitigate the existing system using enhanced LOPA. Again, converting the BPCS interlock to an SIF by adding a relay or Programmable Electronic System provides complete independence for the Overfill Protection IPL. Enhanced LOPA can credit the SIF as risk reduction for the human error initiating event and for failure of any component in the BPCS level control loop.

Table 2B estimates the frequency and risk of the mitigated scenario. Again, a risk reduction factor of 32 is used for SIF-1 pending selection of the logic solver and a test interval. This risk estimate from enhanced LOPA is equal to Table 1B results from standard LOPA because the mitigation eliminated the interdependence of the BPCS.

TABLE 2B – Estimated Risk Using Enhanced LOPA for Mitigated System

Layer Category	Description	Initiating Event Categories				
		BPCS Sensor	BPCS Logic Solver	BPCS Final Element	Other	
Initiating Events	BPCS Loop LIC101 fails closed	5×10^{-2} /yr	2×10^{-2} /yr	3×10^{-2} /yr		
	Human error, closes valve				1×10^{-1} /yr	
RRF of Non-SIS IPLs						
RRF of SIS	SIF-1 Overfill Protection(gen SIL 1)	32	32	32	32	
	Frequency subtotals	1.6×10^{-3} /yr	6.3×10^{-4} /yr	9.4×10^{-4} /yr	3.1×10^{-3} /yr	6.3×10^{-3} /yr
	Total Risk (@ $20,000 per event)					$125 /yr

Table 3 compares estimated risks for the LOPA methods above. Standard LOPA yields an apparent incentive three (3) times the incentive estimated by the enhanced LOPA method. A Risk Benefit Evaluation (RBE) (Garcia and Lewis, 1998) would probably justify the proposed mitigation when the standard LOPA method is used. However, enhanced LOPA is more realistic and the smaller incentive would probably not justify the proposed mitigation; capital could probably be spent more effectively on other safety and health issues provided that the existing system meets required risk criteria. An intermediate mitigation case of adding only the panel alarm (LAHH102) may be justified using enhanced LOPA.

TABLE 3 – Estimated Risk Comparison for Example Application

Method	Existing	Mitigated	Apparent Incentive
Standard LOPA	$2200 /yr	$125 /yr	$2075 /yr
Enhanced LOPA	$ 760 /yr	$125 /yr	$ 635 /yr

Finally, this example is very specific. Other examples may not give similar results.

IMPLEMENTATION IN MS ACCESS

Enhanced LOPA has been implemented in an FMC-proprietary application of Microsoft Access named SISET (Safety Instrumented Systems Evaluation Tool) to conduct SIS evaluations. A spreadsheet format such as Microsoft Excel could have provided similar capability. FMC's criteria for maximum individual risk are included along with the structure to conduct Risk Benefit Evaluations (RBE).

Default values are provided in the database template for a number of initiating event frequencies categorized as sensors, logic solvers, final elements, and others. Similarly, default risk reduction factors are provided for non-SIS IPLs and default instrument failure rates are provided to model SIF architectures. Users are allowed to add initiating events, IPLs, and SIF instruments to their database plus users may modify corresponding values according to local experience. Documentation is maintained throughout. User-specified test intervals can be used for each SIF. SIF architecture options include combinations that can yield SIL 1 (typically a single sensor, a simple logic solver, and single final element) through to SIL 3 (such as 2-out-of-3 voting sensors, a fault tolerant logic solver, and two final elements).

Providing the enhanced LOPA method in this structured but flexible software tool allows realistic analysis of moderately complex systems. Occassionally a complex system is analyzed that is beyond the capability of the enhanced LOPA methodology so QRA by Event Tree Analysis/Fault Tree Analysis may be required for a few cases.

CONCLUSIONS

Modifying the standard LOPA methodology to separate initiating events into four categories — sensor, logic solver, final element, and others — enhances the utility of LOPA. Categorization allows identification of each independent layer of protection that applies specifically to each initiating event failure category so that moderate interdependencies can be more realistically modeled. The calculations are readily implemented in conventional spreadsheet or database software applications.

Strengths of this enhanced LOPA method include:

- Larger number of leaders can be trained (and retained) compared to QRA
- Minimum training required to get active participation from plant operations, maintenance, and engineering personnel compared to QRA
- More realistic results compared to standard LOPA

ACKNOWLEDGEMENTS

The author gratefully recognizes the contributions of the FMC working group that developed the enhanced LOPA technique as part of an SIS evaluation methodology.

REFERENCES

CCPS, *Layer of Protection Analysis Simplified Process Risk Assessment*, 2001, New York, American Institute of Chemical Engineers, Center for Chemical Process Safety.

Fryman, C. (1996), "Managing HAZOP Recommendations Using an Action Classification Scheme," *AIChE Spring National Meeting*, New Orleans, February 25-29, 1996. New York: American Institute of Chemical Engineers.

Garcia, A. A., and D. E. Lewis (1998), "Safety Instrumented System Design Using Risk-Benefit Evaluation," *International Conference and Workshop on Reliability and Risk Management*, San Antonio, TX, September. New York: American Institute of Chemical Engineers.

ISA (1996), *ANSI/ISA-84.01-1996: Application of Safety Instrumented Systems for the Process Industries*, Research Triangle Park, NC: Instrument Society of America.

Rothschild, "Fault Tree and Layer of Protection Hybrid Risk Analysis", *Process Safety Progress* (Vol.23, No.3), September 2004, pp 185 – 190.

Selection and Evaluation of Release Scenarios for an LNG Import Terminal

Jeffrey D. Marx and John B. Cornwell
Quest Consultants Inc.
Norman, OK
www.questconsult.com

Prepared for Presentation at
American Institute of Chemical Engineers
2005 Spring National Meeting
Process Plant Safety Symposium
Atlanta, Georgia
April 11-13, 2005

UNPUBLISHED

ABSTRACT

In most process hazard safety studies, the magnitude and scope of the study are defined by a set of release scenarios that are selected for inclusion in the analysis. Therefore, it is important to carefully evaluate which scenarios are included. Although typically chosen through a hazard identification process, the list of scenarios is increasingly driven by regulatory requirements, corporate requests, or by public perceptions. In the LNG industry, public perception and regulatory pressure are changing the release selection process.

The list of potential release scenarios should be constrained to those that are both possible and credible, yet the list of scenarios often is expanded by outside influences to include impossible or non-credible events. Many of the events that normally would not have been considered in an analysis (e.g., releases caused by terrorist acts) must now be included. This paper will consider the release selection process for a proposed LNG import terminal, demonstrating some of the methods used to evaluate each scenario. The end product is a defensible consequence analysis that evaluates the plausibility of each event, removes impossible events, and presents a realistic set of modeling results.

INTRODUCTION

The LNG industry has experienced substantial growth in the past few years, most notably in the number of proposed import terminals for the U.S.A. This increased activity is aimed at alleviating natural gas shortages in the U.S., but it has also created a significant interest in the safety aspects of LNG. Because LNG is not well understood by the general public, the potential dangers posed by releases of LNG have caused a considerable amount of opposition to several proposed projects. This opposition is not always technically-based. The argument against import terminals has often been an emotionally charged "We don't want it here!"

LNG import terminals are simple facilities compared to much of the petrochemical industry. LNG is unloaded from ships into large, insulated tanks, and later heated to transform it back into gas for delivery into natural gas pipelines. In addition, the LNG industry has a very good safety record, due, in part, to the many regulations surrounding its transport, storage, and handling. What is widely recognized as the major impediment to public acceptance is the large volume of LNG contained in tank ships and land-based storage tanks. As the volume of LNG in a tank ship or storage tank increases, the economics generally become more favorable, so the size of tank ships and storage tanks continue to increase. Large spills from LNG tank ships or insulated storage tanks have become the focal point (and even a scare tactic) of groups opposing an LNG project. Add to this a perceived threat of terrorism, and the problem is further compounded.

Anytime there is a proposed LNG import terminal (or expansion of an existing one), some type of safety study must be done. In a safety study, release scenarios are chosen and evaluated to demonstrate the potential magnitude of the hazards associated with the facility, to show that mitigation measures can effectively limit the magnitude of hazards, or to demonstrate that the risk posed by the facility is acceptable.

EXAMPLE LNG IMPORT TERMINAL

For the purposes of this paper, consider the following proposed LNG import terminal:

Receives LNG from ships at one unloading dock
Dock connected by jetty to land-based terminal
Two 160,000 m^3 insulated storage tanks, full-containment construction
LNG vaporization equipment for peak send-out of 800 mmscfd
- In-tank transfer pumps
- Booster pumps
- Vaporizers
- Gas metering and pipeline connection

Miscellaneous utilities

RELEASE SELECTION

The selection of release scenarios is the first step in any consequence or risk analysis performed for a petrochemical facility. Because of the various siting issues associated with an LNG import facility, there are several ways in which release scenarios can be selected. For the purposes of this discussion, three such methods will be examined. These methods each have their place in different types of studies, but can produce quite different results. The three methods evaluated are listed below, and described in the following sections.

1) **Regulatory** releases required by codes or standards
2) **Credible** accidental releases
3) **Possible** all potential releases, including intentional ones

Method 1 - Regulatory Scenarios

Regulatory scenarios are those required or prescribed by an LNG code or standard. Evaluation of such releases is often required by law. For this discussion, it is assumed that the facility will comply with the siting provisions of Title 49 of the U.S. Code of Federal Regulations, Part 193 (49 CFR 193): *Liquefied Natural Gas Facilities – Federal Safety Standards*.

The scenarios involved in a regulatory compliance siting study are often much different than would be chosen in other studies. LNG is the only petrochemical industry that requires specific release scenarios to be evaluated. The siting provisions in 49 CFR 193 (as applicable to an import terminal) are based on the following items:

1) Siting provisions are in place to minimize the possibility that the hazards associated with an accidental release of LNG might endanger adjacent property or onsite equipment.
2) An impoundment for each LNG storage tank must be provided. The impoundment must hold 110% of the tank's maximum liquid capacity.

3) An impoundment must be provided for every process, vaporization, or transfer area, and must be sized to hold the maximum volume of LNG that could be discharged during a specific scenario referred to as a design spill. Design spills for process, vaporization, or transfer areas are defined as "the flow from any single accidental leakage source", for 10 minutes. In specific instances, the spill duration can be shorter if sufficient surveillance and shutdown provisions exist.
4) An additional design spill requires the examination of a spill from the LNG tank itself. This scenario will vary based on how the tank's piping penetrates the shell, and must also be impounded.

The definition of these release scenarios and subsequent consequence modeling is often considered a "backwards risk analysis" because it determines acceptability based on prescribed scenarios. If the consequences of the prescribed scenarios do not cause unacceptable impacts (generally, this means they remain onsite), then the site is deemed acceptable. The prescribed scenarios are major release events that are not expected to occur in the lifetime of the facility. Put another way, they are low-probability events. The logic behind this approach is that if the consequences of low-probability events do not cause unacceptable offsite impacts, then the risk imposed by the facility is low.

The specific scenarios for the example LNG import terminal, as required by code, are given in Table 1.

Table 1. Regulatory Release Scenarios for the Example Facility

Case	Description
A1	Failure of the primary LNG storage tank, spilling contents into the impoundment (i.e., the outer concrete shell of the full-containment tank).
A2	Ten-minute spill from an LNG storage tank pumpout line into an impoundment.
A3	Spill at the LNG unloading dock during an unloading operation. Typically modeled as a one-minute spill from a failed unloading arm, into an impoundment.
A4	Spill from the process/vaporization area at the maximum vaporization rate into the process area impoundment.

Method 2 - Credible Scenarios

The "credible" label instantly leads to the question:

What is credible?

Credible events can be defined as those that have some <u>reasonable</u> probability of occurring during a defined time period. Probabilities are often derived from the historical record of the industry (and similar industries) and can be stated in numerical or qualitative terms. One example of how qualitative descriptions can be associated with numerical values is presented in Table 1, which is taken from the European LNG code [EN1473, 1997].

Table 2. Probability Ranges

Description	Probability Class
Frequent or quasi-certain event.	more than 10^{-2}/year
Possible but not very frequent event.	10^{-2} up to 10^{-4}/year
Rare Event.	10^{-4} up to 10^{-6}/year
Extremely rare event.	10^{-6} up to 10^{-8}/year
Improbable event.	less than 10^{-8}/year
Event of non-quantifiable probability (falling of meteorite, attempt on life or property, etc.).	Unknown

In general, releases that have the possibility to occur, but are considered to have a very low probability of occurrence, are not considered in an accidental analysis. These events may be defined as incredible. An example would be the failure of an LNG storage tank because it was hit by a meteorite. The probability is not zero, it is simply very, very low. When this approach is taken, some large consequence/low probability events are identified as incredible events and excluded from an accidental analysis.

The purpose of the accidental release case selection methodology is typically to define the maximum credible release scenarios that might result in an adverse impact on the public. The methodology is developed in six steps:

- Review of available documentation (process flow diagrams (PFDs), heat and material balance sheets, process description, plot plan, etc.)
- Review of available safety studies
- Review of accident history in similar facilities
- Development of release scenarios
- Screening of release scenarios using consequence analysis
- Final selection of release cases

The analysis begins with a general review of the process. Written descriptions of the processes are studied to determine the physical and chemical transformations occurring and the general flow of material in the facility. After the process features are known, process flow diagrams (PFDs) are reviewed and compared to the written descriptions. The material balances are reviewed for each major line to determine the exact nature of the material within the line or vessel. Each of the major flow lines is taken individually and evaluated to determine the potential for producing a significant impact if a release occurred. Factors such as line sizes, flow rates, vessel inventories, and the potential for aerosol formation are considered in the selection of release scenarios.

Available safety studies, including preliminary safety studies, are reviewed to determine if all potential credible releases have been adequately identified. Any potential releases identified in these work products are added to the list of releases to be evaluated. Historical incidents in the industry are compared to the operations at the facility to determine if similar releases might occur. In addition, historical failure rate data for the various types of equipment in the facility are reviewed to determine where the most likely release sources may be. Often, a screening-level analysis is performed to determine which release scenarios have the greatest potential for off-site impact.

The final step in this analysis forms a condensed list of unique, credible release scenarios. This list typically does not cover all foreseeable credible releases; some events may be excluded because they lack the potential for off-site impacts. Several credible release scenarios identified for the example LNG import facility that might result in off-site impacts are given in Table 3. The releases shown in Table 3 are accidental events, primarily releases that result from a failure of a piece of process equipment.

Table 3. Credible Release Scenarios

Case	Description
B1	Release from an LNG unloading arm during tank ship unloading
B2	Release from transfer piping between LNG tank ship and LNG storage tank
B3	Release from the suction line to one high pressure booster pump
B4	Release from the header leading to the vaporizers
B5	Release from the outlet line of one vaporizer
B6	Release from the sales gas line after the metering station
B7	Collision/allision/grounding of LNG tank ship sufficient to release LNG onto the water

The one credible accidental release scenario that does not involve a failure of process equipment involves an LNG tank ship (B7). A release of LNG from an LNG tank ship, either when the ship is approaching the LNG terminal or when it is berthed at the terminal, may have the potential to adversely impact persons and property outside the boundaries of the proposed LNG terminal. This incident could involve a collision between an LNG ship and another ship, an allision by the LNG ship with a fixed object (such as a mooring dolphin, pier, or breakwater structure), or a grounding of the LNG ship. In any of these three scenarios, it is assumed that the impact has sufficient strength to penetrate the outer hull, inner hull, and LNG storage tank, releasing LNG onto the water.

Method 3 - Possible Scenarios

Possible events are simply ones that can occur. Inclusion in this category is not an issue of probability or credibility, simply possibility. To evaluate a potential scenario, one question is asked:

Is this scenario possible?

A list of possible events should include both accidental and intentional scenarios. Selecting releases that could be intentionally created by one or more persons requires a different approach than that used to define accidental releases. An accidental release is defined as one that:

- has occurred in an LNG facility, or
- has occurred in an industry using similar types of equipment, or
- could reasonably be argued that it could occur in an LNG facility.

Conversely, selecting a representative set of possible releases of LNG or natural gas requires a broader, more creative approach. If the scenario can be hypothesized, with a reasonable set of contributing events, then it can be included. Events that are physically impossible are excluded. Determination of which events are impossible will often require an engineering analysis. Because of this premise for release identification, the focus is different than that done for the credible scenario analysis. A major motivating factor behind this approach is the perceived threat of acts of terrorism or sabotage. Although the methods of terrorism/sabotage are difficult to define (making the release identification process a very creative one), their inclusion has become an important part of determining possible release scenarios. In order to define a representative list of intentionally created releases, the following factors must be considered.

- The primary target of the action (based on inventory, accessibility, location, etc.)
- The obstacles to overcome in order to effect a release from the target

When discussing intentional acts, it may not be possible to assign a numerical value to the probability of a specific intentional act. In light of this, the discussion changes from credible/incredible to possible/impossible. It is important to recognize the distinction between selecting accidental release scenarios and selecting intentional release scenarios. This is best summarized by the following statements.

- Accidental release scenarios are based on credible events, but not incredible events.
- Intentional release scenarios are based on possible events, but not impossible events.

The events presented in Table 4 have been identified as possible release scenarios that could occur in or near the example LNG import terminal.

The six intentional events listed in Table 4 (C1 through C6) cover a range of potential impacts, as well as a varying degree of difficulty in regard to effecting a release of LNG or natural gas. These six scenarios are not intended to include all possible intentional (terrorist or sabotage) events. They do represent a range of events, all of which are possible to some degree, but many of which are extremely unlikely simply due to obstacles that would need to be overcome by the perpetrators.

Table 4. Possible Release Scenarios

Case	Description
B1	Release from an LNG unloading arm during an unloading operation
B2	Release from transfer piping between LNG tank ship and LNG storage tank
B3	Release from the suction line to one high pressure booster pump
B4	Release from the header to the vaporizers
B5	Release from the outlet line of one vaporizer
B6	Release from the sales gas line after the metering station
B7	Collision/allision/grounding of LNG tank ship sufficient to release LNG onto the water
C1	Hijacking and crashing of a commercial jet airliner into an LNG storage tank
C2	Firing of a rocket-propelled grenade into an LNG storage tank
C3	Hijacking and crashing of a commercial jet airliner into an LNG tank ship
C4	Detonation of a boat bomb near a docked LNG tank ship
C5	Firing of a rocket-propelled grenade into an LNG tank ship
C6	Intentional ramming of an LNG ship by another large ship; Hijacking and intentional damage to an LNG ship
E1	Earthquake-induced failure of an LNG storage tank

As seen in Table 4, generating a list of possible scenarios expands the list of credible scenarios into one that includes a range of intentional and low-probability releases. The list of possible events is separated into three categories: credible events (B1 through B7), terrorist-induced events (C1 through C6), and an earthquake event (E1). The earthquake event is included in Table 4 because the probability of an earthquake whose magnitude is sufficient to fail a full-containment LNG storage tank is expected to be a very low probability event, and thus, not credible.

Before an intentional act resulting in a release of LNG is added to the list of release scenarios, some type of engineering analysis must be conducted. This analysis must evaluate the entire series of events that are necessary for the end result (a release of LNG) to occur. Table 5 presents, as an example, a minimum chain of events that would be necessary for terrorists to successfully cause a release from an LNG storage tank after hijacking a commercial aircraft (scenario C1). At each step in the chain of events, several obstacles to that step are listed. If any one of the obstacles is realized, the end event (release from the LNG tank) will not happen. Although the end event is possible, the many potential obstacles make the outcome highly unlikely.

Table 5. Chain of Events for a Terrorist-Hijacked Aircraft Crashing into an LNG Tank

Event	Potential Obstacles
Terrorists successfully board aircraft	Airline security procedures Airport security checks Flight delayed or cancelled
Terrorists commandeer the aircraft	Post-September 11, 2001 additional in-flight security measures (e.g., Air Marshals, cockpit doors) Passenger/crew intervention
Terrorists choose the LNG terminal as a target, successfully pilot the aircraft and navigate toward the LNG terminal	"Better" targets may be available U.S. Air Force intercepts aircraft – forced to land or crash Inability of terrorists to pilot or navigate aircraft Passenger/crew intervention
Terrorists successfully hit the LNG tank with aircraft	Aircraft flown too low and crashes Aircraft misses LNG tanks (small target compared to a building such as the World Trade Center or Pentagon) Aircraft hits the top of the LNG tank; no liquid release
LNG storage tank is hit by one of the aircraft's engines	Engines may miss the storage tank Most of the aircraft is too "soft" to penetrate a concrete LNG storage tank
LNG storage tank is ruptured by jet engine	Engine hits at obtuse angle, fails to penetrate Aircraft speed is too low; insufficient momentum to penetrate
LNG is released	

CONCLUSIONS

Each of the three methods of release scenario identification provides a unique look into the potential release event at the LNG facility. Each set of scenarios has its own application and meaning.

Regulatory scenarios are analyzed for the purpose of checking compliance with specific requirements. When the consequence analysis corresponding to these release scenarios is complete, the results simply provide an evaluation of whether a site is acceptable. The release scenarios are not defined to be caused by a specific event, nor are they expected to occur during the lifetime of the facility. The consequence analysis for these scenarios examines fire radiation and vapor dispersion impacts using a specific set of modeling conditions. Thus, the results offer a narrow look at the potential impacts associated with releases in the LNG facility.

Credible release scenarios offer an expanded look into accidental releases of LNG or natural gas. These scenarios are identified as potential major release events that could occur at the LNG terminal. Most involve a process equipment failure or leak. Consequence modeling for these release scenarios may involve vapor dispersion (flash fires), torch fires, pool fires, or vapor cloud explosions. The analyst can use a variety of release sizes (e.g., ruptures, leaks) or other variable modeling conditions to demonstrate the range of potential impacts following a release.

When the list of release scenarios is further expanded to include possible terrorist-induced events, the analysis must be structured differently. It is no longer sufficient to simply assume that a release can occur based on historical data, as is the case with most accidental releases. With intentionally caused releases, a chain of events must be analyzed, where each step in the process receives scrutiny and/or engineering analysis. Once this step is taken with intentional releases, the consequence analysis is similar to that for accidental releases.

These three release analysis methods produce differing descriptions of the potential impacts of an LNG import terminal. Depending on the project's needs and requirements, one or more of these approaches may be used when designing or evaluating an LNG terminal. It is important to remember that the public's perception of the hazards (or even risk) posed by the facility can be altered by which analysis is performed and how the results are presented.

REFERENCES

49 CFR 193 (May 2000), *Liquefied Natural Gas Facilities: Federal Safety Standards*. U.S. Department of Transportation, Research and Special Programs Administration, Washington, D.C., 2000.

EN 1473 (1997), *Installation and Equipment for Liquefied Natural Gas - Design of Onshore Installations*. European Standard EN 1473; British Standard BS EN 1473; British Standards Board, 1997.

CRITERIA FOR THE ESCALATION OF FIRES AND EXPLOSIONS

Valerio Cozzani
Dipartimento di Ingegneria Chimica, Mineraria e delle Tecnologie Ambientali, Università di Bologna, via Risorgimento 2, 40136 Bologna, Italy. e-mail: valerio.cozzani@mail.ing.unibo.it

Gianfilippo Gubinelli
Dipartimento di Ingegneria Chimica, Chimica Industriale e Scienza dei Materiali, Università degli Studi di Pisa, via Diotisalvi 2, 56126 Pisa, Italy. e-mail: g.gubinelli@ing.unipi.it

Ernesto Salzano
Istituto di Ricerche sulla Combustione, Consiglio Nazionale delle Ricerche, via Diocleziano 328, 80125 Napoli, Italy. e-mail: salzano@irc.na.cnr.it. *corresponding author*

Prepared for Presentation at
American Institute of Chemical Engineers
2005 Spring National Meeting
Process Plant Safety Symposium
Atlanta, Georgia
April 11-13, 2005

December 2004

UNPUBLISHED

ABSTRACT

The escalation of fires and explosions in an industrial environment, also known as domino effects, may cause extremely severe accidental events. Nevertheless, conventional quantitative risk analysis (QRA) of process plants or industrial areas rarely includes these effects, or limits analysis to the use of threshold criteria for structural damage of equipment as directly linked to the possibility of escalation. Moreover, due to the great uncertainties in the technical literature, very conservative values are adopted to assess the possibility of propagation, thus leading to possible over-predictions of risk. The present study focuses on revision and improvement of the criteria for escalation assessment, based on updated models for fires and explosions and on recent advances in the prediction of damage to process equipment due to different escalation vectors, i.e. heat radiation, overpressure, fragment projection. Specific threshold values are identified for the primary scenarios usually considered in the risk assessment of industrial sites.

INTRODUCTION

The AIChE-CCPS Guidelines for Chemical Process Quantitative Risk Analysis in the process industries defines a domino accidental event as "an incident which starts in one item and may affect nearby items by thermal, blast, or fragment impact, by an increase in consequences or in failure frequencies" [1]. The development of reliable criteria for the possibility of propagation of industrial accidents to nearby equipment is very important for risk assessment in industrial areas, for land use planning, and for inclusion of domino effects in Quantitative Risk Analysis (QRA). However, in spite of the clear definition given above, no agreement is present in the literature on even which kinds of events should be included in domino effect analysis. Hence, further insights are still needed to assess this category of accidents.

A screening of technical literature and case histories evidences that accidents including "domino effects" have the following common elements:

- a primary accidental scenario is followed by the spatial propagation of one or more physical effects (the "escalation vectors")
- the intensity is able to produce structural damage on one or several equipment items
- one or more secondary events may result, involving the damaged equipment items (i.e. fire, explosion, toxic dispersion)
- a relevant "domino effect" takes place if the overall severity of the secondary accidental event is higher or at least comparable to that of the primary event

As a conclusion, in order to take place, a domino effect should associate "propagation" of the primary event to an "escalation" of the overall scenario.

Domino effects can be evaluated using threshold values for the escalation vectors, hence without introducing any information on the primary scenario which has produced the physical effect of concern. Alternatively, escalation criteria may be defined considering particular features of the primary scenario producing the escalation vectors, and taking into account relevant parameters of target equipment, thereby influencing the credibility of escalation.

Table 1 summarizes an historical analysis of more than 100 domino accidents described in the MHIDAS database [2]. Full details on the definition and characteristics of primary scenarios are reported in the literature (see references [3-6]). As shown in Table 1, only three possible escalation vectors have to be considered in the analysis of domino effects: thermal radiation (and/or fire impingement), overpressure, and fragment projection.

In the case of equipment exposed to stationary flames (e.g. pool fires and jet fires); the potential escalation effects due to thermal radiation and to fire impingement are substantially analogous. However, it must be remarked that flame impingement may also trigger further secondary scenarios due to the ignition of combustible materials.

Toxic releases are not considered as an escalation vector in the present approach.

Table 1. Domino Effect Scenarios

Primary Scenario	Human Vulnerability	Escalation Vector	Possible Secondary Scenario
Pool fire	Thermal radiation/fire impingement	Thermal radiation/fire impingement	Jet fire, pool fire, tank fire, BLEVE, toxic release
Jet fire	Thermal radiation/fire impingement	Thermal radiation/fire impingement	Jet fire, pool fire, tank fire, BLEVE, toxic release
Fireball	Thermal radiation/fire impingement	Thermal radiation/fire impingement	Pool fire, tank fire
Flash fire	Fire impingement	Fire impingement	Tank fire
Physical explosion	Fragments, overpressure	Fragments, overpressure	Any
Confined explosion (vapor, gas and dust)	Fragments, overpressure	Fragments, overpressure	Any
Unconfined and partially confined vapor cloud explosion	Overpressure	Overpressure, fire impingement	Any
Toxic release	Toxic concentration	Unlikely	Any

In the case of the overpressures following primary vessel failure, and depending on the substance released, further vulnerability and escalation vectors (e.g. pool fire, fireball, toxic release) should be considered. Note as well that toxic releases may also be generated by the combustion of some substances in fires.

Table 2 shows the generic threshold values for escalation caused by blast wave damage, taking into account the different characteristics of categories of process equipment [7-9]. The table also reports threshold values estimated for equipment damage from the analysis of damage models.

Table 2. Overpressure Threshold Values (kPa) for Structural Damage and Domino Effects

Threshold Type	Substance Hazard (in target vessel)	Target Equipment			
		atmospheric	*pressurized*	*elongated*	*Small*
Damage	All	5	35	17	12
Escalation	Flammable	22	17	31	Unlikely
	Toxic	22	17	16	37

The results in Table 2 do not take into account the features of the primary scenario (e.g., the threshold values are suggested for any type of blast wave). However, the peak overpressures decay with very different rates during the propagation of a blast wave, depending on the mode of energy generation and the amount of energy developed at the source point. A more detailed escalation analysis may involve identification of differences among specific primary explosion scenarios.

Similar considerations may also apply for the escalation caused by thermal radiation and fragment impacts. Indeed, with reference to thermal radiation, the possibility of escalation depends mainly on the duration of the primary scenario, which has to be compared with time to failure of the secondary vessel. For instance, results of an extensive study lead to the following conclusions [10]:

- for a representative set of atmospheric target vessels, times to failure are typically higher than 30 minutes for radiation intensities lower than $10kW/m^2$
- for a representative set of pressurized vessels, times to failure are typically higher than 30 minutes for radiation intensities lower than $30kW/m^2$

However, for events having lower durations (e.g., fireballs and flash fires), only the escalation caused by ignition due to direct fire impingement should be considered (e.g. in the case of floating roof tanks).

Finally, for fragments, the only possible generic threshold is the maximum credible flight distance of a fragment. This is estimated at 250 to 3000m, depending on the shape of the fragment and on the primary scenario [11].

The use of generic thresholds that do not take into account specific characteristics and parameters of the primary scenario leads to escalation criteria that may be greatly over conservative. In the following sections, the primary scenario is considered in addressing fire and explosion escalation.

ESCALATION CRITERIA FOR PRIMARY SCENARIOS

Categories of Accidental Scenarios

Table 1 summarizes the different categories of primary scenarios considered in the present study. The classification given in the table is that widely used in current practice, and follows the guidelines for QRA of process and chemical plants given by CCPS and TNO [1, 12]. However, within the general categories of accidental scenarios defined in the guidelines, some specific definitions are added to take advantage of recent research in the prediction of physical effects.

Explosion phenomena include many scenarios, starting from accidental release of gas, liquid (vapor) or dusts within equipment (confined or partially confined), within congested environment (industrial areas or offshore), or in open atmosphere. The primary scenarios which can be related to the explosion can be classified as:

- unconfined and partially confined gas and vapour gas explosions (VCE)
- confined explosions (including vented explosions or runaway reactions)
- mechanical explosions due to failure of equipment
- point-source explosion of solid explosives

One particular event is a Boiling Liquid Expanding Vapor Explosion (BLEVE), which includes fragment projection and also an unconfined explosion due to the rapid liquid evaporation to atmospheric pressure. When flammable substances are considered, this scenario is also often followed by ignition and a very destructive fire scenario--a fireball.

Other fire scenarios of concern in the QRA framework are the ignition of vapors over a flammable liquid pool (resulting in a pool fire), of a pressurized gas or liquefied gas jet (jet fire), or of a premixed gas or vapor cloud (flash fire). Fires involving solid materials are usually of less concern in the analysis of escalation possibility in a process plant, since these scenarios typically have higher characteristic times and lower damage ranges.

Full details of the characteristics and modeling of the scenarios listed in Table 1 are available in several reference publications [1, 12]. Herein, only the aspects relevant in the analysis of escalation possibility will be discussed. A short discussion will focus on analysis of domino effects. In this framework, it must be remarked that the use of complex tools such as Computational Fluid Dynamics for fires and explosions and the numerical structural analysis for equipment damage is by far out of reach for the purposes of QRA. Thus, in the analysis of accidental phenomena, overly conservative and over-simplifying assumptions are usually adopted. These assumptions are obviously retained in the analysis of escalation, and affect the results obtained.

Vapor Cloud Explosions (VCEs)

In the framework of domino effects assessment, and more generally when QRA is of concern, the overpressure and impulse with respect to distance (generated by any type of explosion) are estimated assuming that the actual blast wave may be compared to the ideal blast wave produced by an equivalent charge of one or more solid "point" explosions. The value of the maximum peak overpressure as a function of distance is thus estimated from the well-known TNT-diagram or from similar, more advanced plots such as those provided within the Multi-Energy (ME) method [13] and the Baker & Strehlow (BS) method [14]. Hence, peak overpressure $P°$ and impulse $I°$ can be determined on the basis of the total available energy of explosion, using an energy (or mass) scaled distance $\underline{R}$:

$$\underline{R} = \frac{r}{\sqrt[3]{\frac{E_{exp}}{P_{atm}}}} \quad (1)$$

where r is the effective distance from the explosion centre or boundary (m), P_{atm} the atmospheric pressure (10^5Pa), E_{exp} the explosion energy (J) which can be alternatively converted to the equivalent mass of the reference explosive. These approaches are certainly acceptable in the far-field, while more accurate analyses are necessary when near-field effects are analyzed.

In the specific case of VCEs, assuming that the detonation regime can be ruled out in practical conditions, the main difficulties are in the definition of the initial explosion strength (or more specifically in the peak pressure reached by the flame propagation in the premixed flame) and in the pressure-time profile. Indeed, as for any complex transient, scale dependent, reactive phenomenon, the prediction of the peak overpressure (and the total duration of blast wave) has to be simplified, thus introducing relevant uncertainties in the analysis. Nevertheless, the ME and BS methods do take into account the geometry of the environment and relate the peak pressure to simple characteristic geometrical and reactivity parameters, in order to allow a preliminary assessment of the hazard related to the explosion. The peak overpressure in the ME method is thus related to an initial source strength factor F, that ranges between 1 (acoustic explosion) and 10 (detonation, or TNT explosion) or, in the BS method, to the acceleration of flame with respect to the speed of sound (the flame Mach number, M_f). Similar approaches are also given for the total impulse of the blast wave. Specific guidance for the application of these methods is reported elsewhere [15].

When VCEs are considered, the total duration of explosion may range from a few milliseconds to hundreds of milliseconds or even more in the case of very low Mach deflagrations (i.e. when slow laminar flames characterize the entire phenomenon). These times are compared to characteristic response times of equipment. Hence, peak overpressures are not sufficiently exhaustive for the prediction of the interaction effects of the blast wave with the equipment, since the impulse and the dynamic factor become relevant in the analysis of the effects on the structure. However, these factors are highly dependent on the cloud dispersion, on the explosion history and on interaction of the blast wave with targets due to shape, relative direction of wave, reflections, and other complex phenomena. These are very uncertain and almost unpredictable factors in the safety analysis of a complex layout. Furthermore, assessment based on static overpressures generally gives conservative results for many types of explosions [6]. Therefore, the maximum peak static overpressure will be the only parameter used to assess the possibility of escalation.

With reference to Table 2, peak overpressures higher than 16 kPa are needed for escalation involving elongated target vessels containing toxic substances, whereas slightly higher overpressures are needed in the case of pressurized equipment and of atmospheric tanks (17 kPa and 22 kPa, respectively). These overpressures are reached in the far field only for an explosion strength factor F ≥ 5 in the ME method (for F = 5 at a scaled distance lower than 0.8) or for a flame Mach number $M_f \geq 0.35$ in the BS method. Hence, it has to be excluded from escalation analysis (at least in the far field) for any explosion with very low strength, independent of the total energy of explosion. On the other hand, blast wave curves collapse to the maximum strength for values of scaled distance $\underline{r} \geq 1.40$ in the ME method and $\underline{r} \geq 1$ in the BS method. As a conclusion, for high strength explosions, the safety distance is independent of the initial explosion source.

Table 3. Minimum Scaled Safety Distances for Peak Static Overpressure of 16kPa and Corresponding Actual Safety Distances for Propane

ME		BS		Actual distance (m)							
Strength Factor	Scaled Distance	Mach Flame	Scaled Distance	**Cylindrical Vessel** **2 m³**		**Cylindrical Vessel** **20 m³**		**Cylindrical Vessel** **200 m³**		**Spherical Vessel** **2000 m³**	
				ME	BS	ME	BS	ME	BS	ME	BS
< 5	-	< 0.35	-	Never reached, independent of total energy							
~ 5	0.8	~ 0.35	0.5	9.3	3.2	20.0	6.8	42.7	14.7	91.3	31.7
> 5	2.0	> 0.35	2.0	23.3		49.9		106.7		228.2	

Starting from these considerations, Table 3 reports the safety scaled distances calculated for the VCE. For the sake of comparison, and in order to understand the relation between the scaled safety distances and the actual safety distances, the table also reports the calculated actual safety distances following a VCE generated by catastrophic failure of several equipment items (three different pressurized cylindrical storage tanks and a spherical pressurized vessel containing propane). In the assessment of VCE overpressures, all the released propane is considered to participate in the VCE.

Safety distances reported in Table 3 double by increasing by an order of magnitude the total energy, which is related to equipment volume. It is important to note that the actual and scaled safety distances reported in Table 3 have to be considered as threshold values for escalation following loss of containment (domino effects) and not for structural damage.

Besides damage caused by the blast wave, combustion of the vapor cloud may also cause damage due to thermal radiation and flame impingement. However, little information is given in the literature on the combined effects of overpressure and heat. Nevertheless, if escalation effects are of concern, the effect of thermal radiation is certainly negligible (as compared to that of the blast wave, at least in the far field), due to the very low duration of the combustion phenomenon. On the other hand, escalation can be caused by the ignition of vapors in the vicinity of vulnerable targets, such as in the case of floating roof tanks.

Confined Explosions

Explosions of equipment due to internal combustion of gases, vapors, or dust are destructive for the equipment involved, and for a narrow area around it. Venting, introduced for mitigation purposes, may further expand the area of concern, since the release of pressurized material through the vent section may produce a blast wave that can travel outside of the equipment. Also, external explosions due to unburnt gases released after venting have been observed. However, the main problem with overpressures generated by confined explosions relates to the safety of the equipment item. In this case, whereas the total energy involved in the production of a blast wave is generally low and the intensity of the blast wave (at least in the far field) is negligible. Indeed, a number of experimental tests and a theoretical analysis reported in the work of Forcier and Zalosh [16] allowed us to define conservatively a threshold distance value of 20m for escalation effects when considering the blast wave propagating outside of equipment after the internal explosion of gas or dust. This value is obtained for a maximum reduced pressure (the maximum pressure reached in the vented vessel) lower than 2 bar, a vessel volume lower than 1000 m^3, and having considered conservatively a vent section $A_v = V^{2/3}$, where V is the vessel volume.

Jet fires (from vents or from pipelines), fireballs and other phenomena may follow the release of products from failed equipment after the explosion. Moreover, if the equipment fails because the venting devices are not able to mitigate the internal pressure, a mechanical explosion may take place due to the rapid depressurization of hot combustion products to the atmosphere. The possibility of escalation due to these secondary effects of confined explosions is discussed in detail in the following section.

Mechanical Explosions

In the chemical and process industries, the failure of vessels containing a compressed gas phase is a common accident. A number of scenarios are included in this category of accidental events. The common feature of all these events is the mechanical failure of an equipment item, followed by the sudden expansion of the compressed gas phase, resulting in the generation of a blast wave. The internal pressure rise may be caused by gas or liquid overfilling of a vessel, by an unvented confined explosion, by a runaway reaction that causes the formation of gaseous products, or by a temperature increase in a vessel containing a non-pressurized liquid at the boiling point (e.g. the heating of a LNG vessel due to an external fire, to the failure of the cryogenic system, or to the sudden mixing with a liquid stream at higher temperature).

The case of failure of a vessel containing a pressurized liquefied gas and subsequent ignition (commonly indicated as BLEVE in the literature [1,12]), in which the blast wave following the failure of equipment is caused by the sum of the expansion energy of the vapor and of that of the flashing liquid, is discussed separately in the following section.

In the case of a mechanical explosion of a vessel, two escalation vectors may be identified: the blast wave following the failure of the vessel, and the fragments that may be generated in the vessel failure. The energy released at the moment of vessel failure is the sum of several components: the energy of vessel fracture, the energy associated with fragment formation and projection, the energy dissipated as heat in the environment, and the energy of the blast wave. The latter is evaluated using a number of models available in the open literature (e.g. the Brode equation). In the far field, the blast waves produced by bursting vessels are similar to blast waves caused by high explosives. Thus, the TNT-equivalence model or similar may be used to estimate the peak overpressure with respect to energy-scaled distance. In the near field, the specific factors of the scenario should be taken into account. The method of Baker et al. [4] introduces pressure curves which evaluate the peak overpressure with respect to energy-scaled distance taking into account geometrical effects and considering that the maximum (initial) pressure of the blast wave is lower than the internal pressure of the vessel at failure time. Using this model, the safety distances for escalation effects may be conservatively evaluated on the basis of the peak static overpressure thresholds listed in table 2 for different equipment types. The TNT-equivalence curves of overpressure with respect to scaled distance may be used to evaluate the safety distances in the far field. The effective distance is obviously equipment specific and depends mainly on the failure pressure of the vessel. Table 4 reports the actual safety distances calculated for equipment with different volumes containing compressed propane gas, obtained assuming a failure pressure of 20 bar. The blast wave energy is estimated directly by the Brode equation.

Table 4. Actual Safety Distances (m) for Escalation Due to Propane Explosions

Cylindrical, Horizontal Vessel $2\ m^3$	Cylindrical, Horizontal Vessel $20\ m^3$	Cylindrical, Horizontal Vessel $200\ m^3$	Spherical Vessel $2000\ m^3$
11.5	24.8	53.4	115.0

Fragmentation of the equipment due to internal pressures may also cause the generation of missiles that may result in escalation scenarios. The possibility of escalation due to fragment impact requires two conditions: the target distance must be within the maximum reasonable projection distance, and the impact must be followed by a loss of containment at the target vessel. Since the second condition is usually assumed to be verified in a conservative approach to missile damage estimate [11], the only possibility to obtain an escalation criterion for fragment damage is to assume a threshold based on the maximum projection distance. The flight distance of a fragment is mainly dependent on the initial fragment velocity, on the drag factor of the fragment, and on the initial vertical angle of projection. To assess the influence of these factors on maximum credible projection distance and on impact probabilities, a number of case studies are analyzed. A representative set of fragment geometries (40) are defined, based upon data from past accidents [6, 17]. The friction factors of the selected fragment geometries range between $1 \cdot 10^{-4}\ m^{-1}$ and $1 \cdot 10^{-2}\ m^{-1}$. A credible range of initial projection velocities are evaluated following the AIChE Guidelines [6]. The range of projection distances is thus estimated using the model of Baker et al. [4]. The maximum theoretical projection distance in the case of mechanical explosion is about 3000m. Although this value might be overestimated, projection distances up to 900m were observed in fragmentation accidents [17]. These results confirm the observation, often reported in the literature, that no safety distance of practical use may be identified for fragment projection on a deterministic basis.

Therefore, a probabilistic approach may be useful at least for a preliminary assessment of the credibility of the escalation events. The impact probability of a fragment on a given target is estimated using the methodology proposed by Gubinelli et al. [11]. Uniform distributions are assumed for: i) horizontal projection angles; ii) initial velocities (between zero and a maximum credible value, assumed at 190m/s); and iii) fragment friction factors (assumed at between $1 \cdot 10^{-4}$ m^{-1} and $1 \cdot 10^{-2}$ m^{-1}). The mean probability to hit a secondary target is thus calculated as a function of distance and target size. To obtain conservative data, vertical cylindrical storage tanks of diameters between 2.7 and 41.2m are assumed as secondary targets. The results, reported in Figure 1, evidence that the estimated impact probability is lower than $3 \cdot 10^{-2}$ at 100m and lower than $5 \cdot 10^{-3}$ at 300m. These figures should be multiplied by a mean number of fragments generated in the accident that might be assumed to be of about 10. Thus, it may be concluded that:

- no practical safety distance for escalation due to fragment projection in a mechanical explosion may be obtained (calculated values are of about 3000m)
- conservative values for the impact probability of a fragment may be estimated to be $3 \cdot 10^{-1}$ at 100m and of $5 \cdot 10^{-2}$ at 300m

Quite obviously, the mechanical failure of equipment can be followed by further secondary effects depending on the substance contained, as for confined explosions. As in the case of confined explosions, the possibility of escalation due to these secondary effects should be separately assessed considering the specific scenarios.

Boiling Liquid Expanding Vapor Explosions (BLEVEs)

The explosion of a vessel containing a pressurized liquefied gas is usually defined as a BLEVE even if the scenario is possibly composed of three phenomena: i) the mechanical failure of vessel (that may also result in the projection of fragments); ii) the blast wave produced by both the rapid expansion of vapor and the expansion (flash) of the evaporating liquid; and iii) the fireball produced if the substance involved is flammable. The latter will be separately considered in the following text.

It must be remarked that the BLEVE, i.e. the explosive evaporation of a liquid that produces a blast wave, requires specific thermodynamic conditions that are seldom verified during an accidental event. More often, the failure of a vessel involved in an external fire evolves into a simple loss of containment, resulting in a jet fire and/or in a fireball, rather than in a blast wave due to liquid explosion [18]. Moreover, as recently stated by Van den Berg et al. [19], a blast wave is formed only if the catastrophic disintegration of the entire vessel takes place in a limited time frame, with the rapid release of the liquid from the vessel. The analysis of past accidents evidences that this failure mechanism is quite unlikely. Nevertheless, scaled nomographs are provided to evaluate the positive phase duration and the peak overpressure as a function of scaled distance [19]. The scaling factor used in the plots is the mass of exploding liquid, in analogy with the TNT models. As an alternative, the TNT model [6] may be used as well to estimate the blast wave effects, provided that the energy of expanding liquid and vapor are known or may be calculated.

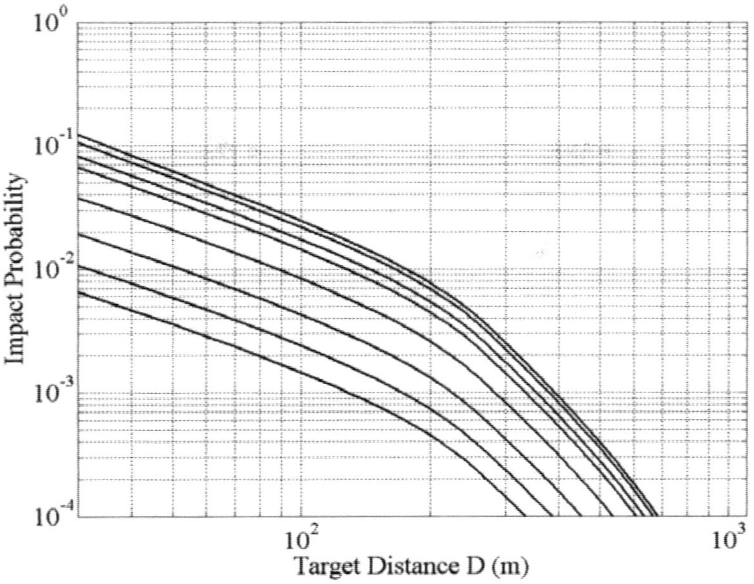

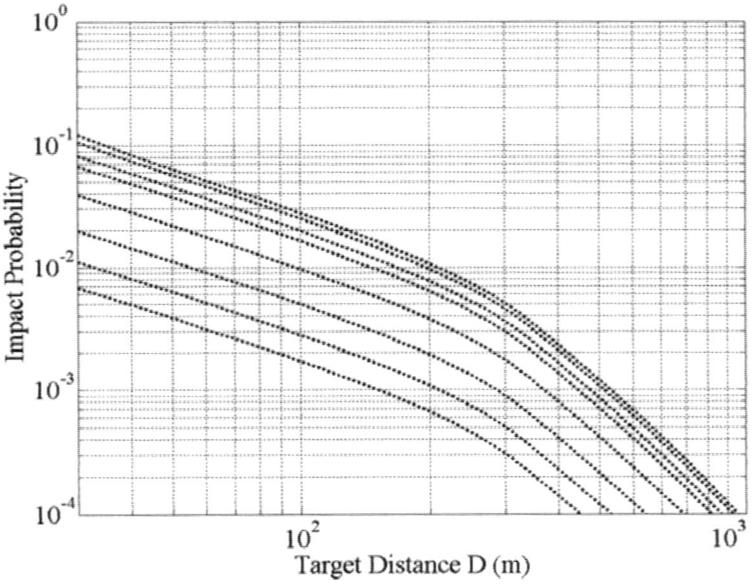

Figure 1a and Figure 1b. Mean Values of Impact Probability of a Single Fragment on a Cone Roof Atmospheric Tank from BLEVEs (Figure 1a) and Mechanical Explosions (Figure 1b)

Table 5 reports the actual safety distances for escalation due to blast wave, calculated for BLEVE of the same vessels considered in Tables 3 and 4. The vessels are assumed to contain liquefied propane, with a liquid filling level of 50% or 80%, and are assumed to fail at 20bar. No geometrical corrections were introduced in the calculations. Table 5 also reports the safety distances calculated considering the total energy obtained by adding liquid and vapor explosion energy. It is worth noting that the vapor contribution is approximately negligible for a filling level of 50%.

Table 5 - Effective Safety Distances for Escalation Due to Blast Waves Following a BLEVE.

	Fill Level	Vessel volume			
		Cylindrical Vessel $2\ m^3$	Cylindrical Vessel $20\ m^3$	Cylindrical Vessel $200\ m^3$	Spherical Vessel $2000\ m^3$
Liquid Explosion	50%	7.68	16.55	35.67	76.84
	80%	9.22	19.85	42.78	92.16
Liquid + Vapor Explosion	50%	8.11	17.46	37.62	81.06
	80%	9.28	19.99	43.07	92.79

A comparison with Table 3 evidences that the safety distances due to BLEVE blast waves are lower than the typical safety distances given for vapor cloud explosions, although the values in Table 5 are sufficiently high to cause interactions with adjacent equipment.

A further escalation vector that may be caused by a BLEVE is the projection of missiles following the failure of the primary vessel. This event is similar to the mechanical explosion, but both the fracture mechanism and the energy fraction transferred to the fragments are different, resulting in somehow different figures for maximum projection distances and impact probability. A maximum projection rate of 120m/s may be estimated from the study of Holden and Reeves [17]. This results in a maximum projection distance of 1350m for the representative set of fragments considered in the present study. Figure 1 reports the corresponding impact probabilities estimated following the approach discussed above. It may be concluded that:

- no practical safety distance for escalation due to fragment projection in a BLEVE may be obtained (calculated values are of about 1350m)
- conservative values for the impact probability of a fragment (obtained assuming a mean number of 10 fragments generated) may be estimated at $2.5 \cdot 10^{-1}$ at 100m and of $2.5 \cdot 10^{-2}$ at 300m.

Point-Source Explosions

For the specific case of blast waves generated by condensed phase explosions or high explosives (or more generally by point-source explosions), the TNT-equivalence models may be used to obtain the peak static overpressure as a function of the scaled distance. The total equivalent mass of reference explosive is the only parameter needed for the estimation of the safety distances for escalation. Hence, the effective distances may be calculated knowing the type of explosive (i.e. the explosion energy per unit mass) and the total mass of explosive involved in the scenario.

Flash Fires

Flash fires involve flame propagation in a gas or vapour cloud without the production of a blast wave, due to a relatively low congestion/confinement of the scenario, or more generally to the low reactivity of flammable mixture (e.g. stratified mixture or non-homogeneous fuel-air clouds or fuel-air mixture with average concentration close to the lower or upper flammability level). The flash fire phenomenon is characterized by low speed flames, and the typical duration is of the order of a few seconds, even for very large clouds.

Therefore, these events have a duration which is always much lower than the "time to failure" of vessels due to thermal radiation. As a conclusion, flash fires are not likely to result in the damage of a secondary vessel and in an escalation.

A single exception should be considered: the ignition of flammable gases or vapors that may be in contact with the flame. The only credible escalation scenario is thus the tank fire that may involve floating roof tanks comprised within the flammable cloud originating the flash-fire.

Fireballs

Following sudden failure of a vessel containing liquefied gas, if the substance contained is flammable and ignition takes place, the lift of vapor in the atmosphere gives rise to mixing with air and to a slow laminar combustion. This does not produce a blast wave, but can be quite destructive due to the high intensity of the thermal radiation associated with the combustion. However, the duration of a fireball is usually less than 60s, thus the possibility of escalation due to radiation damage of target equipment should be carefully assessed.

In evaluating escalation, two different situations should be considered: i) flame engulfment, if target vessels are within the flame radius; and ii) radiation from a distant source, for target vessels at distances larger than the flame radius. In the first case, escalation may be caused by radiation damage or ignition caused by flame impingement. Radiation intensities to be considered are those at the cloud border, and the time of exposure is that of the fireball duration. In the second case, the escalation possibility is due to radiation damage. Radiation intensity depends on the distance of the target vessel from the cloud border and on the view factor. Radiation intensities at ground level are conservatively adopted in the following assessment, and distances are calculated from the projection of the cloud border at ground level.

Table 6 reports the geometric characteristics of a set of atmospheric and pressurized vessels used for the specific assessment of escalation possibility due to radiation generated in fireballs. Four scenarios are considered, consisting of fireballs involving a mass of propane of 130, 52, 26, and 3 tonnes, respectively. The duration of the fireball and the thermal radiation at ground level as a function of distance (from the ground projection of the cloud border) are calculated using standard methods [12]. The conservative value of "time to failure" for the reference vessels is thus calculated by standard models [10, 18], without taking into account any protection or mitigation (e.g., no thermal insulation is considered). Table 7 then reports the values obtained from these calculations.

Table 6. Cylindrical Storage Vessels Geometries for Assessment of Escalation Possibility Due to Thermal Radiation

Design Pressure (MPa)	Vessel Type	Volume (m^3)	Diameter (mm)	Height or Length (mm)	Thickness (mm)
0.1	Vertical Axis	25	2700	4500	5
0.1	Vertical Axis	250	6700	7500	5
0.1	Vertical Axis	500	7800	11000	6
0.1	Vertical Axis	1000	15000	6000	7
0.1	Vertical Axis	2500	16000	13000	7
0.1	Vertical Axis	5200	25000	11000	10
2.5	Horizontal Axis	5	1600	3000	17
2.5	Horizontal Axis	10	1600	4500	17
2.5	Horizontal Axis	25	2300	7000	24
2.5	Horizontal Axis	50	2700	10000	29
2.5	Horizontal Axis	100	2800	18000	30
2.5	Horizontal Axis	250	3800	24000	40

Table 7. Ranges of "Times to Failure" (s) Compared with the Duration of Fireball Scenarios

Vessel type	Duration	Engulfment	5m	10m	25m	50m
atmospheric	6-19	26-38	93-134	96-156	129-243	142-443
pressurized	6-19	142-406	525-1606	550-1936	648-3020	819-5191

Although the results in Table 7 are obtained for the specific case studies defined, these can be considered sufficiently representative to draw some general conclusions. In all of the fireball scenarios, even in the case of full engulfment in flames, the duration of the event is lower (by about an order of magnitude) than the time to failure of any of the pressurized vessels considered in the case studies. It must be remarked that no thermal insulation is considered. Thus, the "times to failure" reported in Table 7 should be considered as extremely conservative values in the case of pressurized vessels. This allows us to conclude on a general basis that the escalation caused by fireball radiation seems unlikely in the case of pressurized vessels.

On the other hand, in the case of atmospheric vessels, the calculated times to failure are comparable to the duration of the fireball in the case of full engulfment. Moreover, in the case of flame engulfment of floating roof storage tanks, the escalation may be caused by ignition of flammable vapors above the roof seal or by failure of the roof seal.

Thus, even if the escalation due to fireball radiation involving atmospheric vessels seems to be credible only for a limited number of scenarios, a specific assessment is necessary if the separation distances between the estimated boundary of the fireball and the secondary target are lower than 25m.

Jet Fires

A jet fire is a turbulent flame following the ignition of a flammable gas or flashing liquid released during the loss of containment from a pressurized vessel. Jet fires are among the more frequent causes of escalation, due to the quite high frequencies of occurrence and to the high damage radius.

Escalation may follow a jet fire due to direct flame impingement or to radiation from a distant source. Escalation caused from flame impingement is a well known cause of domino effects. Although several active and passive mitigation measures are possible (thermal insulation, water deluges, etc.), experience indicates that in the case of flame impingement escalation should always be considered possible, and no threshold may be defined. Recent experimental work confirms this assumption, evidencing that even in the presence of water deluges and thermal insulation, hot spots may cause BLEVEs and mechanical explosions of vessels exposed to jet fire impingement [20].

Outside of the flame zone, less clear data are reported in the literature. Damage due to distant radiation may cause vessel failure, although higher times to failure are expected, and more time is available for active mitigation. Following the approach discussed in the case of fireballs, a specific assessment is carried out to shed some light on this point. Radiation intensities at ground level are calculated for four propane jet fires (from a release diameter of 10, 30, 50 and 80mm respectively and an internal pressure equal to the vapor pressure of propane at 25°C). The radiation intensity is estimated at given distances (5, 10, 25 and 50m) from the flame envelope in the jet direction. The time to failure of the reference set of storage vessels defined in Table 6 is estimated. No mitigation is taken into account and, in particular, no thermal insulation is considered.

The calculated "times to failure" are reported in Table 8. As shown in the table, in the case of atmospheric vessels, escalation is possible in the absence of direct flame impingement. Only at distances higher than 50m from the flame envelope is the time to failure sufficiently high to allow a high success probability for active mitigation. Thus, also considering that usually no thermal insulation is used on atmospheric tanks, this may be assumed as the safety distance in the case of jet fires. It must be remarked, however, that this distance is evaluated from the flame envelope in the jet direction, thus a conservative assessment of the flame shape is needed to estimate the correct value of the safety distance.

Table 8. Ranges of "Time to Failure" (s) in the Case of Radiation from Jet Fires

Vessel Type	5m	10m	25m	50m
Atmospheric	142 - >7200	195- >7200	271 - >7200	>7200
Pressurized	819 - >7200	1065 - >7200	1409 - >7200	>7200

In the case of pressurized vessels, the data in Table 8 evidence that the escalation may not be excluded, although the minimum value calculated for the time to failure is about 13 minutes. It must be remarked that these are very conservative estimates of the time to failure, since pressurized vessels usually have passive fire protection (thermal insulation) as well as active protection (water deluge) that may further raise the value of the required time to failure. Thus, taking into account the time required for emergency management and active mitigation by emergency teams, the escalation involving pressurized vessels is scarcely credible in the case of distant source radiation from jet fires. In the case of pressurized vessels, the escalation is definitely not credible for distances higher than 25m from the flame envelope in the jet direction. This conservative value of safety distance may be further reduced taking into account the passive mitigation due to thermal insulation.

Therefore, the following conclusions may be drawn on the basis of the results obtained:

- escalation from jet fires is always possible in the case of flame impingement (within the flame envelope)
- for a representative set of atmospheric non-protected vessels exposed to distant source radiation, a safety distance from the flame surface can be assumed to be of 50m from the flame envelope in the jet direction
- for a representative set of pressurized non-protected vessels exposed to distant source radiation, a safety distance from the flame surface can be assumed to be of 25m from the flame envelope in the jet direction. This value may be further reduced taking into account the effect of thermal insulation.

Pool Fires

A pool fire is a flame involving the vapors generated from a pool of a flammable liquid, formed as a consequence of loss of containment (LOC) from atmospheric vessels or from instantaneous LOC from pressurized vessels (in this case, only the liquid residue after the flash and the cloud entrainment may form a boiling pool). Further details on pool fires are provided elsewhere [1, 12].

Also, in the case of pool fires, two different scenarios may be identified if escalation possibility is considered. A target vessel may be inside the pool, thus being fully engulfed in the flame, or may be outside of the pool and receive radiation energy from the flame.

As in the case of jet fire, it is well known that full engulfment in a pool fire may result in escalation due to the failure of the target vessel (usually resulting in a BLEVE or a fireball). Thus, within the pool range the escalation should always be considered possible.

On the other hand, it is less clear if escalation should be considered as possible in the case of target vessels receiving distant source radiation from the flame. In order to carry out a specific assessment, the same procedure described for jet fires is applied. Three pool diameters (20, 30 and 50m) and three substances (acetone, benzene, and methanol) are considered. The results of the assessment are reported in Table 9.

Table 9 - Ranges of "Time to Failure" (s) for Vessels in the Case of Radiation from Pool Fires

Vessel Type	**5m**	**10m**	**25m**	**50m**
atmospheric	96 - >7200	130 - >7200	360 - >7200	>7200
pressurized	550 - >7200	751 - >7200	1786 - >7200	>7200

Also, in this case, it is clear that escalation may be considered possible for atmospheric vessels at distances lower than 50m from the pool border. In the case of pressurized vessels, a conservative safety distance of 15m may be assumed, although this value may be further reduced taking into account the effect of thermal insulation.

As in the case of jet fires, some articulated criteria may be thus defined concerning the credibility of escalation caused by pool fires:

- escalation from pool fires is always possible in the case of engulfment (thus for target vessels inside the pool border)
- for a representative set of atmospheric non-protected vessels exposed to distant source radiation, a safety distance from the flame surface can be assumed to be of 50m from the pool border
- for a representative set of pressurized non-protected vessels exposed to distant source radiation, a safety distance from the flame surface can be assumed to be of 15m from the pool border. This value may be further reduced taking into account the effect of thermal insulation.

CONCLUSIONS

The possibility of escalation of primary scenarios resulting in domino effects is assessed for typical accidental scenarios considered in QRA. Damage models for the assessment of damage probability are used to estimate the possibility of escalation in a number of representative case studies. The results obtained allow the development of detailed criteria to assess the possibility of escalation and to evaluate safety distances for the prevention of domino effects.

REFERENCES

[1] Center for Chemical Process Safety of the American Institute of Chemical Engineers, Guidelines for Chemical Process Quantitative Risk Analysis. AIChE: New York, 1989.

[2] MHIDAS, Major Hazard Incident Data Service, AEA Technology plc, Major Hazards Assessment Unit, Health and Safety Executive, UK, 2001.

[3] Lees, F.P., Loss Prevention in the Process Industries (Second Edition). Butterworth-Heinemann: Oxford, 1996.

[4] Baker, W.E., Cox, P.A., Westine, P.S., Kulesz, J.J., "Explosion Hazards and Evaluation." Elsevier: Amsterdam, 1983.

[5] Henrych, J., The Dynamics Of Explosion and Its Use, Developments in Civil Engineering. Vol. 1. Elsevier: Amsterdam, 1979.

[6] Center for Chemical Process Safety of the American Institute of Chemical Engineers, Guidelines for Evaluating the Characteristics of Vapor Cloud Explosions, Flash Fires and BLEVEs. AIChE: New York, 1994.

[7] Cozzani, V., Salzano, E., "Blast Wave Damage to Process Equipment as a Trigger of Domino Effects." 19th Annual CCPS International Conference, AIChE: New York, 2004.

[8] Cozzani, V., Salzano, E., "The Quantitative Assessment of Domino Effects Caused by Overpressure, Part I: Probit Models." Journal of Hazardous Materials, 2004, 107:67.

[9] Cozzani, V., Salzano, E., "Threshold Values for Domino Effects Caused by Blast Wave Interaction with Process Equipment." J. Loss Prev. Process Industries, 2004, 17:437.

[10] Cozzani, V., Gubinelli, G., Russo, G., Salzano, E., Zanelli, S., "An Assessment of the Escalation Potential in Domino Scenarios." Proceedings of the 11th International Symposium on Loss Prevention in the Process Industries, PCHE: Prague, 2004.

[11] Gubinelli, G., Zanelli, S., Cozzani, V., "A Simplified Model for the Assessment of the Impact Probability of Fragments." Journal of Hazardous Materials, 2004, 116:175.

[12] Van Den Bosh, C.J.H., Weterings, R.A.P.M., Methods for the Calculation of Physical Effects (Yellow Book). Committee for the Prevention of Disasters, The Hague (NL), 1997.

[13] Van den Berg, A. C., "The Multi-Energy Method - a Framework for Vapor Cloud Explosion Blast Prediction." Journal of Hazardous Materials, 1985, 12:1.

[14] Tang, M.J., Baker, Q.A., "A New Set of Blast Curves from Vapour Cloud Explosion." Process Safety Progress, AIChE: 1999; 18:235-240.

[15] Van den Berg, A.C., Eggen, J.B.M.M, "GAME, Guidance for the Application of the Multi-Energy method." International Symposium on Hazards, Prevention and Mitigation of Industrial Explosion, 2nd International Specialist Meeting on Fuel-Air Explosions, Bergen, Norway. 1996; 8:23-34.

[16] Forcier, T., Zalosh, R., "External Pressures Generated by Vented Gas and Dust Explosions." J. Loss Prev. Process Industries, 2000, 13:411.

[17] Holden, P.L., Reeves, A.B., "Fragment Hazards from Failures of Pressurized Liquefied Gas Vessels." IChemE Symposium Series No. 93 (1985) 205.

[18] Salzano, E., Picozzi, B., Vaccaro, S., Ciambelli, P., "Hazard of Pressurized Tanks Involved in Fires." Ind.Eng.Chem.Res., 2003;42:1804.

[19] Van den Berg, A.C., van der Voort, M.M., Weerheijm, J., Versloot, N.H.A., "Expansion-Controlled Evaporation a Safe Approach to BLEVE Blast." J. of Loss Prev. Process Industries, 2004;17:6;389-395.

[20] Shirvill, L.C., "Efficacy of Water Spray Protection Against Propane and Butane Jet Fires Impinging on LPG Storage Tanks." J. Loss Prev. Process Industries, 2004;17:111.

Paper 96d

Quantitative Risk Analysis for Hazardous Materials Transportation

Yuanhua Qiao, Nir Keren, William J. Rogers, and M. Sam Mannan
Mary Kay O'Connor Process Safety Center
Chemical Engineering Department
Texas A&M University
College Station, TX 77843-3122 USA

Prepared for Presentation at
American Institute of Chemical Engineers
2005 Spring National Meeting
Process Plant Safety Symposium
Atlanta, Georgia
April 11-13, 2005

February 2005

UNPUBLISHED

QUANTITATIVE RISK ANALYSIS FOR HAZARDOUS MATERIALS TRANSPORTATION

Yuanhua Qiao, Nir Keren, William J. Rogers, and M. Sam Mannan*
Mary Kay O'Connor Process Safety Center
Chemical Engineering Department
Texas A&M University
College Station, TX 77843-3122 USA

ABSTRACT

Historical evidence has shown that incidents due to hazardous material releases during transportation can lead to consequences as severe as those created by releases at fixed plants. In this paper, a procedure is developed to quantify and evaluate individual risk resulting from transportation of hazardous materials.

A transportation route is represented by a sum of risk source units or links. For each link an incident frequency is evaluated using a mathematical model based on a Department of Public Safety incident database. The release scenario probability for an incident is assessed by statistical inference and event tree analysis. CANARY, a consequence assessment package, is used to model consequences. Then individual risk is evaluated using a numerical procedure, which incorporates analysis of hazardous materials, meteorological conditions, and other contributing effects. In this manner a risk evaluation of a transportation network, including fixed installations such as process or storage plants, provides criteria for route selection in hazardous material transportation.

INTRODUCTION

There are roughly 300 million hazardous materials shipments in the U.S. each year [1]. The large number of transportation activities has raised attention from the public and agencies such as the Department of Transportation (DOT) concerning potential risks and the necessity of effective safeguards. DOT has set standards for hazardous material transportation to meet minimum safety levels [2, 3], but these regulations mainly concern hardware and procedures. Therefore, compliance with DOT regulations does not guarantee the safest routes.

How can the best (safest) routes be selected? The designing of hazardous material transport involves comparisons of alternatives in terms of the risk as well as an analysis of tradeoffs between cost and risk. This design approach is possible to the extent that risks of transport alternatives can be quantified. Not only does quantification of the associated risks facilitate transport policy decisions, but it also provides a framework for communicating the risks implied by these policies to the public. Transportation Risk Analysis (TRA) produces quantitative estimates of the potential consequences and the associated frequencies of these

*Corresponding author. Tel.: +1-979-862-3985; fax: +1-979-458-1493.
E-mail address: mannan@tamu.edu (M. Sam Mannan).

consequences. Without the benefits of TRA, judgments about route selection must be made subjectively, and the status quo is often accepted without analysis.

The aim of this paper is to present a framework to assess the risks of hazardous material transportation and to illustrate its potential use in policy-making situations. A numerical procedure is developed step by step to measure the transportation risk. Each transportation route is represented by a sum of risk source units or links, which are formed from connected network points that have the same transportation characteristics. Risk is defined in terms of the incident frequency and the magnitude of the incident consequence. For each link the incident frequency is evaluated by considering various parameters including the type of vehicle, material characteristics, road conditions, and environmental and human factors. The consequences resulting from a transportation incident at each point is assessed by CANARY, which is a commercially available consequence analysis software program. The individual risk within an impact area of the network is then determined by integrating risks created on that impact area by each link of the network, and the risk due to a single link is determined from the sum of the risks created at each point within the link.

BACKGROUND

Risk assessment is a sequential process, beginning with an understanding of the level of involvement, the frequency and type of incident, and the consequence for a given incident. The way those components are defined and determined depends on the data available and the purpose of the risk assessment. In the past two decades, attention has been focused on risk analysis of hazardous materials within transportation networks, and the techniques of quantitative risk analysis initially developed for fixed plants have been extended to transportation risk analysis.

In 1987, Abkowitz and Cheng attempted to measure the risk of hazardous material transportation by summing the cost of fatalities, major injuries, minor injuries, and damage to property [4].

In 1989, Ang et al. suggested a general framework for risk analysis in transportation that decomposed the problem into three separate stages: determination of an undesirable event, estimation of the level of potential exposure, and assessment of the magnitude of consequences [5]. While it is possible to perform a detailed risk assessment for any particular incident location, for routing purposes risks must be estimated for the infinite number of points on a transportation network. A shortcut is employed in this paper to draw a band of fixed width around each link and to use the number of persons living within this band as the link risk consequence. This assumes that all persons within the band will be impacted equally and no one outside of the band will be impacted at all.

In 1993 Purdy estimated the impact to humans from flammable substances and toxic gases [6]. The entire population that may be affected by a hazardous material incident is considered in their model, including motorists on a road where an incident occurs, travelers on trains, and people who live near the transportation route.

In 1995, Erkut and Verter proposed that a path between a given origin-destination pair can be represented by a set of road segments, where the road characteristics are uniform within each segment [7]. The risk imposed on an individual due to a hazardous material shipment can be estimated as the probability of an incident during transport multiplied by the probability of the individual experiencing the consequence as a result of the incident. Erkut and Verter also proposed to employ Geographic Information Systems (GIS) to manage territorial information during the risk analysis of transportation network. GIS is a system of computer software, hardware, data, and personnel to help manipulate, analyze, and present information that is tied to a spatial location. GIS is also employed in several other transportation risk analysis papers for the same purpose [8, 9].

The most common risk measurements used to perform a quantitative risk analysis for a major incident hazard plant, i.e., individual risk and societal risk, are also employed in transportation risk analysis. Individual risk reflects the frequency of a specific health effect at a geographical location. Its predictions are used to answer the question: how does the risk to an individual vary with location? Societal risk represents the overall risk associated with an activity to a particular population. The most common societal risk presentation technique is an F-N curve, which is a graph of number of fatalities (N) on the abscissa and frequency (F) of N or more fatalities on the ordinate [10]. In 1999, Leonelli et al. proposes a detailed procedure to evaluate individual and societal risk for a transportation network [11]. Individual risk is also measured for the transportation process in [12] and [13]. Individual risk and societal risk are powerful tools for the measurement of transportation hazard impact [11, 14].

The proposition of viewing a transportation route as a set of road segments or a set of links, and the employment of individual risk and societal risk as measurements of transportation risk are two prominent achievements in the development of transportation risk analysis methodology. Transportation routes are represented as a network in which each route is divided into various links, with nodes connecting adjacent links. Then the risk resulting from the entire route is the sum of risks of the links, and the risk associated with transportation activities can consequently be determined in a more detailed way and more accurately using a numerical methodology. Following the analysis of transportation risk, transportation route strategies are based on the network models to make possible a more systematic method for transportation route selection [11, 12, 13, 14].

However, the numerical methodology is difficult to employ directly by decision or policy makers, because only the methodology is proposed without input data, or because the methodology is too complicated to obtain available data as input. For example, for most of the methodologies, incident frequency and conditional release probability data are assumed to be available, but the acquisition of the required data requires considerable effort. The methodology should be developed according to the availability of input data, otherwise the meaning of the methodology is in doubt. In this paper, each risk component, including frequency, release scenario, consequence, and resultant fatality, is assessed based on the available data/database. The individual risk in the transportation network is then assessed by a numerical methodology, step by step.

METHODOLOGY

Preliminary Network Analysis

A transportation route can be represented as a risk source consisting of links and nodes over which risk is summed, since a release can occur at each of its points. This representation of the route in terms of links and nodes is shown in Figure 1. Each link in the network must have uniform properties at all its points, which can be accomplished by adding nodes to the network.

Since the primary objective of risk analysis is to facilitate the route selection, the risk analysis is performed for one kind of hazardous material shipment so that the risk is assumed to result from only the shipped hazardous material. After the risk imposed from that hazardous material is assessed, the overall risk resulting from other hazardous materials can be determined by summation.

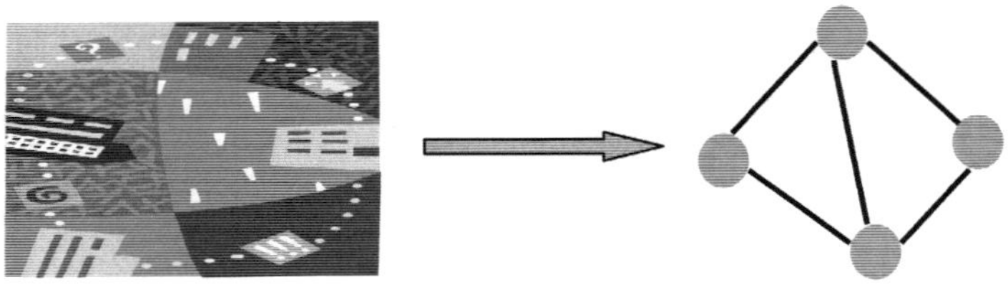

Figure 1. Transformation from Real Transportation Roads to Transportation Network

A transportation route can be considered to be composed of links and nodes. When performing the transportation risk analysis each link must be characterized by certain properties. For a generic link l of a set N_{link} of links, the uniform properties include:

- The transportation network type that the link belongs to, that is, road network, rail, inland waterway, or pipeline. This paper will focus on the road network, but the other transportation types can be assessed in similar way.
- The physical condition of the link including road type and road conditions. For road type, lane number, divided or undivided should be considered because those characteristics will affect incident probability and potential consequences.
- Vehicle type. Both hazardous materials and transporting vehicle type should be included under this category. The vehicle type is selected according to the transported hazardous materials. Title 49, Code of Federal Regulations specifies the type of containers for each type of hazardous materials.
- The yearly shipment data including the types of hazardous material and shipping frequencies.
- The incident frequency expressed in events/(vehicle-mile-year).
- Release probability or the probability of having a release once an incident has occurred.

After characterization of the links, the risk source is defined. The next step is to define the possible consequence and the impact area.

- Environmental condition. After a release occurs in a link, the potential consequences will be characterized not only by the risk source, but also by the environment conditions, which includes temperature, pressure, humidity, wind speed, wind stability, and other parameters used in dispersion modeling.

- Population distribution, which is an input to the societal risk can be simplified as two types or zones, where people may be considered uniformly distributed, and aggregation centers, where people are clustered.

After the transportation network is defined by the risk source and impact characterization, the risk associated with transportation activities can be quantified step by step as follows:

1) Assess the incident frequency for each link;

2) Analyze release scenarios following an incident;

3) Evaluate final outcomes to which each release scenario can lead, e.g., a toxic cloud, explosive cloud, pool flame, or jet-fire.

4) Calculate the individual risk based on the frequency data, release scenario analysis, and the evaluation of the consequences. The transportation route will be viewed as a risk source composed of links and nodes. The overall risk is determined from the sum of the risks created at each link of the route.

This quantitative approach is described below.

Frequency Analysis

Detailed analyses of several publicly available databases by many workers make possible specification of incident probabilities on a per-mile basis as a function of road type, truck type, and population density. One of the most detailed analysis of such data is that conducted by Harwood and Russell [15]. On the basis of computerized data files from California, Illinois, and Michigan, Harwood and Russell calculate incident probabilities by using the number of reported incidents and the total number of truck-miles traveled. Road type, truck type, and population density are considered as parameters affecting the incident frequency in Harwood and Russell's study. In the report of Argonne National Laboratory [16], the Harwood and Russell statistics are divided according to two road categories – interstates and non-interstates (state highway) – and three population density categories – urban, suburban, and rural.

The assessments of transportation incident frequency are not at the same level of accuracy as those of the consequence measurement, which consider the effect of various contributing parameters. Transportation incident frequency varies with the nature of road, truck, environmental, and driver conditions. In this paper the evaluation of frequency incorporates most of those important contributing factors rather than considering only road type and population density categories. For each link the incident frequency is integrated together with fatality to assess the annual individual risk and societal risk rather than viewing it as a constant.

Transportation incident frequency is expressed in events/(vehicle-mile-year). Both the event/year data and the exposure data, which is the number of miles traveled by all vehicles, must be available to determine the incident rate. The Department of Public Safety (DPS) incident database was found to be a good source from which to derive the event/year data, because more of the required parameters are recorded and also because this database can be easily grouped based on road type, making easier the assessment of incident rate for a specific road. The exposure data are obtained from the 1993 Commodity Flow Survey (CFS), and 1997 CFS [17], which are the records of state transportation institutes. These two types of data do not correspond well with each other because they are collected by independent agencies. Incident data are based on police reports, and exposure data are based on questionnaires or traffic counts. This correspondence problem results in the inability to aggregate these two kinds of databases by the affecting factors. Some assumptions are made and numerical models are employed to evaluate the transportation frequency under various conditions. Details of the process are explained in Qiao, et al [18].

The transportation frequency is assessed by incorporating effects of numerous parameters describing the nature of the truck configuration, operation, environment, and road conditions, and it is a function of *Lane # (x_1), Weather (x_2), Surface condition (x_3), Alignment (x_4), Month of year (x_5), Day of week (x_6), Hour of day (x_7), Lane Divided/undivided (y_1), Container capacity (y_3), Container type (y_4), Driver experience (y_5), and Driver status (y_6).* For each link, the *Lane # (x_1), Surface condition (x_3), Alignment (x_4), and Lane Divided/undivided (y_1)* is the same during the entire year; the *Container capacity (y_3)* and *Container type (y_4)* are defined prior to the calculation of transportation risk; and the effects from driver status and driver experience are not considered in this paper. So, for each lane, the transportation frequency is a function of *Weather (x_2), Month of year (x_5), Day of week (x_6),* and *Hour of day (x_7).* Thus the incident frequency is expressed as $P_{incident}(w, m, d, h)$, where w, m, d, and h are the weather, month, day, and hour, respectively.

Release Scenario Analysis

The conditional release probability has a significant influence on overall risk, because it typically addresses the relative likelihood of different sizes of releases having substantially different potential consequences. Given the occurrence of an incident involving a truck carrying hazardous material, this conditional probability of release is a function of vehicle characteristics, shipping conditions, and the nature of the incident, which is hard to characterize from a mechanical standpoint. Statistical inference is the alternative way to derive the release scenarios because some incident data are available. This inference is also the way to assess the incident frequency, or given that an incident occurs, the conditional probability of puncture, for example, is the number of incidents resulting in puncture divided by the number of overall incidents. Rhyne once studied the incident force types and force magnitudes to evaluate conditional failure probabilities for a spectrum of releases sizes [19]. Some incident distribution data are employed to determine the fraction for various forces, including impact force, crush force, and puncture force.

Statistical inference methodology presumes that sufficient historical data are available, and that past observations can adequately be used to infer future expectations. Compared with the incident distribution data of the early 1970s employed by Rhyne [19] and the truck data for the mid-1980s studied by Harwood and Russell [15], the Hazardous Materials Information System (HMIS) database, established in 1971, is more up to date and reliable for estimation purposes.

Event tree analysis, in which an initial event leads to subsequent events, is applied along with the HMIS database inference to provide the conditional frequency of each incident scenario of interest. The HMIS database was disaggregated for each hazardous material and each type of container. For each of those representative and popular hazardous materials, including ammonia, chlorine, LNG, LPG, and gasoline, data is analyzed along with an event tree analysis, so that the probability of three kinds of release scenarios can be assessed, that is leak, puncture, and rupture (which represent the release sizes). In the same way, the conditional probabilities of potential consequence scenarios are evaluated. The release scenario is varied with hazardous material and container type using the event tree evaluation method. So, for each link in the transportation network, the probabilities of various outcomes are viewed as a function of only a consequence scenario, $P_{scenario}(s)$, where s is scenario, and the number of possible consequence scenarios is $N_{scenario}$. One example of this analysis is shown in Figure 2. This methodology will be developed further in a subsequent paper.

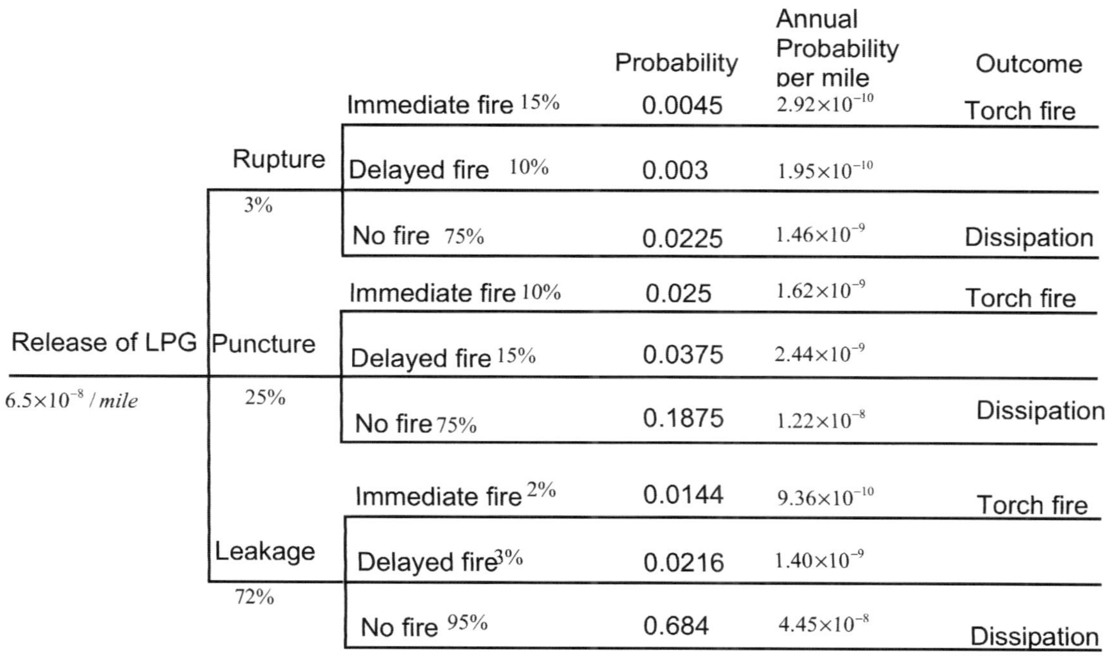

Figure 2. Release Scenario Event Tree Analysis

Consequence Analysis

The estimated consequences of an incident involving a shipment of hazardous materials depends on a variety of factors including the amount released, toxicity of the chemical, health effects from the released materials, population and environmental characteristics adjacent to the incident site, and the weather conditions at the time of the incident. The interplay among these factors produces the risk spectrum for transporting hazardous materials.

The commercially available consequence analysis package CANARY has been used in this paper to assess the transportation incident consequence. CANARY is based on physical property and consequence models that are available in the open literature. Given a release at any point risk source of the transportation route, most of the potential hazards can be modeled using CANARY, including toxic vapor cloud, flammable vapor cloud, unconfined vapor cloud explosion, pool fire, torch fire, flare, BLEVE, and confined explosion.

Number of Fatalities

A personal consequence scenario could be minor injury, major injury, or fatality. In this paper, the probability of incurring death is the basis for assessing the individual risk. From CANARY the incident consequence can be characterized, e.g., as overpressure, toxic vapor density, or radiant heat flux for different consequence scenarios. Combining these effects with proper exposure times the corresponding received doses are obtained, and doses are converted into fatalities through "probit equations". In this way, the fatality resulting from any point risk source in the route is assessed, and the fatality resulting from each point risk source in the route is determined for each point in the impact area. Since the number of fatalities is decided by the incident consequence, and the consequence depends on the nature of the hazardous material, release characteristics, environmental and weather conditions, the fatality will be affected by the same parameters. After the hazardous material and release scenario are defined, the number of fatalities becomes $F_{atality}(w, s)$, the function of only weather, w, and the consequence scenario, s, since the environmental conditions can be assumed to be constant for a link.

Individual Risk Analysis

For the risk resulting from hazardous materials transportation, the individual risk at point $P(x, y)$ is the sum of all risk sources within all links in the transportation route. Translating this concept in transportation risk analysis means that the risk of a single link is determined from the sum of the risks created by all vehicles traveling on that link, and that the risk values of all links are summed to obtain the overall risk of the entire transportation route.

Assume vehicle V, containing hazardous materials and existing in point $Q(x, y)$ on link l, results in a risk consequence at $P(x, y)$. Then the annual individual risk, expressed in fatalities/year, is a function of $P_{incident}(w, m, d, h)$, $P_{scenario}(s)$, and $F_{atality}(w, s)$. Over an entire year, we assume that the effects to incident frequency from month, week, day, and hour can be averaged, so that $P_{incident}(w, m, d, h)$ becomes $P_{incident}(w)$. The purpose of this assumption is to make $P_{incident}(w)$ match with $F_{atality}(w, s)$ and to simplify the integration process.

Performing the calculation for all meteorological/weather conditions, a weather-averaged risk is obtained for a single final outcome. If this is done for all $N_{scenario}$ consequences occurring in *P(x, y)*, the individual risk at point *P(x, y)* due to the point risk source *Q(x, y)* is defined by:

$$PIR = \int_1^{N_{weather}} \{P_{incident}(w) \times \int_1^{N_{scenario}} [P_{scenario}(s) \times F_{atality}(s, w)] ds\} \bullet dw$$

where $N_{weather}$ is the number of meteorology/weather conditions.

Until now the vehicle is assumed to be standing at $P(x, y)$, as if it is a fixed storage vessel from which a release occurs. By taking into account the fact that the vehicle is moving along the link, successive steps are used to translate the point risk source along the link, or perform a linear integration along the link, so that the link risk value created by V at an impact point *P(x, y)* is evaluated from the link individual risk (LIR) integral shown by the following:

$$LIR = \int_l PIR dt$$

The sum of LIR values extended at all links provides the total individual risk created by the entire transportation route on a point *P(x, y)* of the impact area. Finally, the total individual risk (IR) is expressed by the following sum:

$$IR = \sum_{l=1}^{N_{link}} LIR$$

CONCLUSION AND FUTURE WORK

In this paper a numerical procedure, which allows the coupling of time effectiveness and mathematical accuracy, is developed for transportation individual risk evaluation. Each risk component is assessed according to available data sources. The match between input data and methodology makes it easier not only to understand the risk imposed to the impact area, but also to derive the individual risk data quantitatively. This risk analysis result will serve as one of the most important criteria in route selection.

More work is needed to develop the risk analysis tool, including the societal risk evaluation. In addition, a computer program must be compiled to automate the calculation of individual risk and societal risk from input of the CANARY consequence analysis.

REFERENCES

1 http://hazmat.dot.gov/hmpe_execsum.pdf

2. http://www.nhtsa.dot.gov/cars/rules/

3. http://hazmat.dot.gov/rules.htm

4. Abkowitz, Mark, Der-Ming Cheng, Paul, "Developing a Risk/Cost Framework for Routing Truck Movements of Hazardous Materials," Accident Anal & Prev, 1988, 20 (1), 39-51.

5. Ang, A., Briscoe, J., et al., "Development of a Systems Risk Methodology for Single and Multimodal Transportation Systems, Final Report," Office of University Research, US DOT, Washington, DC, 1989.

6. Purdy, Grant, "Risk Analysis of the Transportation of Dangerous Goods by Road and Rail," Journal of Hazardous Materials, 1993, 33, 229-259.

7. Erkut, Erhan, Verter, Vedat, "A Framework for Hazardous Materials Transport Risk Assessment," Risk Analysis, 1995, 15 (5), 589-601.

8. Bubbico, Roberto, Di Cave, Sergio, Mazzarotta, Barbara, "A GIS-Supported Transportation Risk Analysis Approach," Center for Chemical Process Safety 17th International Conference and Workshop, Jacksonville, Florida, October 8-11, 2002, pp361-374,

9. Bubbico, Roberto, Di Cave, Sergio, Mazzarotta, Barbara, "Risk Analysis for Road and Rail Transport of Hazardous Materials: a GIS Approach," Journal of Loss Prevention in the Process Industries, 2004,17, 483-488.

10. Center for Chemical Process Safety, Guidelines for Chemical Transportation Risk Analysis, New York: American Institute of Chemical Engineers, 1994.

11. Leonelli, Paolo, Bonvicini, Sarah, Spadoni, Gigliola, "New Detailed Numerical Procedures for Calculating Risk Measures in Hazardous Materials Transportation, Journal of Prevention in the Process Industries, 1999,12, 507-515.

12. Vayiokas, Nilolaos, Pitsiava-Latinopoulou, Magda, "Developing a Framework for Assessing Risks Involved in the Transportation of Dangerous Goods," Loss Prevention and Safety Promotion in the Process Industries 11th International Symposium Loss Prevention 2004, Praha Congress Centre, 31 May- 3 June, 2004, pp.4495-4503.

13. Bubbico, Roberto,Di Cave, Sergio, Mazzarotta, Barbara, "Risk Assessment for Land Transportation of Hazardous Materials in Italy," Loss Prevention and Safety Promotion in the Process Industries 11th International Symposium Loss Prevention 2004, Praha Congress Centre, 31 May- 3 June, 2004, pp.4297-4304.

14. Advisory Committee on Dangerous Substances, Major Hazard Aspects of the Transportation of Dangerous Substances, London:HSE, 1991.

15. Harwood, D., Russell, E. Present Practices of Highway Transportation of Hazardous Materials, FHWA/RD-89/013, Federal Highway Administration, DOT, Washington, DC, 1990.

16. Brown, D, Dunn, W, et al, A National Risk Assessment for Selected Hazardous Materials in Transportation, ANL/DIS-01-1, Argonne National Laboratory, Argonne, 2000.

17. http://www.bts.gov/ntda/cfs/desc.html

18. Qiao, Yuanhua, Gentile, Michela, Mannan, M. Sam, "Fuzzy Logic Methodology for Accident Frequency Assessment in Hazardous Materials Transportation," CCPS 19th Annual International Conference & Workshop, June 29 - July 1, Orlando, FL, 2004.

19. Rhyne, William R., Hazardous Materials Transportation Risk Analysis: Quantitative Approaches for Truck and Train, Van Nostrand Reinhold, New York, 1994.

Statistical Analysis of Pressure Vessel Failure

Thomas Mikschl and Ahmad Shafaghi, Ph.D.

ABSG Consulting Inc
300 Commerce Drive, Suite 200, Irvine, CA 92602

Prepared for Presentation at
American Institute of Chemical Engineers
2005 Spring National Meeting
Process Plant Safety Symposium
Atlanta, Georgia
April 11-13, 2005

ABSTRACT

This paper presents a simple method for calculating equipment failure rates using all available information that typically includes both "generic" and application-specific data. The paper presents a general methodology suitable for many applications, and an example focused on the statistical analysis of failure rates for pressure vessels. At the core of the methodology is Bayes' theorem [Kaplan], which has been used in data analysis in many fields including the nuclear power plant probabilistic risk assessment (PRA) studies for the past twenty years.

The Bayesian update calculation is simplified through use of the gamma function to represent the prior distribution. The gamma family is conjugate to Poisson data. That is, updating a gamma prior distribution with the Poisson data produces a posterior that is also a gamma distribution. The parameters of the gamma prior distribution, and the application specific event data, can be easily combined to calculate the parameters of the gamma posterior distribution.

The method consists of two main steps: (1) defining a prior [distribution] for the equipment failure, and (2) upgrading the prior using application-specific data. The outcome is the posterior [distribution] for the equipment failure rate. The choice of the prior represents the analyst's state of knowledge regarding the subject failure rate. The selection of the prior must be made independent of the application specific data to be used in the Bayesian update. The prior distribution may be derived from a single source, or from a collection of available sources. The example presented in this paper develops a prior distribution based on a collection of 12 estimates of pressure vessel failure rates found in the literature.

The example presented in this paper applies to the generic pressure vessel used in the various sectors of the process industries. As a result, the example has not taken into consideration the operating or environment conditions that the equipment may be subjected to. This is, however, by no means a limitation of the methodology.

The remainder of this paper presents a description of the Bayesian updating methodology as applied to this application, followed by an example demonstrating the use of this methodology.

BAYESIAN UPDATING METHODOLOGY

In the context of the Bayesian methodology, *"parameters* [e.g., failure rates] *are treated as measurable quantities whose uncertainty is dependent on the state of knowledge of* [the analyst]. *The state of knowledge is presented in the form of a probability distribution over the range of possible values of that quantity. The probability associated with a particular numerical value of an uncertain but measurable quantity indicates the likelihood that the numerical value is the correct one.*

A key issue in developing state of knowledge for the parameters of failure models is to ensure that the information regarding a parameter [e.g., failure rate], *its relevance, and its value as viewed by the analyst are presented correctly, and that various pieces of information are integrated coherently. Coherence is preserved if the final outcome of the process is consistent with every piece of information used and all assumptions made*" [ABS].

The Bayesian estimation of a failure rate consists of three major steps, as follows [ABS]:

1. Prior belief about the failure rate is quantified, i.e., the prior distribution
2. Failure data are collected and the likelihood function is constructed
3. The posterior distribution is constructed by combining the prior distribution and the likelihood function using Bayesian theorem.

The Bayes' theorem is presented in a functional relationship, as follows:

$$f_{post}(\lambda) \propto likelihood\,(\lambda) * f_{prior}(\lambda) \qquad (1A)$$

or

$$f_{post}(\lambda) \propto \Pr(X = x|\lambda) * f_{prior}(\lambda) \qquad (1B)$$

where,

$f_{post}(\lambda)$ = Posterior distribution of failure rate (λ)
$likelihood\ (\lambda)$ = Likelihood function of failure rate (λ)
$\Pr\ (X = x|\lambda)$ = Likelihood function as function of λ, for given failure event (x)
$f_{prior}(\lambda)$ = Posterior distribution of failure rate (λ)
$\propto$ = Proportional symbol

The prior is the state of the analyst's knowledge and the likelihood function is the evidence obtained from the site. If the evidence were too shallow, then the posterior would resemble the prior. However, as the historical data becomes larger, several patterns emerge [Sandia]:

- The posterior distribution bears less resemblance to the prior distribution, because the data become the dominant factor
- The posterior would be narrower and centered around maximum likelihood estimate, implying less uncertainty in the results

The first question that emerges is how to select a distribution reflecting the analyst's prior knowledge about the failure rate.

The Poisson distribution defines the behavior of the facility failure data (the likelihood function). This is an appropriate distribution for random variables that involve counts or events (such as pressure vessel failure) per unit time. The Poisson distribution is presented below:

$$\Pr(X = x \mid \lambda) = \frac{e^{-\lambda t} (\lambda t)^x}{x!} \qquad (2)$$

where,

x = number of events (failures)

t = time interval

The conjugate family of prior distributions for Poisson data is the family of gamma distributions. That is to say the uncertainty of the failure rate (λ) is defined by the gamma distribution. The gamma distribution and the event data can be combined to result in another gamma distribution. This is the meaning of the conjugate family. That is, when we choose the gamma distribution for the prior, updating it using the Bayes' theorem and Poisson likelihood model, then the posterior distribution is also defined by the gamma distribution.

It is noted here, that although the methodology and sample application presented in this paper focus on the calculation of a time based (i.e., failures per unit time) failure rate, based on a Poisson likelihood function and the conjugate gamma distribution, a similar method applies to the calculation of demand failure rates utilizing the binomial likelihood function and it's conjugate beta distribution.

The gamma distribution for the prior with two parameters of scale factor (α) and the shape factor (β), which is shown below:

$$f_{prior}(\lambda) = \frac{\beta^{\alpha}}{(\alpha-1)!} \lambda^{\alpha-1} e^{-\lambda\beta} \qquad (3A)$$

where,

$(\alpha-1)!$ = (α-1) factorial

Expression 3A can also be presented as:

$$f_{prior}(\lambda) \propto \lambda^{\alpha-1} e^{-\lambda\beta} \qquad (3B)$$

Note that this expression is valid when α is positive, which is the case in our study. This is known as the gamma (α, β) distribution, whose mean and variance are defined as follows:

Mean (λ) = α / β (4)

Variance (λ) = α / β^2 (5)

Using the gamma distribution for the prior and Poisson for the likelihood, the updated distribution is also a gamma distribution. The resulting posterior distribution, which is a combination of the prior gamma distribution and Poisson distribution for the likelihood function, is also the gamma distribution, given in Equation 4, which defines conjugate:

$$f_{post} \propto e^{=\lambda t} \frac{(\lambda t)^x}{x!} \lambda^{\alpha-1} e^{-\lambda\beta} \tag{6A}$$

Simplified further, results in:

$$f_{post} \propto \lambda^{(x+\alpha)-1} e^{-\lambda(t+\beta)} \tag{6B}$$

The posterior gamma distribution parameters are simply calculated by Equation 7 and Equation 8.

$$\alpha_{post} = x + \alpha_{prior} \tag{7}$$

$$\beta_{post} = t + \beta_{prior} \tag{8}$$

Equations 7 and 8 along with Equation 4 (the mean) are the primary tools for the Bayesian update in this study.

The method starts with a prior distribution, which must come from sources independent from the subject facility event history. An appropriate source is "generic" data reported in the literature or used in other studies. The issue with data reported in the literature is that it is mostly incomplete and often presented in terms of "point estimates", which are single values representing the mean (or median) of the failure rate. In conditions of insufficient data (incomplete prior knowledge) about the prior distribution or great uncertainty in the generic data sources, we may use "constrained noninformative priors" described by Atwood. This representation of the prior preserves the mean value of the failure rate estimate and maintains a broad uncertainty range to accommodate the site-specific event data.

Based on Atwood's priors, for Poisson data, the constrained noninformative prior is a gamma distribution with the shape factor and scale factor of the following values:

$$\alpha_{prior} = 1/2 \tag{9}$$

$$\beta_{prior} = \alpha_{prior} / \text{prior mean} \tag{10A}$$

$$= 1/2\ (\text{prior mean}) \tag{10B}$$

Given the best estimate for mean value of failure rate (from generic data), then the parameters of the gamma prior distribution, α_{prior} and β_{prior}, will be calculated according to Equations 9 and 10 A/B.

SAMPLE APPLICATION FOR PRESSURE VESSEL FAILURE RATES

The remainder of this paper focuses on the use of the methodology described above to estimate pressure vessel failure rates for several failure modes.

Prior Distribution

There exist a number of methods for estimating the mean value for the prior distribution, which is dependent on the quality and quantity of generic data and the analyst's preference. In some cases the analyst may have access to a single credible source. However, if multiple credible data sources are available, the analyst may simply choose to select the most reliable, or employ one of a number of methods to merge these data into a single point estimate. These methods range from calculation of a simple arithmetic mean, to the use of sophisticated Bayesian procedures (see [Kaplan] or [Mosleh]).

Generic data sources applicable to our sample application have been collected from the various sources and listed in Table 1. The mean values for disruptive and non-disruptive failure modes of the pressure vessel are listed in Table 1. The definitions of the terms used in Table 1 are given below.

> Disruptive Failure – *"a breaching of the vessel by failure of the shell, head, nozzles or bolting, accompanied by a rapid release of the large volume of the contained pressurized fluid"* [Bush].
>
> Non-disruptive Failure -- *"a condition of crack growth rate or flaw size that is corrected, and which if it had not been corrected, could have reached a critical size and led to disruptive vessel failure" or "a local degradation of the pressure vessel boundary that is localized cracking with or without minor leakage. Such a crack would not reach critical size and lead to disruptive vessel failure"* [Bush].
>
> Range Factor – The range factor implies the level of confidence that the analyst has in the data source. The smaller the range factor, the higher the confidence of the analyst in the data source. For a lognormal distribution, the range factor is equal to the square root of the ratio of the 95th percentile value to the 5th percentile value.

The mean values in Table 1 are taken directly from the referenced data sources. Based on the analyst's level of confidence in each source, a range factor and a probability (weight) are assigned to each source (and corresponding mean failure rates) listed in Table 1. The data reported by Bush has received a small range factor, relative to the other sources, because of the analyst's high level of confidence in this source. On the other hand, data reported in Rijnmond report was assigned a range factor of 9, indicating a lower level of confidence in this data source. The probability weights also represent the analyst's confidence in each data source.

Using the data given in Table 1, we have calculated prior distribution mean values for the pressure vessel failure modes using the various methods, including the first stage of Kaplan's methodology (utilizing the mean values and range factors presented in Table 1). The calculated parameters of the resultant distributions for the pressure vessel failure rates are presented in Table 2. Mean values for the prior distributions have also been calculated and presented in Table 3 based on a simple average of the mean values, based on a geometric mean, and based on a weighted average (using the probabilities assigned to the data sources).

Table 1. Generic Data on Pressure Vessel Failure Modes and Assigned Range Factors and Probabilities

Source	Disruptive [per year]	Non-Disruptive [per year]	Assigned Range Factor	Assigned Probability
Savannah River Site [Blanton]	3.33E-04	3.24E-03	7	0.05
EEI-TVA [Bush]	3.00E-04	1.70E-03	4	0.08
EEI Boiler Drum [Bush]	1.40E-04	2.00E-04	4	0.08
Chemical	5.48E-05	1.05E-04	3	0.11
UK Steam Drum Sample [Bush]	5.00E-05	6.00E-04	3	0.11
IRS-TUW [Bush]	4.50E-05	6.00E-04	3	0.11
NBBPV [Bush]	3.50E-05	-**	3	0.11
UK-Smith& Warwick [Bush]	3.20E-05	2.60E-04	4	0.08
CCPS	9.55E-06	5.57E-05	5	0.07
German LWR Study Group [Bush]	8.80E-06	-**	4	0.08
ABMA [Bush]	4.20E-06	-**	4	0.08
Rijnmond	1.00E-06	1.00E-05	9	0.04

(**) No frequencies reported for Non-Disruptive failures. Assigned probability weights were normalized for calculation of prior distributions.

Table 2. Calculated Prior Distribution Parameters for Pressure Vessel Failure Modes Using Bayesian Method

Failure Mode	Mean [per year]	Median [per year]	5th Percentile [per year]	95th Percentile [per year]
Disruptive	6.7E-05	5.1E-05	1.5E-05	1.1E-04
Non-Disruptive	6.0E-04	5.7E-04	3.3E-04	1.8E-03

Table 3. Calculated Prior Distribution Mean Values for Pressure Vessel Failure Modes (alternate methods)

Method of calculation	Disruptive	Non-Disruptive
Arithmetic Mean (Average)	8.4E-05	7.5E-04
Geometric Mean	3.1E-05	4.1E-03
Weighted Average (used in sample calculation)	7.7E-05	4.8E-04
Bayesian (Values from Table 2)	6.7E-05	6.0E-04

Facility Specific Data

A hypothetical case is used to demonstrate the data analysis method presented in this paper. Consider Company XYZ, a worldwide gas and oil firm, has collected pressure vessel failure data for their facilities for the past 15 years. The number of pressure vessels in operation is 187. The history has shown zero disruptive and 7 non-disruptive failures for the past 15 years among the 187-pressure vessel population.

Data Update

The evidence given in the previous section will be used to update the generic data (prior). For the following example, the prior distributions calculated as a weighted average of the reference sources (see Table 3) will be used.

Using the constrained non-informative gamma distribution, the parameters of the prior distribution for the disruptive pressure vessel failure mode are calculated as follows:

$$\alpha_{Prior}^{Dis} = 0.5$$

$$\beta_{Prior}^{Dis} = 0.5 /7.65E\text{-}05$$

$$= 6{,}536$$

The posterior gamma distribution parameters are calculated:

$$\alpha_{Post}^{Dis} = 0+0.5$$
$$= 0.5$$
$$\beta_{Post}^{Dis} = 15*187+6{,}536$$
$$= 9{,}341$$

The posterior mean is then calculated as:

λ_{Post}^{Dis} = 0.5/9,341
= 5.4E-05 (events per year; disruptive failure mode)

Following the same steps for non-disruptive failure mode, we would get the following mean value for posterior distribution of the non-disruptive failure mode:

λ_{Post}^{N-Dis} = 2.0E-03 (events per year; non-disruptive failure mode)

Table 4 presents calculated distribution parameters of disruptive and non-disruptive pressure vessel failure modes for the example.

Table 4. Prior and Posterior Parameters for Pressure Vessel Failure Modes

Gamma Distribution Parameters	Disruptive		Non-Disruptive	
	Prior	Posterior	Prior	Posterior
α	0.5	0.5	0.5	7.5
β	6,536	9,341	1,048	3,853
λ	7.7E-05	5.4E-05	4.8E-04	2.0E-03
5th Percentile	4.4E-05	3.1E-5	2.1E-05	9.2E-04
95th Percentile	2.7E-04	1.9E-4	1.7E-03	3.2E-03

REFERENCES

ABS Consulting, Methodology for One-stage and Two-stage Bayesian Updating.

Atwood, Corwin L., Constrained Noninformative Priors, Idaho National Engineering Lab, Idaho Falls, Idaho 83415, October 1994.

Blanton, C.H. & Eide, S. A., Savannah River Site, Generic Data Base Development (U), June 30, 1993, WSSRC-TR-93-262.

Bush, S.H., Statistics of Pressure Vessel and Piping Failures, Journal of Pressure Vessel Technology, August 1988, Vol. 110/225-233.

CCPS of AIChE, Guidelines for Process Equipment Reliability Data, 1989.

Chemical Data, Confidential source

Handbook of Parameter Estimation for Probabilistic Risk Assessment, Sandia National Lab, NUREG/CR-6823, SAN2003-3348P.

Kaplan, S., On a “Two-Stage” Bayesian Procedure for Determining Failure Rates from Experimental Data.

Mosleh, Ali et al, Development of Risk and Reliability Data Base for Chemical Facilities, AIChE 1986 Summer National Meeting, Boston, MA 1986.

Rijnmond, Risk Analysis of Six Potentially Hazardous Industrial Objects in the Rincon Area, A Pilot Study, D Reidel Publishing Company, 1981.

Atlantis Case Study: Getting the Most from Your HAZOP

Donnie Carter, BP
Mary G. Gerschefski, W.S. Atkins Inc
Kevin D. Watson, P.E., formerly of W. S. Atkins Inc.

Prepared for Presentation at
American Institute of Chemical Engineers
2005 Spring National Meeting
Process Plant Safety Symposium
Atlanta, Georgia
April 11-13, 2005

December 2004

UNPUBLISHED

SYNOPSIS

Hazard and Operability (HAZOP) analysis is a proven technique for identifying design deficiencies that could lead to undesirable consequences. When integrated into the design stage as part of an overall Hazard Management Program, this risk assessment technique can improve a facility's safety and operability. A flexible and functional tool, the HAZOP study has become the preferred means of hazard identification in the chemical process industries.

Although widely popular and adaptable, the HAZOP, like any tool, is only effective when correctly implemented. Unfortunately, many projects apply the HAZOP technique inefficiently, and these poorly executed HAZOPs can result in late and costly design changes, multiple HAZOP sessions, and/or overlooked hazards. By nature, HAZOPs for large-scale projects can be particularly challenging because of the high volume of design information, the long project duration, and the numerous internal and external interfaces that must be managed. Additionally, opportunities are often missed to integrate other process hazard analysis (PHA) techniques, such as Layer of Protection Analysis (LOPA) or Safety Integrity Level (SIL) Assessment into the typical HAZOP review.

In contrast, the Atlantis HAZOP case study presented here illustrates how the full benefits of the HAZOP technique can be realized when the design stage HAZOP analysis is strategically planned and fully integrated into the project schedule. Key success factors from the Atlantis HAZOP experience include:

- Scheduling the HAZOP sessions to match design development
- Predefining nodes and populating worksheets as part of the pre-session preparation
- Including dedicated and experienced operators as HAZOP team members
- Using a dedicated, qualified scribe and fully documenting each scenario
- Predefining repeatable acceptance criteria
- Implementing an auditable recommendation close-out process that defines required approval levels
- Implementing a process for information transfer to Operations

Furthermore, the Atlantis HAZOP expanded the traditional methodology by incorporating LOPA principles as a means of evaluating safeguards and proposed changes. This approach enabled the project team to determine the adequacy of in-place and recommended hazard mitigations and identify instances where more rigorous evaluation was required.

This paper details the lessons learned from the successful implementation of the Atlantis HAZOP and recommends key actions for achieving high quality and efficient HAZOP analysis in the design stage. Implementation of these recommendations by other projects will ensure that future HAZOP studies provide cost effective solutions for the improvement of overall facility safety and operability.

ATLANTIS PROJECT AND HAZOP OVERVIEW

The Atlantis Development is located approximately 150 miles offshore in the Southern Green Canyon (GC) area of the Gulf of Mexico (GoM) and is one of the largest deepwater developments worldwide. The development area straddles the Sigsbee escarpment in water depths ranging from 4,200 to 7,200 feet. BP and BHP Billiton are joint venture partners in the Atlantis Development. Production is scheduled to commence from Atlantis in 2006.

The host facility for Atlantis is a production and quarters package semi-submersible platform designed to process up to 200 MBOPD and 180 MMSCFD of gas. The platform uses a four-column hull with permanent taut leg mooring (12 line wire/chain held by suction piles). Subsea production and test wells will be tied back to the platform via steel catenary risers.

The Atlantis HAZOP studies took place during the execution phase of the project and were divided to cover the subsea and topsides systems separately. The Atlantis HAZOP targeted identifying and assessing hazards that could lead to personnel injury, environmental damage, or equipment damage, with causes including equipment failure, human error, or design error. (Purely operability issues were not addressed unless the concern was one that would prevent the facility from operating as intended.) The methodology of the HAZOP process applied is presented schematically in Figure 1 below.

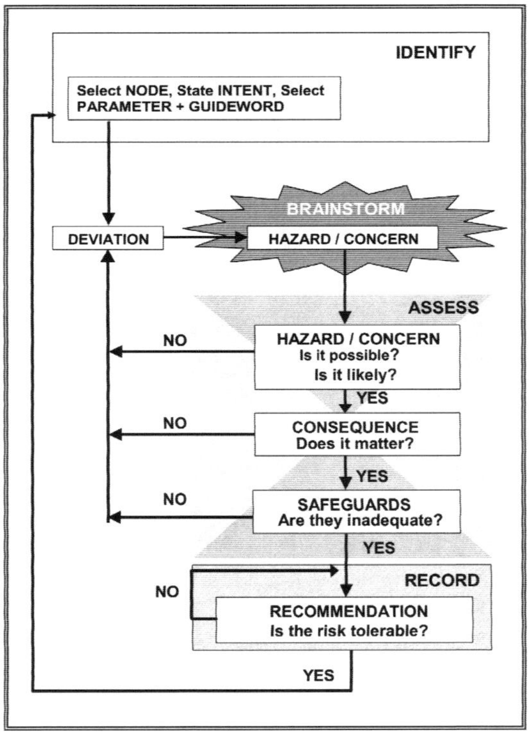

Figure 1. Methodology of the Applied HAZOP Process

The discussion that follows outlines some of the key factors that allowed Atlantis to maximize the benefits of the design HAZOP reviews conducted. In particular, emphasis will be placed on the areas of:

- HAZOP Timing
- Preparation
- Participation
- Documentation
- Incorporation of Layer of Protection Analysis (LOPA) Principles
- Action Item Close-Out
- Transfer of Information to Operations

HAZOP TIMING

One of the most important factors of a high-quality design HAZOP is determining the appropriate time to schedule the review. For most projects, two considerations should be weighed when scheduling the HAZOP, namely the need to:

- Minimize impact on cost and schedule from implementing late design changes or needing to conduct redundant HAZOP reviews
- Maximize thoroughness and completeness of the review

In general, the level of detail available on the P&IDs at the time of the HAZOP review is an indicator for how complete the review will be. A HAZOP review conducted fairly early in the design stage will highlight major design concerns and areas where applying inherently safer design principles should reduce risk to personnel, the environment, or the asset. The drawback of this approach is that early in design, many details of the process are not known or may be subject to change, requiring follow-up HAZOP reviews of the same material to ensure completeness and accuracy of the analysis.

An alternate approach is to conduct the design HAZOP review just prior to the P&IDs being "Issued For Construction," effectively serving as a validation of the design. Although the scope of a HAZOP conducted at this point would be extremely thorough, by this stage in the project there is little opportunity to implement recommended changes resulting from the HAZOP review without incurring significant cost or schedule impacts.

To gain the benefits of both of these approaches (i.e., maximize scope of review and minimize cost of review and implementation), the Atlantis Project combined elements from both philosophies. The combination of these activities enabled the project team to identify high-risk issues early on, while planning for the more detailed HAZOP review at a stage where completeness of the analysis could be assured.

Atlantis first performed a high level PHA (in the form of a checklist review) early in the design stage. The results of the checklist review were in turn integrated into the Operations review conducted prior to issuing the drawings as "Approved for Design." The HAZOP review then followed once the design was sufficiently mature, i.e., when sufficient vendor data and design detail was available to ensure a complete and thorough review. The results of the HAZOP review were then incorporated into the "Approved for Construction" P&IDs.

To further optimize the HAZOP review process, the Atlantis HAZOP sessions were phased to match the P&ID issue dates for each of the major systems rather than holding a single review session once all of the P&IDs were completed. This approach allowed for scheduling Operations and Engineering resources for several short reviews, rather than one long marathon review, improving overall productivity and reducing fatigue on the team members. While this approach requires more planning in coordinating schedules and logistics, the extra effort is well worth the benefits gained. In addition to a better quality review in terms of productivity and level of completeness, the team members' project availability is less disrupted when the HAZOP is made up of several short sessions, as opposed to a single HAZOP lasting several weeks.

PREPARATION

Effective HAZOP preparation will allow the team to maximize their time spent identifying hazards, while reducing the opportunity to overlook hazards. The following preparation activities were considered as key success factors for the Atlantis HAZOP, and they are each described in more detail below.

- Node Definition
- Inclusion of Interface Issues
- Defined Acceptance Criteria
- Consequence Modeling Information

Node Definition

It is a common HAZOP analysis practice to divide equipment and piping shown on P&IDs into specific process systems, referred to as study nodes. This practice allows the HAZOP team to examine hazards on a node-by-node basis, a process which provides for methodical evaluation of issues within each node and between nodes in turn. Although useful for the overall success of the HAZOP, attempting to identify and delineate study node boundaries is not an effective use of HAZOP session time. It is preferable for the study leader to identify study nodes prior to the session, to allow for the most productive use of the team's time.

The Atlantis HAZOP incorporated this strategy, enabling the team leader to distribute an agenda and the node sequence to be studied to all team members before the start of the study. Team members had an opportunity to become familiar with the systems prior the review, and also to gather any additional information or personnel that would be beneficial when that node was reviewed. On each day of the HAZOP study, the team arrived knowing the plan for that day, which resulted in a positive and effective start to the study. The use of an established agenda benchmarked by node also allowed the progress of the review to be measured against the plan.

Node size also affects the overall effectiveness of a study. Nodes can vary from the size of a purge line to the size of an entire compressor train. The benefit of small nodes is that they help ensure all details are thoroughly reviewed, while larger nodes tend to keep teams on a tight schedule. Conversely, the rigor of using small nodes tends to make reviews much longer, while analyzing overly large nodes could cause the team to lose focus and potentially overlook hazards.

The nodes defined for the Atlantis HAZOP were based on unit operation, such that a node typically included the unit operation under review (e.g. single vessel, exchanger, compressor) and the lines connected to that piece of equipment. This was considered an optimal node size because the detail and focus of the review was retained while keeping the HAZOP on a schedule of about 2 hours per node.

Inclusion of Interface Issues

Interface issues can be a struggle for most projects as they typically have no real owner and can easily be overlooked during the HAZOP. However, interface hazards are real and demand the same attention as other parts of the facility.

In order to thoroughly address interface issues, the Atlantis HAZOP study process included specific nodes for the purpose of identifying and discussing interface hazards. Representatives from all of the involved design teams participated in the review of the interface nodes. This allowed the HAZOP team to focus on items that may have been missed when independently reviewing other process nodes. In some instances, the HAZOP also provided the first opportunity for interface discussions to take place, opening the window for future conversations as the solutions to hazardous issues were resolved.

Another approach to addressing interfaces is to hold specific interface HAZOPs once the individual HAZOPs have been completed. This option was considered to be less efficient with an increased risk of late design changes, and was therefore rejected for the Atlantis study.

Defined Risk Assessment Criteria

Most HAZOP techniques used in industry include a risk ranking step, which provides the HAZOP team with an understanding of what level of risk the hazard presents. With this knowledge, the team can determine if further safeguarding measures are required to eliminate or control the risk to a tolerable level. In order to achieve this step effectively, well defined risk assessment criteria must be established by the owner/operator of the facility prior to the HAZOP.

The Atlantis risk assessment criteria were defined in terms of severity of consequence, likelihood (or frequency), and overall risk categories. The project applied a 4 x 5 risk ranking matrix (see Figure 2).

Risk Assessment Matrix					
Hazard Severity Category	Frequency Level (Likelihood)				
	A	B	C	D	E
I					
II					
III					
IV					

Figure 2. Risk Ranking Matrix for Atlantis HAZOP
(Note: Risk Ranking Criteria have been excluded from the figure.)

To support evaluation of hazard scenarios, the risk matrix included clear definitions to differentiate severity of consequence and likelihood aspects of risk. (Ambiguous terms such as "severe consequence" were not used in order to avoid lengthy discussions regarding how the terms should be defined and potential inconsistencies between HAZOP teams.) The HAZOP team was also provided with failure rate data and consequence information to assist in hazard evaluation. Finally, the overall risk categories (red, yellow, green) were defined in terms of expected actions, such that the highest risk region stipulated that engineered remedial action was required to reduce the risk, and so on.

For an effective HAZOP, it is also important to clearly define the risk ranking process prior to starting the review (i.e., how the risk ranking matrix will be applied for each scenario discussed). For the Atlantis HAZOP, the risk ranking process was carried out as described in the following steps:

1. Once a hazardous scenario was identified, the consequences for that scenario were estimated along the vertical axis of the risk ranking matrix, without taking credit for any identified safeguards.
2. The likelihood (or frequency) of the worst credible consequence was then determined along the horizontal axis of the risk ranking matrix, taking into consideration any safeguards in place.
3. With a known consequence and likelihood, the risk matrix identified the level of risk associated with the particular hazard.

This risk ranking method avoids taking double credit for existing safeguards. It is also crucial that the risk ranking be targeted to a specific event. For instance, if the worst possible consequence was determined to be a severity II, the likelihood was evaluated for the scenario escalating all the way to that particular consequence (not merely for the initiating event), hence the application of safeguards in the likelihood portion of the ranking process.

If a further action was recommended, the scenario was risk ranked a second time, taking credit for the recommendation as a safeguard. Reranking the scenario allowed the team to evaluate how effective that recommendation would be and to demonstrate that the risk had been reduced to a tolerable level, per the HAZOP process illustrated in Figure 1. (If the recommendation was for further study or evaluation, the scenario was not reranked.)

Consequence Modeling Information

Another effective HAZOP preparation technique involves modeling known hazards prior to the HAZOP session. This information can be used during the meeting to ensure all team members understand the consequences of particular events. For example, modeling riser fire events could show that the duration of the fire could last 10 minutes or 10 hours depending on the leak size. These results aid the team in the risk ranking process and in determining the need for additional safeguards.

During the Atlantis HAZOP, information such as manning levels, hydrocarbon inventories, leak rates, peak overpressures, and jet fire lengths and durations for various scenarios provided further insight into the consequences of the scenarios under review. Oil spills were also modeled to better define the environmental consequences. Detailed design information was also available, such as valve capacities and pump deadhead pressures. Having detailed knowledge of the consequences readily available reduced the number of "conduct further study" recommendations made and enabled the team to make more concise recommendations to lower the overall risk.

PARTICIPATION

Committed participation by design engineers, Operations, the facilitator, and the scribe is vital to the success of any HAZOP study. Typically three basic types of participants are required on a HAZOP team:

- A HAZOP leader, or facilitator
- A dedicated scribe
- Experienced team members (engineering and Operations) with the necessary technical skills and experience to perform the review

While inclusion of these participants is standard practice for HAZOP reviews, participation in the Atlantis HAZOP differed from the norm in two areas, namely Operations and the scribe.

Operations and maintenance representatives were assigned full time to the Atlantis Project team during the definition design phase and were dedicated HAZOP team members – a longer and more intensive involvement for Operations personnel than is generally found. These representatives have remained with the Atlantis asset during commissioning and will continue into the operation phase. In addition to adding valuable specific input as to how the facility would be operated and maintained, their participation in the HAZOP reviews helped transfer information to the Operations team as a whole. Furthermore, having the same Operations representatives involved in all of the HAZOP reviews and subsequent close-out of the recommended actions has ensured consistency in the operational requirements.

The second noteworthy difference in participation for the Atlantis HAZOP was the high quality of the scribe. To effectively document the discussions that serve as the basis of the HAZOP study, the scribe should be familiar with process terminology, the HAZOP and risk ranking process, and have good computer, writing and organizational skills. Too often, secretaries or junior engineers without the requisite skills are recruited as scribe, with predictably poor results. Although a secretary may have good computer skills, their lack of familiarity with the technical content can hinder real-time recording, while similarly the engineer may be hampered by poor typing and software skills. Using a professional scribe with all the requisite qualifications allowed the Atlantis HAZOP to flow smoothly with minimal documentation delays and ensured capturing an accurate record of the proceedings, as discussed further below.

DOCUMENTATION

The recording process is an important part of the HAZOP study. There are two general approaches used to document HAZOP reviews. One method is to document scenarios on an exception-only basis, that is, to only record information about a potential scenario if an action or mitigation is recommended. Obviously this abbreviated recording style requires less effort during the session, but it also generates a HAZOP report that may be found lacking. The other approach to documentation is to record the discussion surrounding significant aspects for all hazardous scenarios identified, regardless of whether recommendations are made.

During the Atlantis HAZOP, all hazardous scenarios discussed were recorded, even if after brainstorming the team decided a particular scenario did not warrant further action. Information captured included causes, consequences, safeguards, recommended actions, and any pertinent comments. This recording practice ensured complete documentation of everything reviewed during the HAZOP study.

In the long term, this documentation approach provided a thorough HAZOP analysis report, sufficient to answer the typical questions raised later in the project or during the operating phase, such as “Did the HAZOP team consider … ?” and “Why was this hazard scenario acceptable?” Without full documentation from the HAZOP session, those questions would be unanswerable, as it would be impossible to know if the issue was not included because it was not a significant hazard or because it was simply overlooked.

INCORPORATION OF LAYER OF PROTECTION ANALYSIS PRINCIPLES

Although Layer of Protection Analysis (LOPA) is traditionally handled as a stand-alone study, Atlantis opted to integrate the principles and evaluation techniques of LOPA methodology into the HAZOP study. Seamless integration of the two processes provided consistency between the various project HAZOP sessions and achieved the objectives of LOPA with only a minimal increase in the time required to complete the HAZOP. Together, the simultaneous application of LOPA and HAZOP study techniques proved to be a key factor to the overall success of the Atlantis HAZOP.

LOPA is a semi-quantitative technique used to provide a consistent basis for assessing the adequacy of independent protection layers (IPL) in a design [CCPS]. As illustrated in Figure 3, multiple IPLs may be in place for a given hazard scenario, each of which must be capable of preventing the consequence associated with the hazard. The required number of IPLs depends on the scenario's overall risk and the reliability of each individual IPL to prevent the consequence.

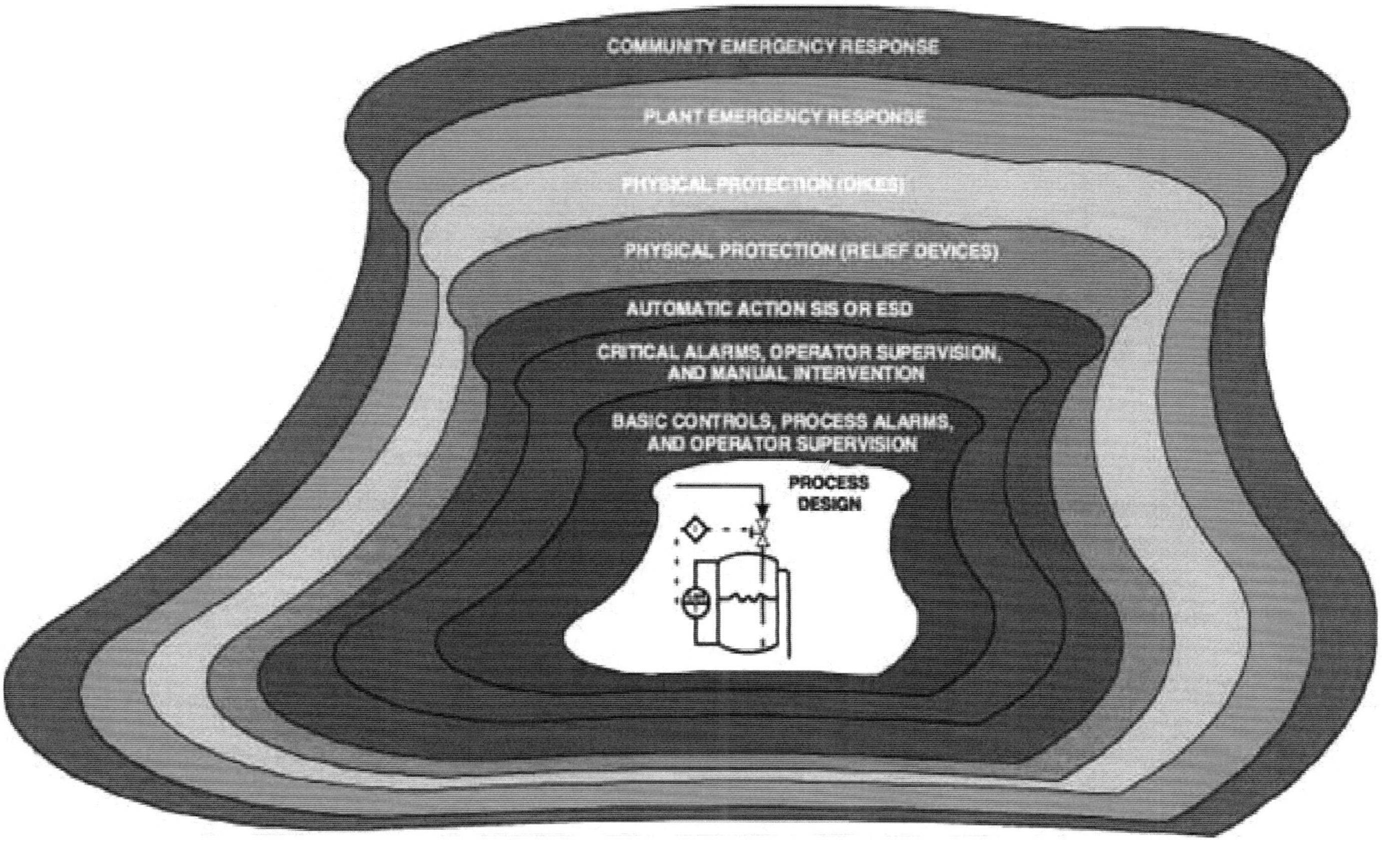

Figure 3. Independent Protection Layers (IPL) for a Process Hazard [CCPS]
(Source: adapted from Figure 2.1 of AIChE/CCPS (2001). Reproduced by permission of AIChE.)

In order to maximize the value of integrating LOPA into the HAZOP methodology, additional preparation work was needed during the HAZOP planning phase, including the following activities:

- Establishing criteria for evaluating the effectiveness and independence of safeguards
- Establishing criteria for qualifying operator action as a safeguard
- Establishing reliability data for typical initiating events and safeguards included in the design

As part of the HAZOP methodology, the team had to identify and then evaluate the adequacy of existing safeguards to protect against the hazard. Unfortunately, a common deficiency noted in previous HAZOPs has been the tendency of teams to generate a long list of protections, some of which are not effective at preventing the specific hazard or consequence being considered. Another typical shortcoming is the failure of a team to recognize that some of the identified safeguards are either not independent of the initiating event or other identified safeguards. Either of these errors can cause a team to take too much credit for existing safeguards, therefore underestimating the true risk associated with the hazard.

To avoid the problem of underestimating risk through excessive safeguarding credit, the Atlantis HAZOP team was provided with the following criteria for evaluating if a safeguard is an IPL:

- The safeguard must be effective in preventing the consequence(s).
- The safeguard must be independent, both from the initiating event and from components of other IPLs already credited for the same scenario. Consideration should be given to common mode failures as well as the design configuration.
- The safeguard must be dependable and auditable.

Similarly, typical HAZOP teams sometimes take too much credit for operator intervention as a safeguard. Many factors may prevent an operator from responding correctly to a hazardous event, including the operator having insufficient information about the hazard, not being available to implement the required action in sufficient time, or simply not having a course of action that could prevent the consequence(s). HAZOP teams have also mistakenly considered multiple alarms as independent levels of protection, disregarding that a single operator or initiating device represent common cause failures.

To improve the evaluation of effectiveness of operator intervention as an IPL, the Atlantis HAZOP team was provided with the following criteria:

- An alarm or other mechanism must clearly identify the problem to the operator.
- There must be a course of action the operator can take to prevent the consequence(s).
- The operator must be trained to take the appropriate action in response to the alarm.
- The operator must have sufficient time available to respond to the alarm and implement the required action (typically a minimum of 15 minutes).
- The operator must be available to implement the appropriate action.

Using the safeguarding criteria listed above to identify IPLs for the particular hazard scenario, the Atlantis HAZOP team then had to determine if the protection provided was adequate through risk ranking the scenario as described previously (see Figure 2). To improve the quality and repeatability of the risk assessment, the Atlantis HAZOP team was provided with failure rate data for typical initiating events (human error, basic control loop failure, pump failure, loss of power, seal failure, etc.). The team was also provided with quantitative Risk Reduction Factors (RRF) for the various safeguards included in the design (basic process control loop, operator intervention, safety instrumented functions (SIF), PSV, emergency support systems, etc.). The RRFs were calculated based on the actual hardware specifications and configurations included in the design, expected testing frequencies, and industry specific reliability data such as OREDA using commercially available software.

If the safeguards identified for a particular hazard scenario were determined to be inadequate, the HAZOP team recommended design changes to either eliminate the hazard or to provide additional protection to reduce the risk. Options for additional protection included adding IPLs or improving the availability of existing IPLs. In several cases, the HAZOP team recommended improving the Safety Integrity Level (SIL) of the existing SIF. The hazard scenario was then reranked, taking into consideration the recommended changes, to determine an after-mitigation risk level. This process also eliminated the need to perform a separate SIL evaluation for every SIF.

ACTION ITEM CLOSE-OUT

To ensure effective management of action items resulting from the HAZOP, the Atlantis Project established a robust action item close-out process. Implementation of the process provided the delivery teams with an auditable and standardized method for tracking, closing, and documenting the recommended actions arising from the Atlantis HAZOP.

The Atlantis Action Item Close-Out Process addressed the following issues:

- Tracking the status of open action items
- Action item closure and approval
- Documentation requirements
- Transfer of action items between delivery teams

Recommendations were entered into a central database that enabled tracking progress to close-out. Data entered included definition of the risk, the recommended action, the risk ranking for the hazard scenario, and the name of the person designated as the risk owner. Status reports were generated to update management on the close-out progress.

Resolution of each recommendation from the Atlantis HAZOP study was documented on a formal close-out form, including record of:

- Any transfer of ownership of the responsibility for action items between delivery teams. (Transferring action items required sign-off on the form by both the initiating and receiving delivery teams.)

- Final resolution of the recommendation and hazard, with approvals and verification documentation attached, as described below.

Authority to close action items was based on conformance with HAZOP Team recommendations and the scenario's risk. The recommended action could only be closed by the responsible delivery team conforming to the action recommendation, or defining the reasons for non-conformance and obtaining appropriate approvals. Additionally, close-out required verification that the recommendation had been incorporated into the appropriate design documentation or operating procedures. This verification was performed by the designated accountable engineer and Operations.

If the recommended action was not implemented, or an alternative solution was applied, additional approvals were required from management and the Process Safety Specialist. Approval requirements were set based on the risk ranking of the hazard scenario, with the action items associated with higher risk scenarios requiring approval by higher level managers.

By clearly recording and maintaining transparency of ownership and progress of recommended action items, the Atlantis Action Item Close-Out Process provided an auditable record of action item closure. The forms tracking close-out of the HAZOP recommendations will ultimately be part of the hand-over documentation provided to Operations.

TRANSFER OF INFORMATION TO OPERATIONS

As noted previously, thorough documentation of the hazard scenarios discussed during the HAZOP sessions is one of the keys to a successful review. The information is essential to resolution of the recommended action items and evaluation of design changes proposed subsequent to completion of the HAZOP. Much of the information collected during the HAZOP analysis is also valuable to the Operations team as part of Operations Readiness planning and during subsequent operation of the facility.

Operations used the information from the Atlantis HAZOP to assist in preparing the Operating and Maintenance Procedures for the facility, for which operator actions and testing assumptions identified as safeguards are of particular importance. Additionally, information from the HAZOP was used to help develop the Safety Critical Equipment List.

The Atlantis Project is also providing Operations with Hazard Management Plans (HMP) for all the major hazards identified during the review. The HMPs document the causes, safeguards, and other information related to the hazards and will be a key resource to Operations, particularly as future modifications are made to the facility. Much of the information for the HMPs is being derived from the HAZOP documentation.

SUMMARY

The HAZOP technique is flexible and effective, and many projects in industry are adapting some version of the methodology to their process hazard reviews. However, the HAZOP process can be misused or applied poorly, leading to increased project, design and review cost, in addition to overlooked hazards. Large-scale projects are particularly susceptible to inefficient HAZOPs due to the long time, many resources and personnel that must coordinated. Additionally, opportunities are often missed to integrate other PHA techniques such as LOPA or SIL assessment into the HAZOP.

This paper discussed the key success factors identified during the Atlantis HAZOP and recommended actions for achieving high quality and efficient design HAZOP analyses.

Applying these key factors during the Atlantis HAZOP reviews resulted in the Atlantis Project realizing the following benefits:

- Use of a consistent approach for identifying and evaluating hazards
- Completion of the HAZOP reviews on schedule
- Issuing of Approved for Design and Approved for Construction P&IDs on schedule
- Zero safety-initiated changes were required post HAZOP
- Limiting the need to re-HAZOP to one short session
- Determining the adequacy of proposed SIL integrity levels
- Successful transfer of key safety information to Operations

Implementation of these key HAZOP success factors, identified through the Atlantis HAZOP case study, by other projects will help ensure that future HAZOP studies provide cost effective solutions that improve overall facility safety and operability.

REFERENCES

Center for Chemical Process Safety, Layer of Protection Analysis: Simplified Process Risk Assessment, American Institute of Chemical Engineers, New York, New York, 2001.

Offshore Reliability Data Handbook (OREDA), 4th Edition, 2002.

Operator Initiated Action As An Independent Protection Layer

Scott Sandler and Dr. Angela Summers, P.E.
SIS-TECH Solutions
12621 Featherwood, Suite 120
Houston, Texas 77034

Prepared for Presentation at
American Institute of Chemical Engineers
2005 Spring National Meeting
Process Plant Safety Symposium
Atlanta, Georgia
April 11-13, 2005

KEYWORDS

Layers of Protection Analysis, LOPA, hazard analysis, protection layers, IEC 61511, ANSI/ISA 84.01-2004, risk mitigation, alarm management, operator response

ABSTRACT

Process alarms are often cited as primary safeguards in mitigating process risk. A process alarm can also be considered an independent protection layer (IPL) if it meets three criteria:

1. it is independent of the initiating cause and other safety functions mitigating the same hazard,
2. it is designed to provide the necessary integrity, and
3. there is sufficient time for the operator to detect that there is a problem, to determine what to do, and to take the action necessary to bring the process within its safe operating limits.

The time it takes the operator to detect the problem and to take action is related to the design of the operator Human-Machine Interface (HMI), the information and training provided to the operator, and the complexity of the required response. The operator action may include shutting down the process.

This paper will discuss the use of Layer of Protection Analysis (LOPA) to identify, design and manage any operator-initiated IPL. The process includes identification of critical process variables; verification of independence of the alarm function; documentation of alarm procedures, and management of change to the alarm function.

INTRODUCTION

Operators have always had a significant role in the operation of chemical plants and refineries. Operations personnel are responsible for monitoring the process and taking appropriate action when a process variable is approaching the boundary of its acceptable operating range. Operator action may be intended to prevent an operational upset, an environmental excursion, or a safety-related hazard.

When a process hazards analysis (PHA) team is assessing the risk associated with a particular deviation, consideration is typically given to safeguards associated with the operator responding to process alarms. Depending on the severity and likelihood of the consequence, operator response to an alarm may be considered acceptable as the only line of defense or as one of several safeguards.

During a PHA using a qualitative methodology, such as HAZOP, multiple process alarms may be cited as safeguards towards mitigation of a process hazard. Little or no consideration is often given to the independence of the alarm in relationship to the initiating cause; the manner in which the alarm appears to the operator relative to the other less critical alarms; or the relative complexity of the required operator actions to bring the system to a safe state. As a result, multiple layers of protection may be assumed when, at most, only one independent protection layer actually exists.

Layer of Protection Analysis (LOPA) is a semi-quantitative method of assessing and managing process risk. Safeguards must satisfy specific criteria to be considered an Independent Protection Layer (IPL). This paper provides a means of using LOPA to identify, design and manage operator-initiated IPLs.

BEGIN WITH LAYER OF PROTECTION ANALYSIS (LOPA)

LOPA is typically employed after a more general Process Hazards Analysis (PHA) technique has been used to identify the hazardous scenarios, including the consequence severity and initiating causes. The consequence severity, assuming no response from the safeguards, and the initiating cause frequency are components of the unmitigated risk, e.g., a catastrophic release is expected to occur once in 100 years. This is compared to the predetermined risk criteria, which are often expressed in terms of a risk matrix, where frequency and consequence severity are the x- and y-axis and the cell coordinates define the required risk reduction. If the unmitigated risk is too high, safeguards that meet the IPL criteria are required to offset the residual risk. It is at this stage of the analysis where alarms are identified as potential operator initiated IPLs.

An introduction to LOPA is provided in Summers and Sandler, "Introduction to Layer of Protection Analysis [Summers]. For a more detailed description of LOPA, refer to CCPS, "Layer of Protection Analysis" [CCPS].

INDEPENDENT PROTECTION LAYER (IPL)

IPLs are a subset of safeguards and may be comprised of instrumented functions, mechanical devices, and operator actions that rely on instrumentation and/or mechanical devices. All IPLs are safeguards; however, not all safeguards are IPLs. For a safeguard to be considered an IPL, it must satisfy the following criteria:

<u>Independence:</u>

An IPL must be independent of the initiating cause and all of the other protection layers, associated with the identified hazard scenario. Independence requires that the performance must not be affected by the failure of another protection layer or by the conditions that caused another protection layer to fail.

Specificity:

The IPL is intended to prevent or fully mitigate the consequences of the identified hazard scenario.

Dependability:

The protection provided by the IPL must reduce the identified risk by at least ten-fold. In terms of availability, the IPL must be at least 90% available. This means that out of ten demands, the safeguard must successfully perform its function and fully mitigate the hazard at least nine times.

Auditability:

The IPL must be designed to allow regular, periodic validation of the protective function.

OPERATOR INITIATED IPLS

Now let's apply the above IPL criteria to a safeguard comprised of a critical alarm with a particular operator response:

Independence:

The indication of the process deviation must be from an independent sensor, with signal communicated to the operator independently from the failure or other protection layers. Ideally, communication should be independent of the basic process control system (BPCS) and displayed on a dedicated separate human-machine interface (HMI). However, the BPCS may be used if the alarm signal is communicated via an independent channel in the BPCS. The required response is carried out by the operator who should be independent of the initiating cause, utilizing final elements that are independent of the initiating cause and other protection layers.

Specificity:

The process alarm that triggers the operator action should be a process variable, which is a clear indication of the hazard scenario. The action to be carried out by the operator must fully mitigate the event by interrupting its propagation.

Dependability:

The operator initiated safeguard must be at least 90% available. Again, this means that in at least nine of ten opportunities, the operator will take the appropriate action to successfully interrupt the propagation of the hazardous event sequence and to bring the process to a safe state. Therefore, the process alarm that triggers the operator action should be clearly indicated, such that it stands out from other alarms that may be activated by the event. Ideally, it should be annunciated on an independent HMI to minimize response time to the alarm.

There must be ample time from the point that the alarm sounds for the operator to carry out the required action(s). The action(s) should be straightforward and documented in a written procedure. The procedure should state the specific steps to be taken to bring the process into control. The operator should be trained regularly on the procedure.

Auditability:

All devices used to generate the alarm and to take action on the process should be periodically tested. This includes the operator who is, in effect, the primary logic solver. The operator should be trained on the response procedure including the alarm set point and the required mitigating actions. In addition, the operator should be tested or drilled periodically to reinforce skills.

CRITICAL ALARM IDENTIFICATION

As discussed earlier, the LOPA team determines the initiating cause frequency and consequence severity, defining the unmitigated risk. Further, the unmitigated risk is compared against the risk criteria to determine if there is a need for IPLs. If the unmitigated risk requires IPLs, the team reviews the instrumented system design and assigns the appropriate IPL credit to any existing safeguards.

As the various safeguards are reviewed, the team identifies process alarms that warn the operator that a deviation has occurred that require manual action as defined by an operating procedure. For most hazard scenarios, only one operator alarm can be claimed as an IPL, since the limiting component is the panel operator. As such, the team should identify the most effective combination of process alarm and final element(s). For example, in an overfill scenario, a high level alarm is generally better than a high pressure alarm, because the process safety time associated with high level allows more time for the operator response. The number of individual final element(s) to be manipulated should be minimized to limit the response complexity.

Next, the team should confirm independence of the sensor and the final element(s) to be utilized in the operator-initiated IPL. These field-mounted devices must be independent of the initiating cause and other IPLs identified to mitigate the hazard scenario. For instance, if the initiating cause is a failure within a temperature control loop, the temperature transmitter within the failed loop cannot be used to provide the critical process alarm, nor can the associated control valve be used to isolate a process stream. An independent sensor and block valve (or other effective final element) is required.

The LOPA team should evaluate the time available for an operator to execute the required action. There should be sufficient time available for the operator to intervene and stop the hazardous event propagation. Human factors, such as operator stress level and location of the final elements, should be considered.

For example, let's look at a plant-wide cooling water failure as the initiating cause; the hazard scenario associated with overpressure of the fractionator column would be realized 5 - 10 minutes after the critical high pressure alarm sounds; the panel operator must scroll through the list of process alarms to recognize the problem, radio to the outside operator, who then must climb to the top deck to close a valve. Given the greater level of stress and the limited time available to react, this is not a realistic combination of alarm and final element towards an operator-initiated IPL.

Typical time limits used in industry to determine whether or not adequate time is available for an operator-initiated IPL include 10 minutes for operator action taken by the control room operator and 30 minutes by the field operator. These assume that the operator action is not overly complex.

The team now evaluates whether the candidate process alarm and its corresponding operator action satisfy the IPL criteria. If so, one order of magnitude risk reduction, or one IPL credit, may be assigned to this safeguard. If not, the team may recommend that the observed design deficiencies be corrected, so that risk reduction can be achieved using the safeguard. This may require the installation of additional hardware.

DESIGN OF OPERATOR INITIATED IPL

At this point, it will be assumed that the required instrumentation and devices are in place and the IPL criteria are satisfied for the operator initiated IPL. The next step is to determine how the alarm is communicated to the panel operator. If the process includes a safety Programmable Logic Controller (PLC), as well as a BPCS, the safety PLC would be the preferred routing, since it would be independent of any causes initiated within the BPCS. Further, a greater level of security and administrative control will be implemented for the safety PLC, which provides greater protection against systematic errors, where changes to the hardware and software configuration result in IPL failure.

If it is not feasible to have the critical process indication and alarm connected directly to the safety PLC, or if an independent PLC is not available, the BPCS may be utilized. As stated above, it must be ensured that the alarm is communicated via an independent channel in the BPCS, providing separation from the initiating cause and other safety functions mitigating the same hazard scenario. Therefore, if the initiating cause is a failure within a control loop in the basic process control system (BPCS), the process alarm should ideally be communicated independently of the BPCS, e.g., via an independent, safety-PLC or hardwired system. If this is not practical, it is possible to achieve the required separation if the alarm is communicated via an independent channel within the BPCS.

This independent channel must also be designed to meet the required integrity, which means that redundancy and enhanced diagnostics must be provided for the I/O modules and main processors. The BPCS outputs must also be configured to fail safe rather than fail-last position. The independent channel must also be managed more rigorously than normal BPCS channels with greater emphasis on security and administrative control of modifications within the BPCS, which will be discussed later in this paper.

Another important consideration is how the alarm condition physically appears to the operator. It is important that this alarm stand out against many other non-critical process alarms that may be sounding simultaneously. To improve the integrity of the operator response, the required operator action may be displayed on the HMI when the operator acknowledges the alarm. Many companies are now using an independent HMI for critical process alarms to make them distinct and to provide enhanced alarm response information. If it is not practical to implement an independent HMI, a different appearance or sound should be utilized. Otherwise, the critical process alarm may go undetected, particularly during an event in which several alarms could be expected, such as a utility failure.

The next step is to completely document the operator initiated IPL specification. The LOPA report and operating procedures generally provide all necessary technical information, as follows:

> <u>Process Hazard Description</u>: A detailed description of the hazardous event that requires mitigation,

> <u>Functionality Description</u>: A description of the operator initiated safety function, including the critical alarm, the final element(s), and the action(s) to be performed by the operator(s),

Associated Tag Numbers: Instrumentation and device numbers,

Communication of Alarm: BPCS, PLC, Hardwired, Independent or shared HMI,

Additional Process Data: Alarm critical set point value, time available for operator response.

An example operator initiated IPL specification is included in Figure 1.

To achieve the required risk reduction, the devices must be periodically function tested. At a minimum, the devices should be tested during each regularly scheduled plant shutdown. Historical performance of field devices may indicate that more frequent testing is required. Without testing, the 90% availability requirement cannot be assured.

In addition to testing the associated instrumentation and devices, the operator's availability must be proven, since the operator is performing the task of the logic solver. The operator action must be documented in a written critical operating procedure. A separate section is recommended to highlight critical procedures within the main operating procedures. If critical procedures are fully integrated into the main operating procedures, special tagging is recommended, so that critical procedures are quickly identified. Critical procedures should be part of the routine operator training with periodic drills to assure integrity of the operator response and to decrease response time.

MANAGEMENT OF OPERATOR INITIATED IPLS

Management of operator initiated IPLs requires maintaining the relevant technical documentation, administrative control of the configuration and alarm set points, and conducting the required audits to verify continued performance.

All supporting documentation for the operator initiated IPLs should be maintained by a designated representative from the technology or operations department and updated when appropriate. The relevant documentation should be made available to all employees that are directly involved with the unit operation, including engineers, operators and management.

The platform in which the critical alarms reside must be maintained with a similar level of security as a safety PLC. Therefore, if the safety critical alarms reside within an independent channel of the BPCS, it must be ensured that this separation is always maintained. This will require more stringent administrative requirements than what is typically placed on the BPCS. Any changes, which require adding or relocating control functions, should be reviewed through the management of change process to ensure that separation of functionality is not inadvertently compromised.

The set points of the critical alarms should also be controlled via the management of change process. Routine changing of set points or complete disabling of safety critical alarms should not be permitted. Otherwise, the operator initiated safeguard would not meet the 90% availability criterion.

The auditing function includes verification that all required documentation is in place, up to date, and readily available to affected personnel. As stated previously, the components of the function need to be validated periodically. The associated devices should be tested at least as frequently as the unit is shut down for maintenance. The operators should undergo regular training on the operator initiated safety functions in their unit. This process should include periodic testing or drilling to help reinforce the concepts. Since the required training and testing can be a significant burden on the operators, it is beneficial to limit the quantity of operator initiated IPLs for a given unit.

CONCLUSIONS

Operator initiated IPLs make up a significant portion of the safeguards in a chemical plant or refinery. A systematic approach was presented to develop scenarios and apply LOPA to determine when safeguards were needed to mitigate the process risk to a tolerable level. The scenarios were explored in enough detail to determine if an operator initiated safeguard would satisfy the IPL criteria. This included examination of the hardware and software involved, as well as whether the operator had a reasonable amount of time to perform the required actions. After the acceptance of a particular critical process alarm and operator action as part of an IPL, the details of the design were discussed. This included the means of alarm communication to the operator as well as proper documentation. Lastly, management of the overall system was discussed with an emphasis on security and administrative control to prevent unapproved disabling of alarms or alterations to set points.

REFERENCES

Center for Chemical Process Safety, "Layers of Protection Analysis: Simplified Process Risk Assessment," American Institute of Chemical Engineers, New York, New York, 2001.

Summers, A. E., Sandler, S. H., "Introduction to Layers of Protection Analysis," presented at ISA EXPO 2003, Houston, Texas.

Figure 1. Operator Initiated SIF Specification

Unit	Resin A101
Area	Utilities/Propylene Refrigerant
Operator Initiated SIF#	3
Process Hazard Description	High level in Propylene Refrigerant Compressor Accumulator could lead to liquid carryover to the Propylene Refrigerant Compressor, with possible mechanical damage to the Compressor, potential injury to personnel, and a release of Propylene vapor to the atmosphere; possible fire.
Functionality Description	On indication of high level (> 20%) in the Propylene Refrigerant Compressor Accumulator, the operator will shut down the compressor from the control room. A single liquid level transmitter connected to the safety system PLC will be used to indicate an accumulator high level condition. The signal will be displayed on a dedicated monitor for critical alarms. A manual pushbutton (HS-901A) routed via the safety system PLC will shutdown the Compressor.
Input	LAHH-901, Accumulator high level alarm
Output	HS-901A, manual push button routed via the safety system PLC.
Alarm Communication	A high level alarm communicated via single liquid level transmitter connected to the safety system PLC will be used to indicate an accumulator high level condition. The signal will be displayed on a dedicated monitor for all critical alarms.
Critical Set Point	20% on LAHH-901
Operator Response	Operator presses manual pushbutton (HS-901A) routed via the safety system PLC to shutdown the compressor.
Time Available for Response	12 minutes from Control Room

Incident Investigation:
The Transmitters Are Leaking in One Unit, But Are Working Fine in Another Unit, Or Why Didn't the First Fixes Work?

A. M. (Art) Dowell, III, P.E.
Rohm and Haas Company
6519 State Highway 225
Deer Park, TX 77536 USA
Phone: 281-228-8258
Fax: 281-228-8159
E-Mail: ADowell@rohmhaas.com

Prepared for Presentation at the
American Institute of Chemical Engineers
2005 Process Plant Safety Symposium
TH004 Incident Investigations and Methods of Interpreting Evidence
Atlanta, GA
April 10-14, 2005

Incident Investigation: The Transmitters Are Leaking in One Unit, But Are Working Fine in Another Unit, Or Why Didn't the First Fixes Work?

A. M. (Art) Dowell, III, P.E.
Rohm and Haas Company
6519 State Highway 225
Deer Park, TX 77536 USA
Phone: 281-228-8258
Fax: 281-228-8159
E-Mail: ADowell@rohmhaas.com

Abstract:

This paper describes an actual incident investigation into unexpected premature failure of sealed diaphragm pressure transmitters in a chemical process. Some non-typical investigation tools from the *CCPS Guidelines for Investigating Chemical Process Incidents, Second Edition* were used.

A typical failure allowed a release of corrosive process fluid with potential injury to personnel. The failures also required a shut down of the process with loss of production. The cause of the problem was elusive; pressure transmitters from the same manufacturer had no problems in a similar process unit – however, component designs differed between the two units.

The investigation included confirmation of materials of construction, photographic and visual analysis of failed components, simulation of assembly of the pressure transmitter in the process and measurement of torque values, and brainstorming of possible failure mechanisms. Several hypotheses were tested using a fact-hypothesis matrix to determine most likely cause scenarios. Similarly, a matrix was used to illustrate which scenarios could be prevented by which corrective actions. The investigation included a root cause analysis tree to confirm cause scenarios.

The paper will discuss the incident investigation process, including diversity of skills on the investigation team, and how each of the tools was used. The paper will also discuss the communication of the findings to operations.

The Incidents

Four premature failures of sealed diaphragm pressure transmitters occurred in a chemical process unit soon after start-up. Each failure allowed a release of corrosive process fluid with potential injury to personnel should they be in the area of the release. Each failure caused a shutdown of the process for repair.

The failures were puzzling since pressure transmitters from the same vendor were working successfully in a similar process unit. A sketch of the diaphragm installation is shown in Figure 1.

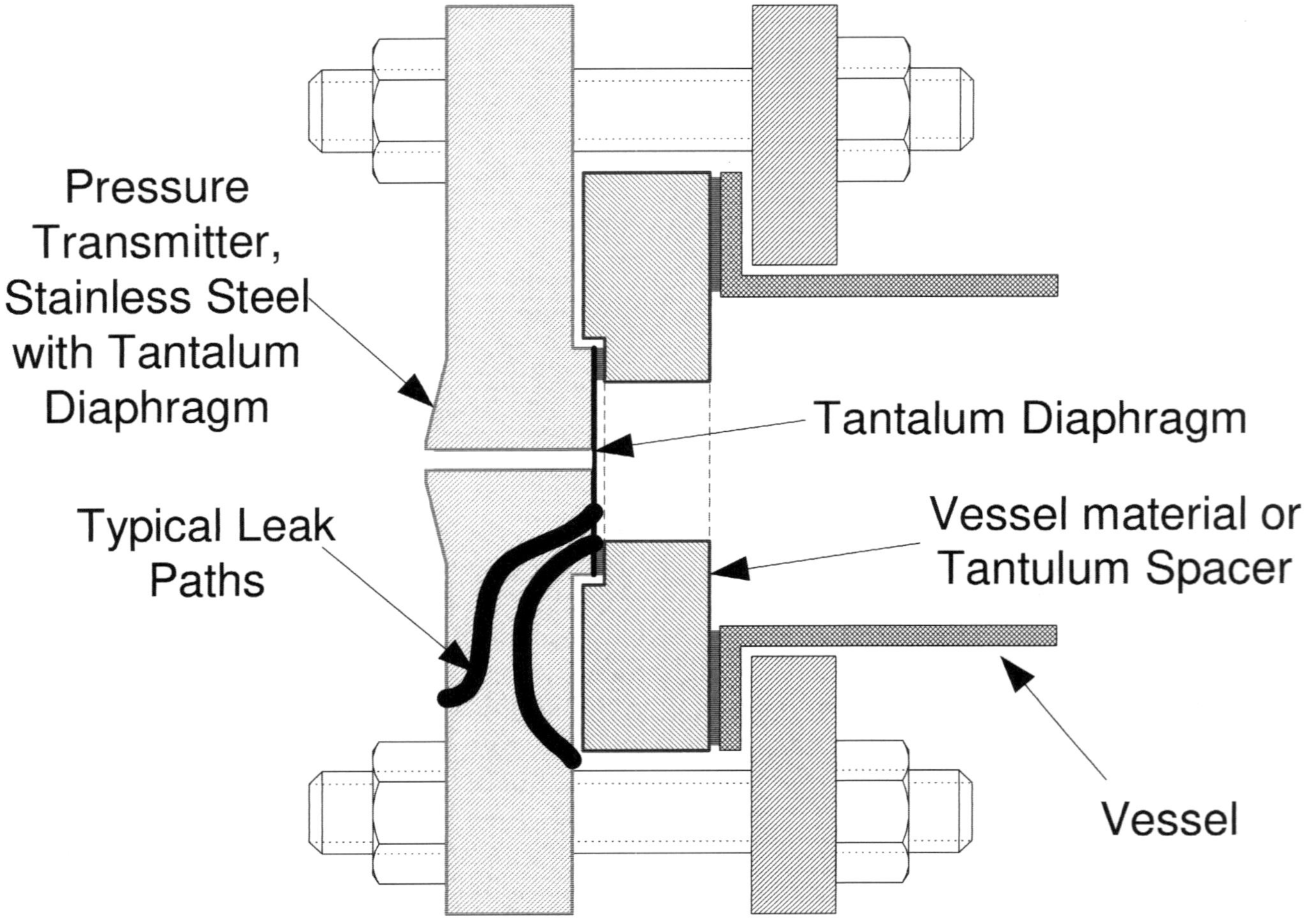

Figure 1: Diaphragm Transmitter Installation Sketch

The diaphragm is a thin corrugated sheet of tantalum brazed to a stainless steel backing plate. The tantalum diaphragm sees the process pressure and transmits it to a fluid inside the stainless steel portion of the pressure transmitter where the pressure can be measured. The transmitter is attached to the process vessel with bolts to a backing flange. There is a spacer between the vessel and the diaphragm. There is a gasket between the diaphragm and the spacer, and another gasket between the spacer and the vessel.

The failures all involved rapid corrosion of the stainless steel backing plate, followed by release of acidic process fluid to the atmosphere.

The first failure was on March 5 after about two months of service. Hydrogen embrittlement of the failed tantalum diaphragm was noted. The transmitter was replaced. There was no indication of embrittlement on two other transmitters in the same service area that were inspected.

The second failure was on May 13 at a different transmitter. Ten+ transmitters were inspected; only three passed visual inspection and pneumatic pressure testing. The transmitters were reinstalled with insulation kits to attempt electrical isolation to avoid galvanic corrosion. However, complete electrical isolation could not be achieved.

The third failure involved three transmitters during restart after the May 13 failure. Spacer misalignment was suspected and guide bolts were added to improve alignment. Torquing values for the transmitters were reduced to avoid overtorquing the flange and crushing the insulating washers. Even so, the washers extruded. Since electrical isolation could not be achieved, the insulating washers were removed completely.

The fourth failure was on June 16. The leak sprayed on the process manual shutoff valve, causing an extended release. Heavy instrument covers were installed to minimize product spray on failure. We reviewed the location of each transmitter to see if there were any important valves, walkways, etc., that would become inaccessible if the transmitter were to leak. Nonessential transmitters were removed. The gasket material was changed based on manufacturer's recommendations and testing.

The Investigations

As each incident occurred it was investigated by maintenance, mechanical engineering, and materials specialists in consultation with the vendor. Control and process engineers and a gasket specialist joined the team later. After the fourth incident, a risk analyst (the author) joined the team. Management charged the team to conduct a thorough root cause analysis to ensure that we understood the failure mechanisms and that we had taken sufficient precautions against each failure mechanisms. We were concerned that the changes made after each incident had not solved the problem.

Time line

First the team developed a detailed timeline or sequence of events as described in CCPS (1). The dates of each failure were listed along with data that had been gathered and conclusions that have been drawn at the time.

Then the team reviewed all the data from the inspections and simulation of transmitter installation.

As each failure occurred, the investigators had confirmed the materials of construction of the transmitter components and the flange and vessel components in contact with the transmitter. The investigators had done detailed photographic and visual analysis of failed components.

After the third incident, the team simulated the assembly of the pressure transmitter in the process, measuring torque values during assembly, and then disassembling the transmitter to inspect for damage.

Next, the team brainstormed possible failure mechanisms. Approximately 8 potential failure mechanisms were identified as hypotheses; several of the failure mechanisms would result in galvanic corrosion through the stainless steel backing of the transmitter, the tantalum diaphragm, and the electrolyte fluid in the process as shown in Figure 2.

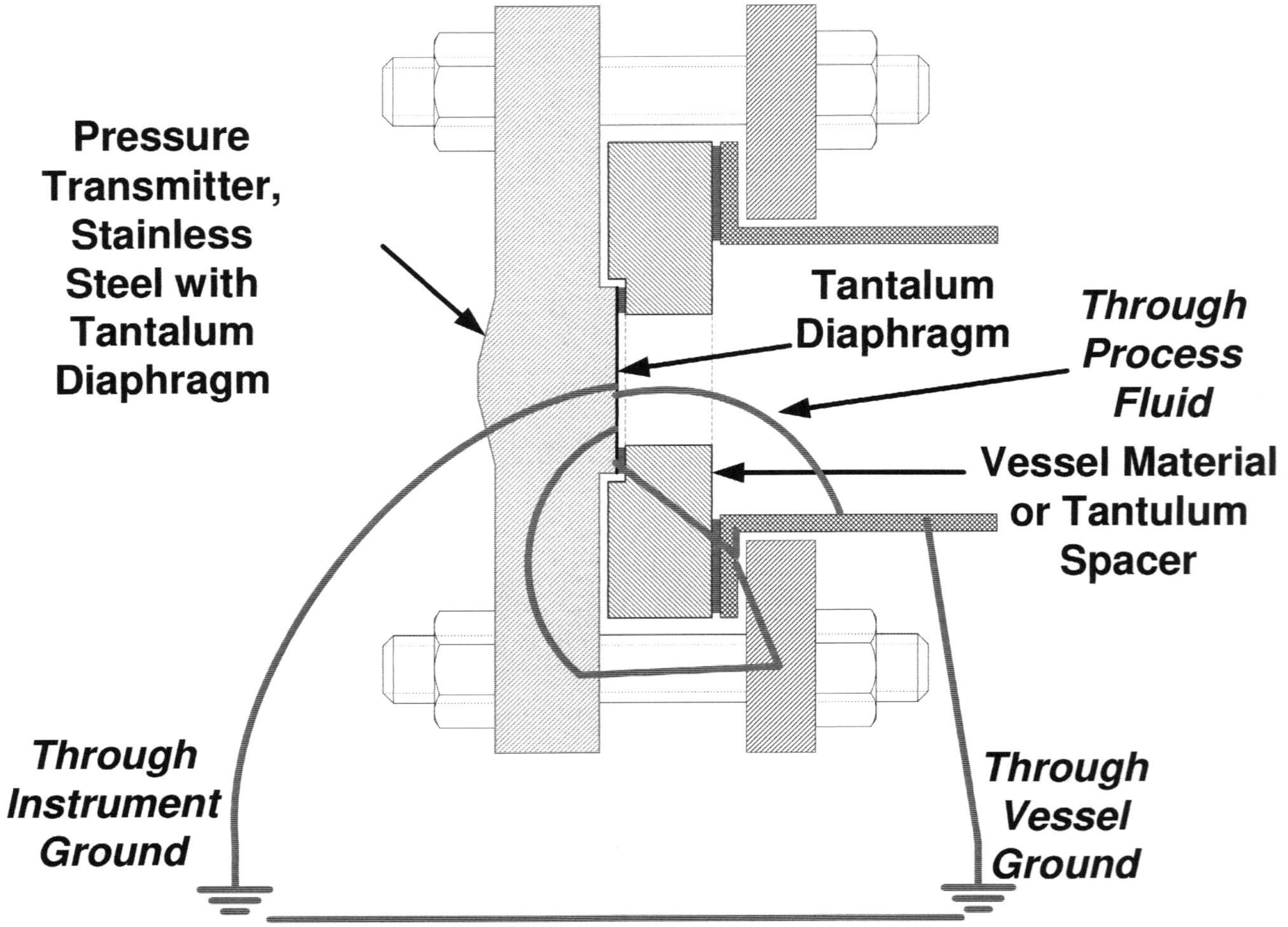

Figure 2: Paths for Galvanic Current Flow

Fact-Hypothesis Matrix

The hypotheses were tested using a fact-hypothesis matrix to determine most likely cause scenarios. (See CCPS (1) for illustration of this technique.) The hypotheses are listed across the top of the matrix. Facts that had been observed during the investigation and conclusions that had been drawn are listed down the side of the matrix. A sample portion of the matrix is shown in Table 1. At the intersection of each fact and each hypothesis, we placed a “Y” if that fact supported that hypothesis. If the fact disapproved the hypothesis, we placed a "N". If we thought the fact might support the hypothesis, we placed a "?".

The facts included the materials of construction of each component, the leak path, the observed conditions of the diaphragm, the gasket materials, the performance of the gasket materials under various torque values, the availability of conductivity paths, the torque values used in assembling the transmitter to the process flange, and the diaphragm vendor. In short, we listed every fact that we could think of that potentially could be a factor in loss of containment through the transmitter. As we filled out the matrix we recognized that we needed more precision and detail in the facts, and we split groups of facts into smaller elements.

Table 1: Sample Fact-Hypothesis Matrix

				Any of these will result in Galvanic Corrosion through SS & tantalum AND electrolyte					
	HYPOTHESES / FACTS	**Galvanic corrosion** through vessel material & tantalum AND electrical conductivity AND electrolyte	Diaphragm **gasket** failed (leaked), allowing a leak path **due to misalignment**	Diaphragm breach caused by mechanical failure, **manufacturing defect**	Diaphragm breach caused by mechanical failure, **particulates**	Diaphragm breach caused by mechanical failure, **diaphragm impacts the spacer**	Diaphragm breach caused by mechanical failure, **installation dent**	Diaphragm breach caused by mechanical failure, **Extruded gasket**	Diaphragm breach caused by mechanical failure, **Warped diaphragm due to overtorque**
1	Diaphragm brittle on failed PT	**Y**		**Y**	**Y**	**Y**	**Y**	**Y**	**Y**
2	Capillary hole enlarged, leak through fill hole								
3	Leak through gasket, bad joint								
4	Leaked through backing plate through corner of plate, not through the gasket	**Y**		**Y**	**Y**	**?**	**Y**	**?**	**Y**
5	Diaphragm size								
6	Diaphragm material	**Y**							
7	Spacer material	**Y**				**Y**			
8	Diaphragm backing flange material	**Y**							
9	Is there H2 embrittlement on the tantalum spacer? **Action**								

Continued for additional facts

We found that the transmitters in one unit had been ordered from one office of the vendor and the transmitters in the other unit had been ordered through another office of the same vendor. While the design principles were the same, the details of construction differed between the two sources of transmitters. The failed transmitters were more difficult to install than those in the other unit; they are heavier and more cumbersome to handle.

We also found that the gasket selection was different between the two units. The initial combination of gasket selection and bolt torque values (or lack of specified torque values) contributed to gasket extrusion and mechanical failure of the diaphragm.

From observation of some of the inspected transmitters, we found mechanical failure of the diaphragm due to handling problems during installation, such as setting the diaphragm down on the floor grating, or hitting it against another object. On some transmitters we observed mechanical failure due to misalignment of the spacer and the two gaskets.

As discussed above, in response to one of the failures, we attempted to electrically isolate the tantalum diaphragm and stainless steel backing plate from the vessel using an isolation kit including washers and spacers for the bolts. We were not successful in achieving electrical isolation.

Initially, we attempted to fill out one matrix for all of the incidents at once. However, it soon became clear that the facts were different for each of the four incidents. We copied, pasted, and edited the matrix in a spreadsheet program to generate four matrices, one for each incident.

Root Cause Analysis Tree

The investigation included a root cause analysis tree to confirm cause scenarios. (See CCPS (1) for a discussion of this methodology.) Sample fragments of the tree are shown in Figures 3-5. There were a total of 20 pages for the full tree.

We started with the top event of potential injury from the release and worked backwards in time, asking the question "What had to happen immediately before this event?" We attempted initially to develop a root cause tree for all of the incidents at once. However as we got into the details at the bottom of the tree, we quickly saw that the causes differed. In fact, the differences in the root causes between the various incidents led to difficulty in being sure that we understood the root causes and that we had appropriate corrective actions.

The construction of the fact-hypothesis matrices and the root cause analysis trees was iterative. As we developed the trees, we questioned the facts and did additional analysis to clarify them. We then corrected the facts in the fact-hypothesis matrices. As we updated the fact-hypothesis matrices, we incorporated those changes into the trees. This iterative loop is described in CCPS (1).

As we developed the tree, we identified each element in the tree as initiating (active) or enabling (passive). For each "AND" gate in the tree, we attempted to make sure that there was only one initiating element and that the rest of the elements were enabling. Where we had galvanic corrosion, all the elements appeared to be enabling; we did not see an obvious initiating element.

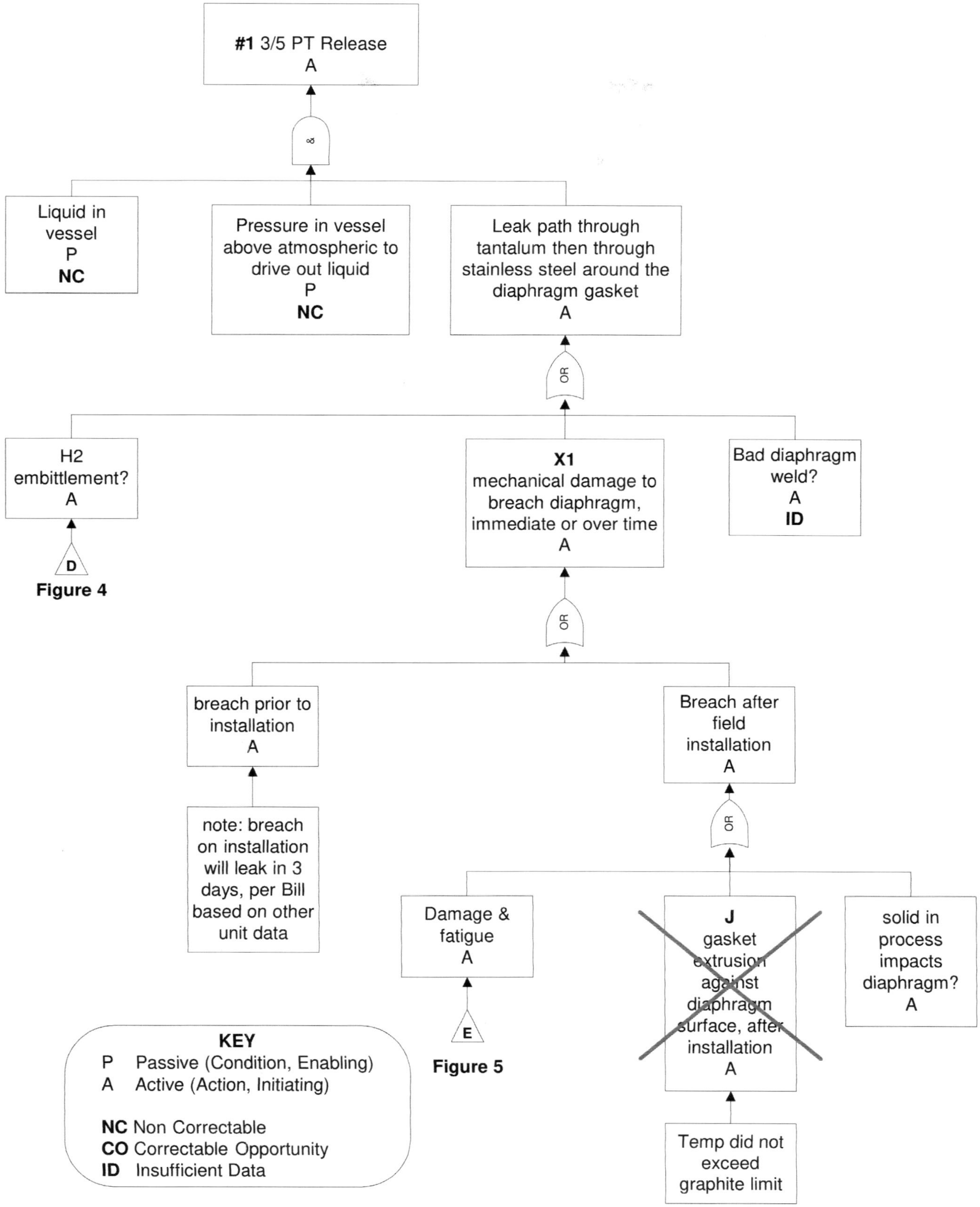

Figure 3: Sample Root Cause Analysis Tree for One Release

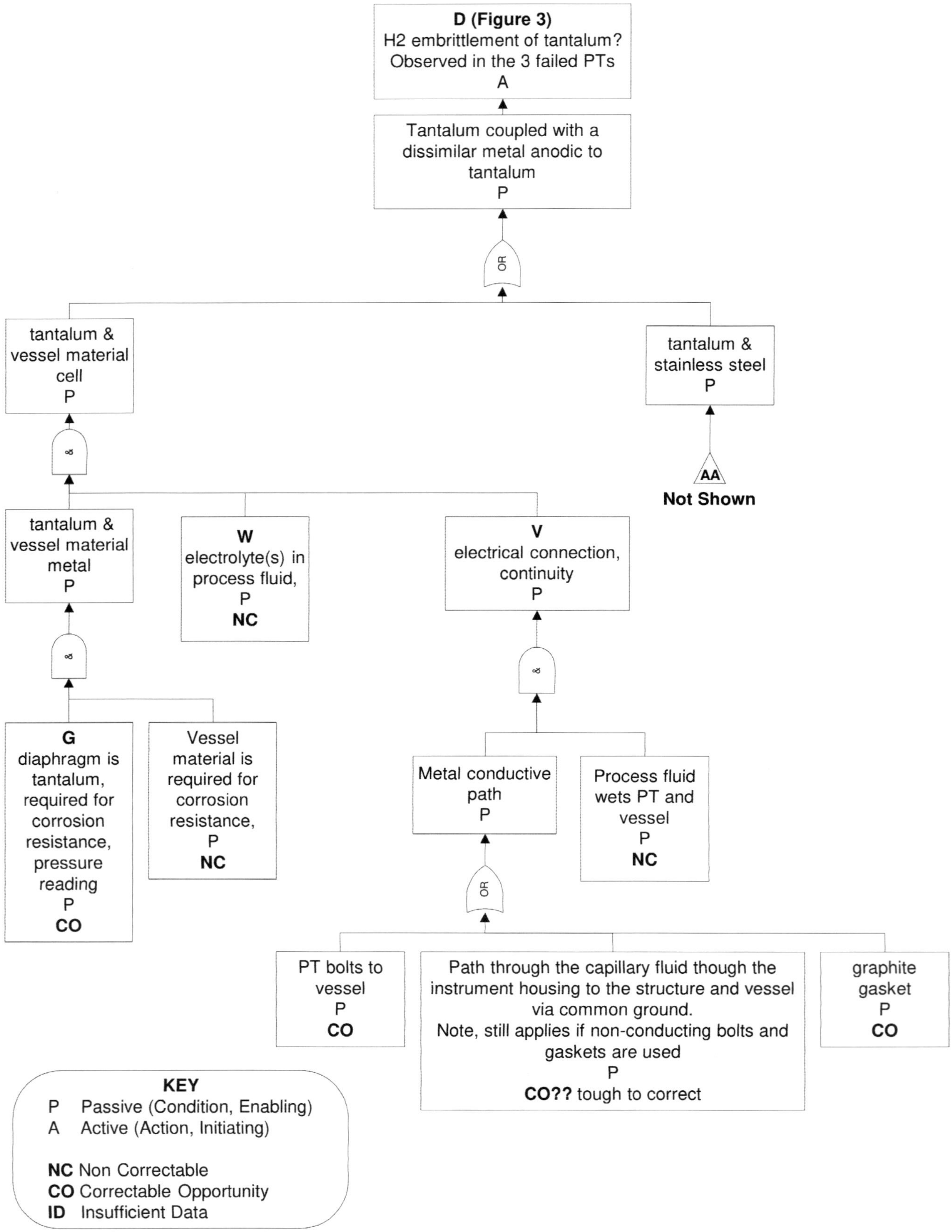

Figure 4: Sample Root Cause Analysis Tree for One Release

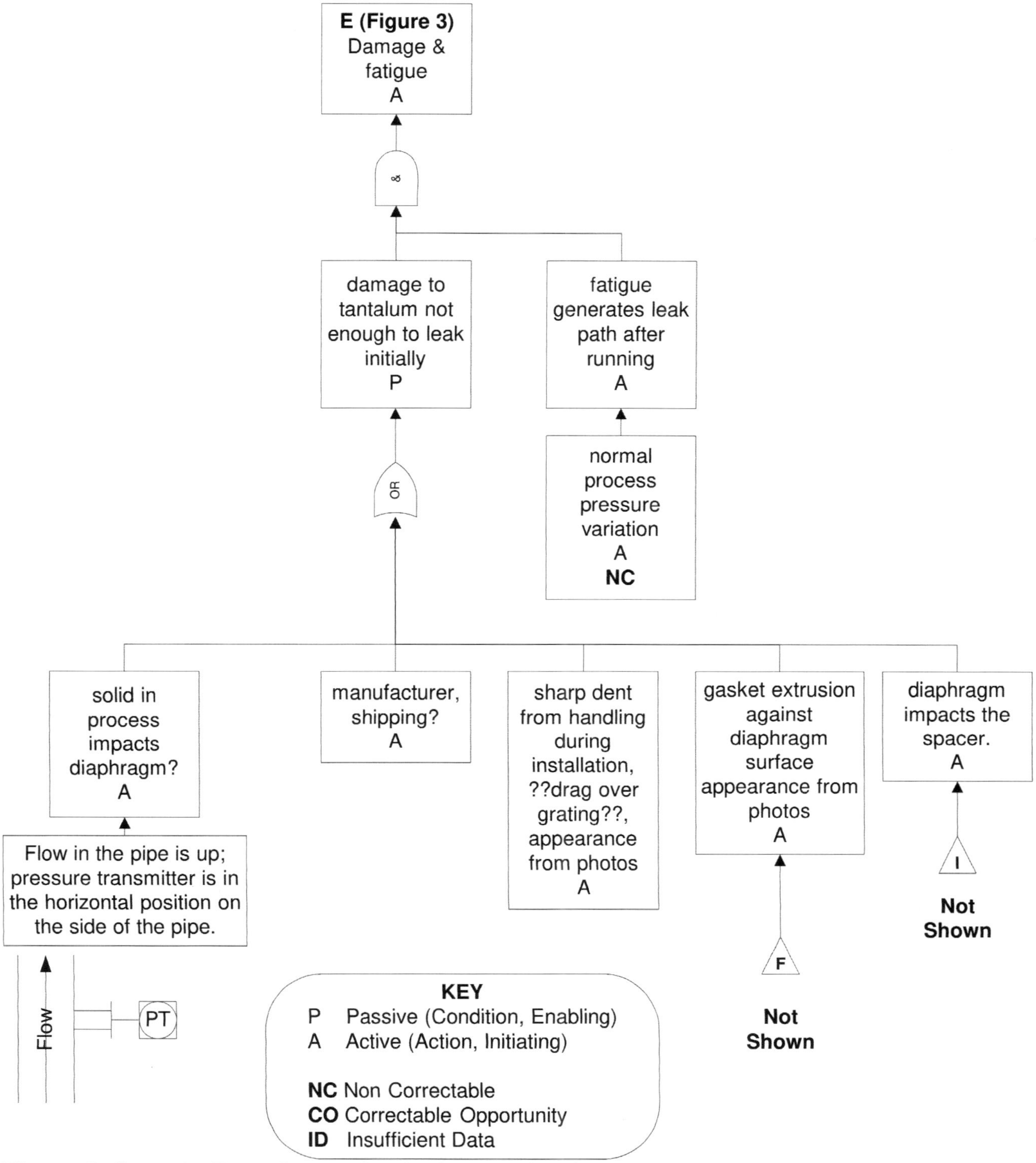

Figure 5: Sample Root Cause Analysis Tree for One Release

For several of the incidents, there was not a clear-cut single root cause. There were several possible paths that could have led to the failure. We identified these in each tree.

Finally, we had generated a fact-hypothesis matrix and a root cause analysis tree for each release. The combination of these two tools gave us confidence that we understood the potential mechanisms and causes of the failures. We grouped the failure paths into five major areas.

Corrective Actions

With the output from the fact-hypothesis matrices and the root cause analysis trees, we reviewed the corrective actions that had been taken up to that point and brainstormed additional corrective actions that might protect against the identified five failure paths.

To avoid mechanical failure from handling, we insured that the handling procedures would protect the transmitter from damage and instituted an inspection before the transmitter was installed.

To avoid mechanical damage from misalignment of the spacer and the gaskets, we installed alignment bolts to secure the spacer, the small gasket, and the stainless steel backing plate assembly. This reduced the difficulty that the technicians faced in attempting to hold the transmitter, the spacer, and the two gaskets in the correct alignment while installing the flange bolts.

To avoid mechanical damage due to gasket extrusion, we changed the specification of the gasket to a better design and established appropriate torque values and a torquing pattern to ensure that the gasket was compressed enough to seal, but was not over-stressed.

To avoid galvanic corrosion, we are considering a more robust diaphragm of another material that is less prone to damage during handling and is impervious to galvanic corrosion.

We used a matrix to illustrate the applicability of corrective actions against the five failure paths. On the left-hand side of the matrix we listed the failure paths. Across the top of the matrix we listed the releases, the configuration in place at the time of the investigation, and future planned configurations. In each cell, we listed any corrective actions that were in place at the time of the release, or that were planned for future states. If there was no corrective action to protect against a particular mechanism during a release or state, the cell showed the color red. If a corrective action was in place that could protect against the mechanism, the cell showed the color green. If we were not sure if a particular corrective action would be sufficient to protect against a mechanism, the cell showed the color yellow. A sample of the protection against failure matrix is shown in Table 2.

This graphic illustration was very helpful in discussing the effectiveness of the corrective actions with the operating personnel. The operators were very concerned about a leak of acidic material that could injure them. Explaining the mechanisms using the fact-hypothesis matrix, the root cause analysis trees, and showing the protection against failure matrix helped them understand the effectiveness of the corrective actions and reassured them that the risk was appropriate.

We also found that the detailed figure of the diaphragm seal arrangement, the fact-hypothesis matrices, the root cause analysis trees, and the protection against failure matrix were very helpful in discussion with the operating management to ensure that the investigation was as thorough and complete as possible, and that appropriate corrective actions were in place or were planned.

Table 2: Protection Against Failure Matrix

		Incident				Today	Future
		#1 3/5	#2 5/13	#3 Release after repair, 5/26	#4 6/16	6/25 In service	Future New Material Diaphragm
1	Mechanical Failure - Particulates	N	N	N	N	N	Y ?**
2	Galvanic Mechanism	N	N	N	N	N	Y
3	Mechanical Failure - Gasket Extrusion	N	N	N	N	Y (Gasket & torque values)	Y (Gasket & torque values)
4	Mechanical Failure - Handling	N	N	N	Y (Inspection)	Y (Inspection)	Y (Inspection)
5	Mechanical Failure - Misalignment	N	N	N	Y (alignment bolts)	Y (alignment bolts)	Y (configuration)
	Key: **N** = no protection was available against this mechanism for this installation **Y** = protection available against this mechanism for this installation **Y?** = protection is likely available, but not has not been proved, against this mechanism for this installation						** Test shows tougher than Tantalum, however, not unbreakable

Lessons Learned

We reinforced prior learnings about incident investigations and learned additional lessons:

1. There is no substitute for subject matter expertise. Our team was blessed with a metallurgist, a gasket reliability engineer, a mechanical engineer, a control engineer, technicians who had installed the transmitters, and a process engineer. Each of these individuals provided knowledge and expertise that was essential to drawing correct conclusions in the investigation. The team also included the EHS representative for the operating unit. This ensured that the investigation team understood the environmental, health, and safety drivers for the investigation.

2. Diversity of personnel on the investigation team is critical to ensure that subject matter expertise is available and to stimulate alternate perspectives when the root causes are not immediately obvious.

3. A person skilled in the incident investigation methodologies is important. This person brings a discipline in recording facts and identifying hypotheses, and in developing root cause analysis trees. He/she also frequently serves as an objective person to ask questions that might not be obvious to those closely associated with the problem.

4. For a complicated system where several failure paths may lead to the undesired event, the fact-hypothesis matrix is a data-based approach to identify which failure mechanisms may be applicable.

5. This incident illustrates the situation when several failure paths are possible and corrective actions only address one or two of the possible paths. Although corrective actions were made failures continued to occur until we identified the additional failure paths and insured corrective actions had been taken.

6. The root cause analysis tree is highly recommended to confirm that all of the required conditions that could lead to the undesired event are captured. For example, for a galvanic cell, there must be dissimilar metals, an electrolyte, and a closed-circuit for electrical current to flow. The root cause analysis tree helped us to identify all of these elements and understand which ones we could take protective actions against.

7. The protection against failure matrix showed which corrective actions were in place for each release, and which were planned for future installation. This matrix was very helpful to assure operations personnel and management that the risk was reduced to an appropriate level and that we had a clear understanding of the mechanisms.

Acknowledgments

To the members of the investigation team, who provided subject matter expertise, who brainstormed mechanisms and corrective actions, who suggested the protection against failure matrix, and who spent their time clarifying the facts, the mechanisms, the matrices and the trees.

Disclaimer

This information was developed solely for Rohm and Haas Company's internal use. We are making this information available without any guarantee that it is appropriate for your purposes as conditions and methods of use of the information are beyond our control. Rohm and Haas Company disclaims any liability for use of this information by persons outside of the Company.

Reference

1. Center for Chemical Process Safety (CCPS). *Guidelines for Investigating Chemical Process Incidents, Second Edition*. New York: American Institute of Chemical Engineers, 2003.

Incident Investigation of an Explosion at a Styrene Plant

Daniel Wiff
NOVA Chemicals, Inc
Moon Township, PA
Wiffd@NOVAChem.com

On June 11, 2003, an explosion occurred in the natural gas fired ethyl benzene reboiler at a styrene plant in Bayport, Texas. The incident occurred while the air damper and registers were being adjusted to stabilize the flame and the excess oxygen content of the stack discharge. Following the rupture of the heater housing, fire erupted in the radiant section and the inlet header. The fire burned for three-and-one half hours as ethyl benzene continued to feed the fire from return piping connected to the ethyl benzene column. The reboiler was completely destroyed.

This paper summarizes how it was determined what evidence to collect; how initial evidence was collected soon after the incident; how the physical evidence was preserved during the 'make safe' and demolition stages; and how evidence was staged and protected after demolition.

1. THE PROCESS:

The natural gas fired heater is a reboiler for an ethyl benzene recovery distillation column. The process feed to the reboiler is from the bottom of the distillation column through a pump, as shown in the simplified process flow (Figure 1). The process flow travels through several tube passes then returns to the distillation column exiting the reboiler at a level above the inlet to the reboiler.

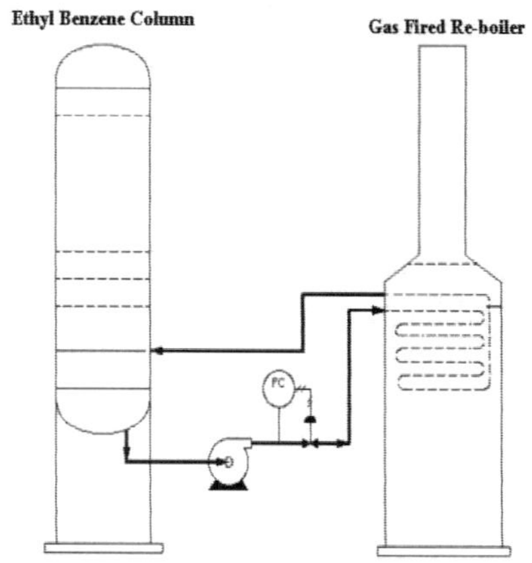

(Figure 1) – A Simplified Process

2. INCIDENT SUMMARY:

On June 11, 2003, an explosion destroyed the natural gas furnace at NOVA Chemicals' Bayport, TX site (Figure 2). The fire was quickly extinguished, with no personal injuries and minimal environmental impact. A full investigation was launched after the incident.

(Figure 2) Explosion at Bayport Facility

Immediately prior to the explosion, an operator noticed flame instability problems in the furnace. The operator began to manually adjust airflow to the burners in an effort to change the burner flame appearance and oxygen content of the stack discharge. During the few minutes in which adjustments were being made, a loud puff was heard, followed by an explosion in the furnace. The internal pressure from the gas explosion caused the radiant section enclosure of the reboiler to burst. This broke the seam bolts, dislodged supports for the convection section overhead, and knocked down several of the vertical tubes from their hangers.

Following the rupture of the heater housing in the radiant section, a fire erupted in the radiant section and at the inlet header. After about one half hour, the convection chamber, plus flue and stack fell (Figure 3). This caused a rupture in the ethyl benzene return line. The fire burned for three-and-one-half hours as ethyl benzene continued to feed the fire from piping connected to the ethyl benzene column.

(Figure 3) Collapsed Convection Chamber

More than 150 fire fighters promptly responded and proceeded to bring the fire under control. Streams of water were strategically applied to cool the support structure so that the convection chamber, flue and stack collapsed into an open area of the plant site. The reboiler was completely destroyed.

3. The Investigation Team:

The formation of the investigation team began soon after the emergency was mitigated. The team included company specialists in the field of gas-fired equipment and metallurgy. The team was initially responsible for identifying what evidence to collect, how to collect it during demolition, and how to preserve and properly stage the evidence.

To follow is the thought process the team used.

4.0 General:

The determination of the root cause of an incident is an important process that seeks to establish appropriate corrective measures and to prevent a recurrence. Therefore, the initial stages of an incident investigation, namely the collection, preservation and traceability of evidence, are crucial in providing for a complete and accurate root cause analysis. It is additionally important to understand that the reckless loss or contamination of evidence at this point will likely result in erroneous conclusions later in the investigation.

4.1 Background Data:

Evidence and data to be collected after an incident includes knowledge of events leading up to the incident itself. All documentation or information that may be necessary for the investigation should be secured as soon after the incident as possible. The documentation can be in physical form such as drawings, equipment specification sheets, maintenance records and operating procedures. Information may also be discovered through interviews or electronically such as process conditions, both historical and at the time of the incident. Additionally, any changes in the design, materials of construction, or operating conditions should be collected.

One commonly overlooked piece of information is meteorological data. For some incidents, wind speed, wind direction, temperature, humidity, as well as amount of precipitation are important.

After securing and making the incident site safe, one of the first steps is to identify and record the placement and appearance of the failed equipment and its components. Photographs and sketches should be made from different angles and elevations. Photographs freeze the incident in time. In the case of this incident, over 1000 pictures were taken. Not all of the photos were used during the investigation, but they were available to the investigation team. Aerial photographs (Figure 4) were used in order to capture the wide damage area and the dispersion of debris.

(Figure 4) Aerial View

Another importance of photographs is the use of reference points. A grid or some other reference scale allows for quantitative measurements of size, distance, location, orientation, and other dimensions.

5.0 Selection and Preservation of Evidence:

The actual position of failed equipment or its components provides valuable information to the incident investigation team, especially if fragmentation has occurred. This is why it is imperative that all parts be left in their original position, if possible, until a failure analyst has arrived and appropriate photographs have been taken.

After the incident at this site, an effort was made to make the facility safe. This required the removal of the heater stack, which lay across street. The stack was removed in order to provide access for emergency vehicles. If it is necessary for removal of any equipment or parts of equipment, then it should be documented, identified, and saved for the failure analyst.

The preservation of evidence, gathering and recording, is a time-sensitive issue. Improper methods of collection or preservation may reduce the value of the evidence. Fragile evidence has a high likelihood of degradation due to exposure. Evidence can become fragile for many reasons including being removed, misplaced, cleaned-up, destroyed, altered, disfigured, dispersed and taken apart. It is important that all personnel understand how to communicate the location of evidence especially when the evidence is located outside of the immediate scene.

5.1 Positional Evidence:

Sometimes it may be more important to know where the evidence is as opposed to what it is. It is essential to document each individual part along with its location and orientation. Positional evidence includes:

- Valve positions relative to the incident
- Controls and switches
- Relief devices
- Tank levels
- Gauges
- Flame, smoke and scorch marks
- Layers and strata debris
- Glass fragments
- Fragmentation debris
- Bulging, bending and bowing of structures
- Location of witnesses
- Indications of components that should be present but are missing
- Melting patterns
- Impact marks

During this incident, heater seam bolts broke and were propelled by sling shot action in four directions. After analysis, it was determined that the distance the bolt pieces traveled was due to the spring action in the paneling and not the pressure from the compression. Other positional evidence provided answers as to how the incident started, where the initial release of ethyl benzene was and where the fire was initiated.

5.2 Corrosion Evidence:

In the case where equipment has been subjected to corrosion, it is necessary to retain not only the equipment, but the corrosion products as well. The equipment should not be cleaned or sandblasted. If cutting is required, the cuts should be made far away from the corroded area to prevent altering the microstructure, introducing contaminants or damaging any residues that may be present. This will provide areas that were not exposed to corrosion, which will be valuable for comparison purposes.

During the investigation of this incident, rust debris was found in the natural gas headers. The headers had to be removed. The cuts were made so as not to damage any evidence.

5.3 Burned or Overheated Evidence:

No attempt should be made to clean any evidence that has suffered fire or overheating damage. The color of scale may indicate the temperature reached during a fire. This may indicate the origin of the fire.

5.4 Buried Artifacts:

Evidence buried as the result of an incident requires excavation of the debris. Great care needs to be taken during the excavation as to not damage any evidence. Photographs should be taken to document the location and condition of the evidence as compared to its surroundings.

5.5 Demolition and Preservation of Evidence:

After a major incident, demolition of the damaged equipment may be necessary. Prior to any demolition, a hazard review should be performed. The purpose of the review is to assure that reasonably foreseeable hazards will be identified so that the risk to people, facilities and the environment can be reduced to an acceptable level. During the demolition hazard review, the investigation team can communicate with the demolition team and identify what evidence needs to be preserved.

For this incident, the burners and associated headers, tubes, the section of the stack that contained the explosion doors, and the valves and instrumentation were among some of the equipment that had to be preserved. All cut items were labeled and match marked. They were then removed from the demolition and transferred to a properly designed containment staging area.

As the investigation progressed, the team was able to reduce the amount of evidence to be preserved to a level that would fit in several trailers. More than a year and a half later, the amount of evidence fills the floor of a room the size of an average office.

6. Conclusion:

The importance of having a company standard for preserving evidence of an incident, as outlined in this paper, cannot be emphasized enough. The standard should include what type of documentation and information should be collected, the importance of photographs, positional relationships, how to maintain the integrity of evidence soon after the emergency has been mitigated and how to stage evidence during demolition. It is equally important to identify how to preserve different types of physical evidence including corroded evidence, overheated evidence, and buried evidence.

This particular facility was fortunate to have experts, who were from another NOVA facility and had a guideline for preserving evidence, involved in the investigation. The guideline has since become an engineering standard for the company.

7. Reference:

1. Capstone Engineering Services, Inc, **"**Failure Analysis Ethyl Benzene Column Reboiler HS-104 for NOVA Chemicals Bayport, Texas**"**, September 15, 2003.

2. NOVA Chemicals, “Guideline for the Preservation of Physical Evidence for Metallurgical Examination”, October 28,2004.

INVESTIGATING CHEMICAL PROCESS ACCIDENTS: EXAMPLES OF GOOD PRACTICES

D.R. Morrison, R.A. Ogle, M.J. Viz, A.R. Carpenter, Y. Su

Exponent, Inc.

Prepared for Presentation at
American Institute of Chemical Engineers
2005 Spring National Meeting
Process Plant Safety Symposium
Atlanta, Georgia
April 11-13, 2005

February, 2005

UNPUBLISHED

INVESTIGATING CHEMICAL PROCESS ACCIDENTS: EXAMPLES OF GOOD PRACTICES

D.R. Morrison, R.A. Ogle, M.J. Viz, A.R. Carpenter, Y. Su

Exponent, Inc.

1 ABSTRACT

The accident investigation process, rooted in the scientific method, relies heavily on the use of evidence. This paper reviews aspects of the investigation process for three large-scale chemical process unit accidents (explosions or fires) with a focus on evidence: its collection, handling, use and analysis. Real incidents provide challenges to the investigation process; some of these challenges involve questions surrounding evidence, such as when is evidence collection complete and how do we know that certain non-physical evidence (e.g., witness statements) is reliable? These case studies provide practical examples of the importance of following good engineering practices throughout the life cycle of the investigation with respect to evidence.

The objective of an accident investigation is to determine what happened so that a recurrence of the accident can be prevented. The investigator of a chemical plant accident is confronted with numerous challenges to the fact-finding process because the accident has destroyed or altered significant sources of evidence such as the physical equipment and the process data. The investigator must often cope with incomplete or contradictory data. The best strategy for dealing with these difficulties is the scientific method: gather facts from numerous independent sources, develop and test alternative hypotheses, and evaluate potential causal factors using sound engineering and scientific principles. The ability to ultimately identify the root causes of a complex chemical plant accident depends heavily on the quality and comprehensiveness of the fact-finding process.

2 INTRODUCTION

The investigation of a chemical process accident can be significantly more difficult than an accident investigation in most other industrial settings. This difficulty arises because a chemical process unit, whether a batch or continuous process, is typically composed of a number of interdependent pieces of equipment that must work together in a precisely coordinated fashion for successful operation. A seemingly trivial fault condition in an apparently minor piece of equipment, such as a valve or pump, can cause catastrophic consequences for the entire unit. In other words, a piece of equipment that may appear to be minor may in fact cause a major process disruption if a few additional, apparently unrelated events occur simultaneously. The high degree of connectivity often introduced by recycle streams or the integration of heat transfer operations can magnify or multiply the consequences of a fault condition.

As with any accident, the ultimate objective of a chemical process accident investigation is to determine the causal factors so that corrective actions can be identified that will prevent a recurrence. The American Institute of Chemical Engineers (AIChE), through the Center for Chemical Process Safety (CCPS), has published a guidance document for the investigation of chemical process incidents (hereafter referred to as the CCPS *Incidents* book)[1]. One of the primary goals of a process accident investigation is to identify the root causes of the accident. However, as we show in Figure 1, before root causes can be identified, a number of other activities must be performed first, including the gathering of evidence. Although the following may seem obvious, we have found from our investigations that our ability to clearly identify root causes with certainty is usually a direct function of the thoroughness of evidence gathering, preservation and analysis tasks. In this paper we examine some of the best practices we have found for evidence gathering. We then illustrate some of the pitfalls to be avoided in the investigation process relying on case studies from our experience. The common theme running through our case studies is the role that robust and methodical evidence gathering has played in avoiding some common mistakes in accident investigation.

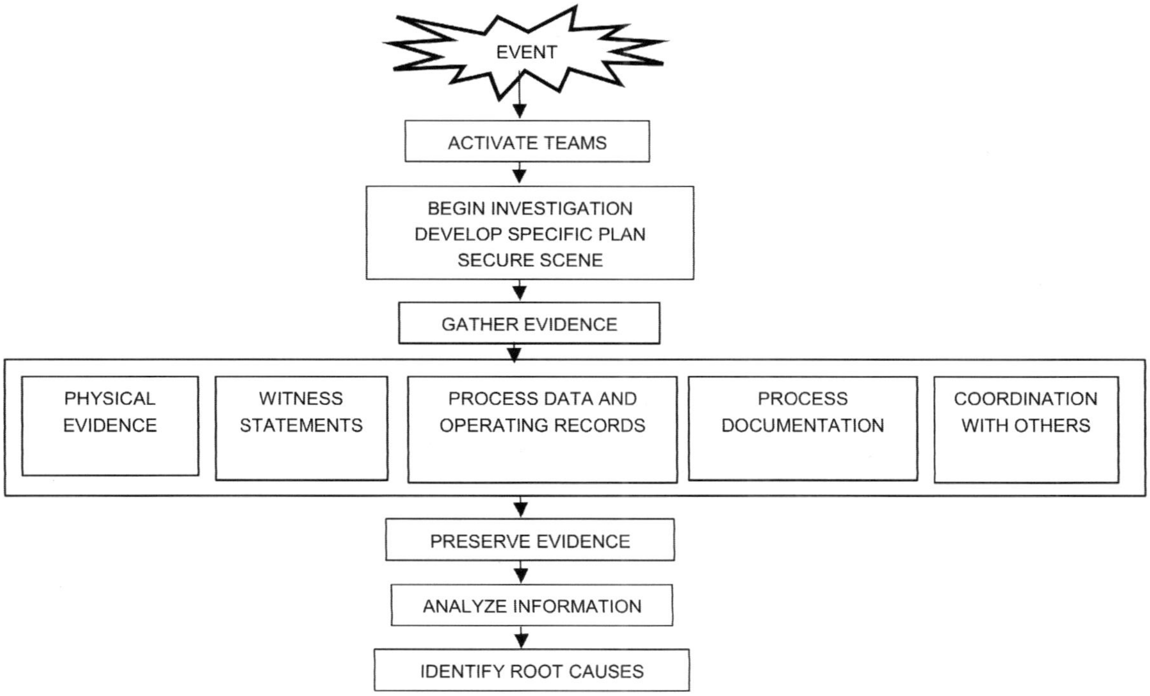

Figure 1. Event response action flow for a chemical process accident (as modified from that presented in the CCPS *Incidents* book [1]).

3 GATHERING EVIDENCE

The collection and management of evidence involved in the investigation of a chemical plant accident is a multi-faceted topic. Rather than provide a comprehensive narrative on all aspects of evidence, we limit our discussion to a few important aspects of evidence collection that are especially important in chemical process investigations. The discussion is organized into four areas: physical evidence, witness statements, process data and operating records, and process documentation.

3.1 Physical Evidence

Physical evidence includes any physical object that may be useful in testing a hypothesis or substantiating a statement initially assumed to be factual. This can include processing equipment, valves and piping, process instrumentation, process controllers, chemical samples, structural elements, explosion and fire debris or any other physical object. To ensure the maximum utility of physical evidence, it must be carefully documented, both prior to and after relocation, and preserved in a manner that establishes its chain of custody. The evidence must be handled in such a way that long after the accident, the question "What is this and where did it come from?" can be answered and any physical changes or alterations to the artifact after location and collection can be readily identified. Guidance on proper evidence handling is available from the National Fire Protection Association (NFPA 921),[2] the American Society of Testing and Materials (ASTM)[3,4] and the CCPS *Incidents* book.

3.1.1 Documentation

One of the most important steps in the identification, collection and preservation of physical evidence is documentation. An artifact or feature can be documented in numerous ways, including sketches, notes, photographs and video. The primary method of documentation and often the most useful is photography.

It is not unusual to accumulate dozens of photographs of the same piece of evidence, and in the case of a complex chemical process investigation with a large accident envelope (i.e., a large physical area effected by the accident) and numerous artifacts, this may generate hundreds or thousands of photographic images. Thus a system of managing the photographs is essential. A protocol should be used to give each photographic image a unique identification label that includes a reference number, name or initials of the photographer and the date the photograph was taken. A well-documented accident site will include photographs of the entire area from multiple angles. To facilitate later review of the photographic record, it is often useful to maintain a log of photographs where a short description of an individual photograph or sequence of photographs is described. Similarly, a site plan drawing showing the orientation of the photographer for a sequence of photographs can be useful.

3.1.2 Artifact Location

Just as capturing the physical condition of the evidence is important, accurately recording the location where evidence was recovered is important as well. This becomes especially so in the case of structural collapse or explosions where the physical evidence has been displaced substantially from its original location. The location of physical evidence can be captured in several ways, including photographs (with scales added), sketches and site maps with measurements, or GPS surveying. Aerial photographs are occasionally helpful especially if similar aerial photographs exist of the location before the accident occurred. It can also be useful to develop a debris map of the physical artifacts showing the starting and ending location in relation to other fixed plant structures. The following figure (Figure 2) shows an artifact debris field map where the starting and ending locations of each piece are connected by vectors.

ARTIFACT MAP		
Artifact #	Weight (lb)	Distance (ft)
CH77161-SY1	280	30
CH77161-SY2	40	160
CH77161-SY3	450	195
CH77161-SY4	500	310
CH77161-SY5	925	800
CH77161-SY6	160	600
CH77161-SY7	120	515
CH77161-SY8	100	760
CH77161-SY9	50	1080
CH77161-SY10	30	2400

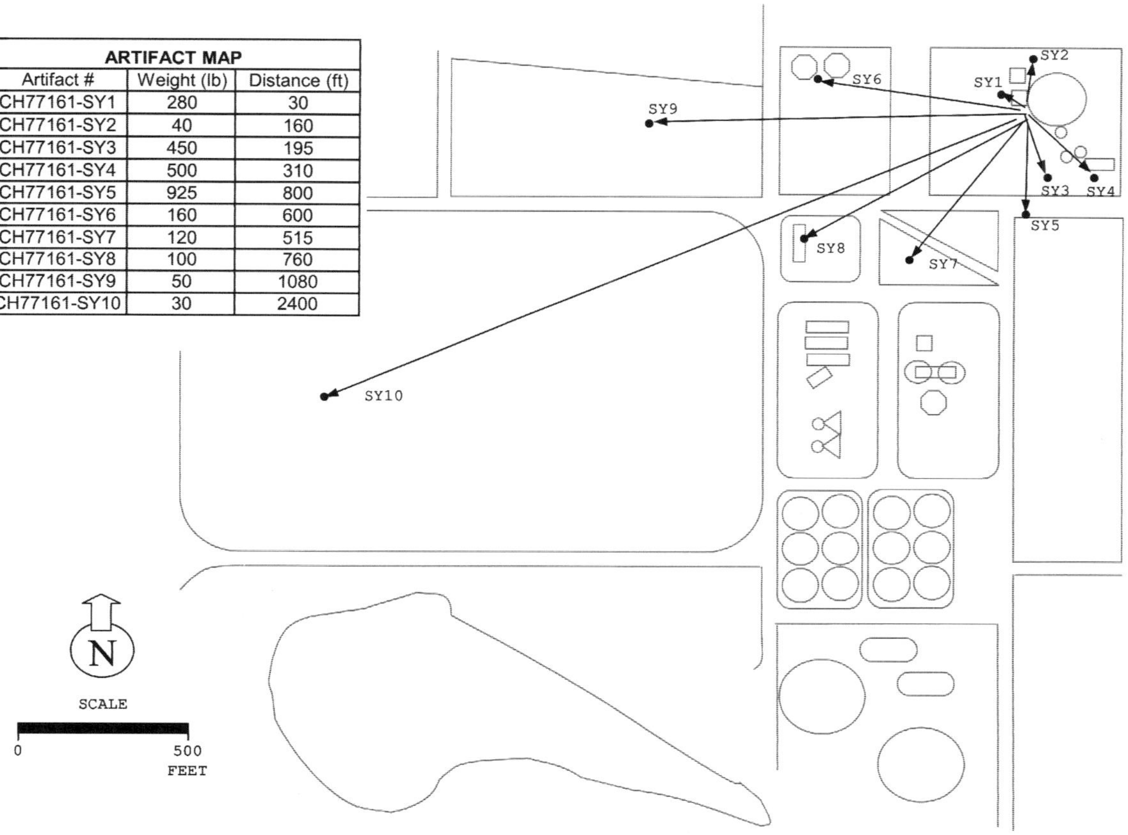

Figure 2. Example of an artifact debris field map.

A debris field map like the one above can be very useful in determining the likely trajectory and minimum energy required to form the debris field. However, the creation of such a map usually requires the coordinated effort of an entire team of field personnel involved in the search for artifacts. The field collection activities of the team must be directed with precision and strict adherence to the evidence collection written protocol. If hundreds or thousands of pieces of debris exist scattered over a large physical area, the potential for errors in identification, labeling and cataloging are significant and must try to be avoided. An additional concern lies in discriminating useful debris from junk, e.g., ejected pipe fragments versus spare pipe spool.

3.1.3 Artifact Identification

All of the recovered physical artifacts need to be uniquely identified and cataloged. Artifacts should be assigned a unique identification, typically an alphanumeric tag unique to that artifact. Additionally, the artifact needs to be identified as to what it was (or what part of a larger whole) and where it fit into the larger facility. This becomes very important in instances where pieces look similar and have similar functions, e.g., differentiating pipes and valves from neighboring legs in a parallel train.

Since the condition and position of valves can be very important in determining the operating status of the facility at the time of the accident, and often look identical, they require careful study. All of the valves in a relevant portion of a facility should be documented in place (where possible) and their position recorded. Table 1 is an example of a valve inventory.

Table 1. Example of a valve inventory table.

Artifact #	Valve #	Type	Method of Activation	Position	Failure Mode
CH09020-AC13	SL-104	Ball	Manual	Open	—
CH09020-AC42	RV-002	Gate	Automatic	Closed	Failed open
CH09020-AC02	TV-023	Gate	Automatic	Open	Failed in last position
CH09020-AC55	CV-004	Gate	Automatic	Open	Failed closed
CH09020-AC08	SL-088	Ball	Manual	Closed	—

3.1.4 Chain of Custody

In addition to documentation and identification, a chain of custody log or form must be developed for each collected artifact. This document should include the date and location where an artifact was collected and where the object is being stored. The chain of custody must be updated whenever the artifact is moved or transferred to another party's control. In addition to confirming the location of an artifact, the chain of custody will help to identify when any modifications were made to the artifact as a result of storage, transportation, examination or testing. Both the NFPA and ASTM offer guidance on chain of custody.

3.1.5 Exemplars

Whenever it is possible, an exemplar of key subject artifacts should also be obtained. An exemplar artifact is an artifact that is identical or substantially similar to a subject artifact but was not directly involved in the accident, whereas, a subject artifact was directly involved. Studying an exemplar can be very useful in determining any abnormalities with a subject artifact and can be useful in refining examination and testing procedures without risk of damaging or altering a subject artifact.

3.2 Witness Statements

Collecting and recording witness observations are especially fragile tasks. The investigator must seek ways to corroborate witness observations. The first way to test witness observations is to compare them with independent objective data such as physical evidence ("The switch was in the 'off' position.") or process data ("The tank level was at 30%."). Another approach, sometimes less satisfactory, is to compare observations from different witnesses. Care must be taken here because witness observations can be both imprecise and inaccurate. A suggested method to compare witness observations is to try to synchronize them with each other by finding a single common event to use as a benchmark ("I heard a loud blast."). Another technique for synchronization is to compare the sequence of events as related by different witnesses. Developing a map of witness locations can also be useful in analyzing their statements.

3.3 Process Data and Operating Records

Chemical process accidents are often the result of a deviation from an expected operating condition. Process data and operating records can be an invaluable source of information for this type of accident. This is especially true for processes with a distributed control system (DCS). First, it is important to classify the type of process event that preceded the accident (e.g., process excursion, normal startup, emergency shutdown). A comparison data set should be obtained, if possible, to enable the investigator to determine how the pre-accident event differed from previous process events.

3.4 Process Documentation

It is important to assemble the relevant process documentation as early in the investigation process as possible. The process documentation is a valuable way to establish a common ground of understanding for the investigation team. It is important to bear in mind, however, that the process documentation describes things as they should be, not necessarily as they are. As an example, detailed plant layout drawings, piping isometrics and P&IDs (process and instrumentation drawings) may not have been regularly updated, revised and re-released by the plant's document control office to show the as-built versus the as-designed configuration. It is helpful to know in advance of the evidence collection and identification activities how closely the configuration drawings match the actual plant layout at the time of the incident. Knowing what major differences exist between these two may save substantial amounts of time looking for process equipment that simply does not exist.

4 PITFALLS IN ACCIDENT ANALYSIS

Some of the pitfalls to be avoided in the investigation process are illustrated with case studies from the authors' experience. Specific details of the case studies have been omitted or altered to protect the identity of all concerned parties. Again, our focus here is on the role good evidence gathering plays in assisting the investigator to avoid common analysis pitfalls.

4.1 Beware of the Single Cause Explanation

A continuous hydrometallurgical unit processed a mineral ore for separation and purification by downstream units. The ore was mixed with chemical reagents in an aqueous solution and the resulting slurry was fed to the reactor with superheated steam at 300°C and 40 atmospheres. The superheated aqueous slurry product was returned to atmospheric pressure and a temperature of 100°C by flashing off steam.

The unit was forced into an emergency shutdown by a power outage. Contrary to the plant's procedures and training, the operator bypassed the reactor drain valve interlock. He then manually controlled the draining of the reactor inventory and allowed the contents to drain so quickly as to overwhelm the mechanical integrity of the downstream pressure vessels. The explosion of the pressure vessels devastated the process unit. Various parties were quick to focus their analysis of the accident in terms of one particular causal factor as a result of considering only the most easily found and readily available physical evidence. In fact, this accident was the result of a tangled web of causal factors.

A safety-instrumented system, which was intended to automatically perform an emergency shutdown and was under study for two years, had yet to be implemented. Pressure relief valves intended to protect these pressure vessels were unavailable because they had been blocked off for emergency maintenance. In subsequent interviews the operator stated that his objective in draining the reactor was to prevent excessive overpressure in the downstream pressure vessels. He focused his attention on the pressure dials mounted at eye level on the control board; these dials corresponded to the strongest pressure vessels. The pressure dials for the weakest pressure vessels were mounted at knee level. The operator thought that the downstream pressure vessels had a uniform pressure rating.

The power outage was caused by an electrical malfunction that occurred in a different process unit. The power distribution control system was designed to isolate different process units in the event of a power distribution failure. A software error prevented the power distribution system from correctly diagnosing the fault and isolating it. The electrical malfunction that triggered the power outage was a ground fault to a 13-kilovolt bus bar in one of the distributed electrical switch houses.

As can be seen, this tangled web of causal factors did not become clear until an exhaustive variety of evidence had been collected, documented and analyzed. We found ourselves returning to certain key pieces of evidence – physical (e.g., pressure relief valves), witness statements (e.g., operator's actions), process data (e.g., pressure and temperature records of downstream pressure vessels), process documentation (e.g., piping and vessel layout drawings) – time after time during the period of our investigative efforts. A consideration of <u>just</u> the evidence most easily identified did not provide a complete understanding of all of the root causes by itself.

4.2 The Initiating Event May Be Distant from the Accident Epicenter

A production facility manufactured an organic intermediate using a nitration process. The process unit was a batch operation with numerous reactors, separation units and storage tanks. The facility generated a number of waste streams including hazardous organic wastes (both liquid and solid), wastewater and a nitrogen-exhaust stream. Over a period of a few years, the manufacturer implemented a number of undocumented process improvements. One of these improvements was to recycle one of the liquid organic waste streams back into the front of the process.

During a routine maintenance shutdown, it became necessary to drain one of the waste holding tanks. The tank was thought to contain wastewater and so it was drained into the wastewater neutralization basin located approximately 100 meters from the production building. Shortly after the transfer was completed, an explosion destroyed the neutralization basin and associated piping. The destruction of the neutralization basin shut down the wastewater treatment facility causing, in turn, the shut down of the remaining three production operations at the facility. The ensuing business interruption eventually resulted in the bankruptcy of the manufacturer.

This case study illustrates why it is often important to broaden, rather than narrow, the scope of the evidence gathering process. During the initial accident investigation, the manufacturer and their insurance company focused their attention on the waste holding tank and the equalization basin. Their conclusion was that a shock-sensitive, explosive organic compound had accumulated in the waste holding tank. This material, determined to be tetranitromethane, was determined to be the cause of the explosion in the basin. The gathering of evidence (physical and process documentation) was thus initially limited in scope. The question left unanswered was, "Where did tetranitromethane come from and why did it accumulate?"

Upon further investigation by other parties, it was determined that the manufacturing process had generated tetranitromethane throughout the eleven-year life of the facility. In the original design of the facility, the tetranitromethane safely exited the process through the wastewater and nitrogen exhaust streams. As a result of the recent process changes (recycle addition), however, the tetranitromethane was captured and accumulated in the waste holding tank. Thus, the explosion was caused by equipment, materials and operator actions 100 meters distant from the explosion epicenter.

This conclusion was based on a reconstruction of the process changes using witness statements and process documentation. The investigation was seriously hindered by three factors:

- The failure of the manufacturer to retain process data and operating records
- The failure to update process documentation, especially process flow diagrams, general arrangement drawings, and process and instrumentation drawings (P&IDs)

- The failure to implement a sampling program for quality control, waste characterization and material balance analysis.

4.3 Detecting Process Abnormalities from Process Measurement Data

Many chemical process accidents are the result of an unexpected departure from normal operating conditions. Process instruments can detect changes in operating conditions that are either not remembered or not readily observed by operators. The archive of process measurement data (hereafter, "process data") from a distributed control system (DCS) is analogous to the "black box" of a commercial airliner. The process data provide a record of events that is independent of the inherent limitations of human witnesses.

Successful use of process data for accident analysis requires an understanding of both the chemical process itself and the instrumentation used. The P&IDs are a useful starting point for understanding the process, but, if possible, they should be corroborated against field observations to verify their accuracy. The standard operating procedures (SOPs) are also helpful if it is confirmed with the operators that the SOPs are indeed followed.

The accident data, i.e., the process data recorded prior to and during the accident, are most useful if a reference set of data is available for comparison. A suitable reference data set is selected by classifying the type of process event that preceded the accident (e.g., process upset, normal shutdown, emergency shutdown) and by determining the time frame that defines the pre-accident period. The accident data is then compared with the reference data to identify departures from normal operation. The evaluation of process data (both accident and reference sets) can benefit from comparison with other independent operating records (e.g., operator logs, batch sheets). Figure 3 illustrates how process data can be related to the normal operation of, as an example, a suspension polymerization reactor, and Figure 4 illustrates the process data from the accident event. A comparison of these two plots shows the departure from normal operation; this departure may not have been fully appreciated by those who witnessed it.

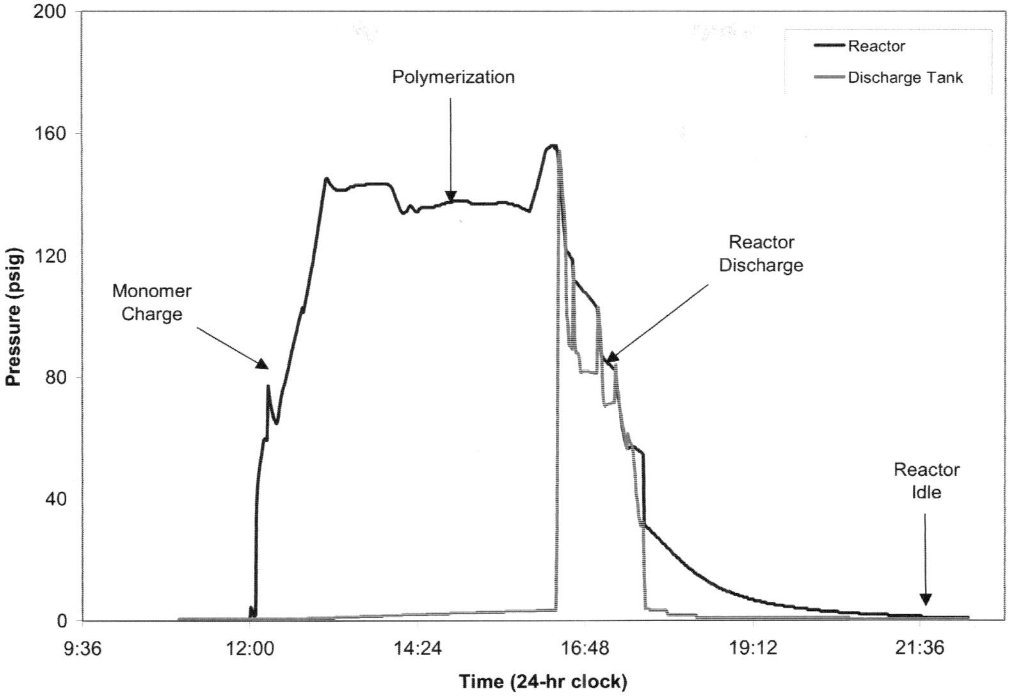

Figure 3. Pressure versus time plot as an example of normal reaction behavior.

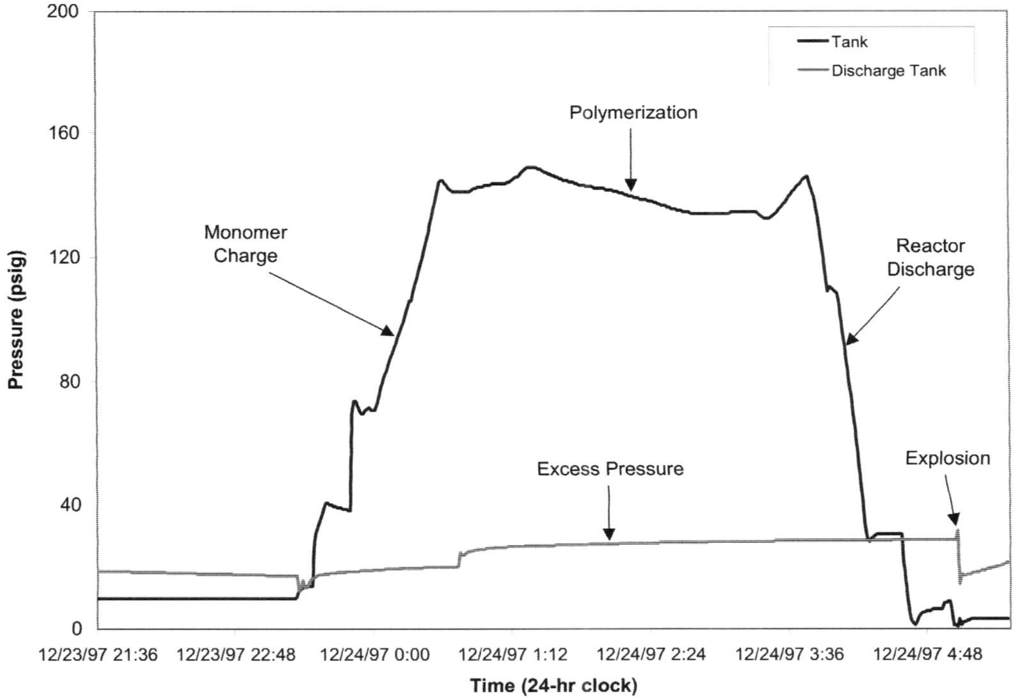

Figure 4. Pressure versus time plot that shows abnormal pressure behavior leading up to the eventual explosion.

The investigator must be cautious about relying uncritically on process data, however. Measurement or interpretation errors can occur. Some potential sources of error in process data are:

- Failure to synchronize clocks: Electronic data sets from different computers will bear the time stamp of the host computer. Different computers rarely have synchronized clocks. Identifying a common event benchmark on the two different data sets and noting the time difference can synchronize the data.

- Inaccurate instrument readings: Check the calibration records for the instrumentation to verify the instruments' precision and accuracy.

- Out-of-range measurements: Beware of zero and full-scale readings. Analyze earlier data trends to establish, if possible, that the instrument reading is off-scale.

- False instrument readings: Spurious process data from one instrument that seem inconsistent with other data sources may indicate a fouled sensor or plugged orifice.

- Sampling frequency too slow to detect transient events: The failure of an instrument to detect a transient event may be due to a slow sampling rate rather than an instrument deficiency.

It is generally easy to test for these potential sources of error. Whether the data can be corrected depends on the nature of the error. The key point here is that process data often will tell a more detailed story of what process deviations occurred prior to the accident. However, these data should be evaluated critically to establish what they truly mean.

5 CONCLUSIONS

The investigation of a large chemical plant accident, such as an explosion or fire, involves the use of the scientific method. Although certain aspects of the scientific method rely on logical inference, other aspects of it depend on a careful consideration of the empirical data. As such, the careful consideration of all varieties of evidence in an accident investigation is critical. Hopefully, our case studies have illustrated in particular detail some of the general concepts the investigator should keep in mind when considering evidence. Evidence gathering should be complete. Incomplete or unnecessarily limited evidence collection may lull the investigator into concluding that a single root cause was at play when in fact the other root causes have been missed due to insufficient evidence collection. As the initiating event may be distant from the epicenter of the accident, it may be wise during an investigation to broaden the evidence-gathering scope rather than limiting it. Broadening the evidence gathering activities also serves to generate more evidence from different sources, which may then be used as a means to corroborate more subjective evidence like witness statements. Insofar as preventing the recurrence of an accident is dependent on finding the root causes of the event, the ability to ultimately identify the root causes of a chemical plant accident depends heavily on the quality and comprehensiveness of the fact-finding process as supported by evidence.

---

[1] CCPS, *Guidelines for Investigating Chemical Process Incidents, Second Edition*, American Institute of Chemical Engineers, Center for Chemical Process Safety, New York (2003).

[2] NFPA 921, *Guide for Fire & Explosion Investigations*, 2004 Ed., National Fire Protection Association, One Batterymarch Park, Quincy, Massachusetts (2004).

[3] ASTM E860, *Standard Practice for Examining and Testing Items that are or may become involved in Litigation.*

[4] ASTM E1188, *Standard Practice for Collection and Presentation of Information and Physical Items by a Technical Investigator.*

Paper (Number)

Investigation Challenges of the Kinston Dust Explosion Accident

Quentin A. Baker
Baker Engineering and Risk Consultants, Inc.
Robert C. Gombar, Sr., Esq.
McDermott Will & Emery LLP
James A. Lastowka, Esq.
McDermott Will & Emery LLP
Michael A. Polcyn
Baker Engineering and Risk Consultants, Inc.

Prepared for Presentation at
American Institute of Chemical Engineers
2005 Spring National Meeting
Process Plant Safety Symposium
Atlanta, Georgia
April 11-13, 2005

January 2005

UNPUBLISHED

Abstract

An explosion occurred at the West Pharmaceutical Services plant in Kinston, NC, on January 29, 2003. The facility manufactured rubber parts for pharmaceutical delivery devices. The explosion occurred in the rubber compounding section of the plant, which suffered extensive damage. There were six fatalities and a number of injuries. The incident received national attention and was the subject of concurrent investigations by West Pharmaceutical Services, numerous federal and state government agencies, and dozens of third parties.

In-depth investigation by West Pharmaceutical revealed that the explosion involved polyethylene dust. Unlike most dust explosions, no residual dust accumulations were found. The investigation posed major technical challenges in light of the extensive damage to the facility and injuries to key witnesses. Investigation challenges were also presented by the large number of parties conducting investigations, combined with initial concerns about a possible airplane crash or terrorist incident. The paper will focus on the investigation challenges including investigation methodology, interpretation of data, management of the investigation site, and management of information and evidence requests.

Background

West Pharmaceutical Services Inc., founded in 1923, is one of the largest manufacturers of closure systems and components for sealing drug vials and pre-filled syringes. The closure systems and components are made of rubber compounds developed by West and other materials. Headquartered in Lionville, Pennsylvania, West has eight manufacturing facilities in North America and ten facilities in Europe and Asia.

West's plant in Kinston, North Carolina, started operation in 1975 and underwent a substantial expansion in the mid-1980s. There were essentially two operations in the plant: rubber compounding and product finishing. The plant was primarily a single story structure, except for the rubber compounding operation (known as the automated compounding system (ACS) or "the Tower" section) that was approximately 60 feet tall. The total plant area was approximately 150,000 square feet. The rubber compounding process operated 24 hours a day, five or six days a week, producing sequential batches of rubber compounds. In January 2003, the workforce at West's Kinston plant included 264 West employees and 35 full-time contract employees.

The ACS had two separate rubber compounding lines, each with a mixer, a mill, and batchoff equipment. Mixer 1, Mill 1, and Batchoff 1 were located at the southern part of the ACS. Mixer 2, Mill 2, and Batchoff 2 were in the northern part of the ACS.

In the rubber compounding process, raw materials were prepared in a separate area known as "the kitchen" on the first floor of the ACS. The materials for each separate batch were placed into a tote and transported by an elevator and a series of conveyors to the mixers located on the second floor of the structure. Other bulk ingredients, such as clay and mineral oil, were fed directly to each of the mixers. The mixers performed the actual compounding process, converting the raw materials through a series of steps into a rubber compound batch.

Once a rubber compound batch was completed in a mixer, the batch was dropped through a chute to the first (or ground) floor of the ACS where a rolling mill smoothed the batch into a sheet. The sheet of rubber was then trimmed into a narrower strip and conveyed to a batchoff machine where the strip was dipped through a vat containing a slurry of very fine polyethylene powder and water. The strip was then air dried, folded, and packaged in containers either for shipment to other West plants or for use in the finishing area of West's Kinston plant.

A suspended ceiling comprised of acoustic tiles hung 10 1/2 feet above the first floor of the ACS. The ceiling covered the entire room where the rubber compounds were rolled, dipped, cooled, folded, and packaged.

There were two separate dust collection systems in the ACS, one for the kitchen area and one for the mixers. The extraction ducts from the kitchen and mixer systems terminated at dust collector units located outdoors.

Because West manufactured products for use in the pharmaceutical industry, the cleanliness of the facility was a very high priority. A cleaning staff worked around the clock vacuuming and wiping up any visible dust accumulations in the work areas.

The Incident

On the day of the incident, operations in the plant were typical and described by some workers as very smooth. According to witness accounts, there were no sights, sounds, or odors that would have indicated a problem. At 1:28 p.m. on January 29, 2003, an explosion occurred abruptly in West's Kinston plant. Workers throughout the plant heard the explosion or, as some described it, a sound like "rolling thunder." Some workers closer to, but outside of the rubber milling area, saw a bright flash and felt a pressure wave that knocked them off their feet. Workers inside the rubber milling area saw flames and were knocked off their feet.

The explosion started a fire that engulfed part of the plant. The entire plant was affected to some extent by the explosion and fire, though the damage was most severe in the ACS or rubber compounding part of the plant, which was essentially destroyed. There were six fatalities and a number of injuries as a result of the explosion and fire.

A Large Number of Investigating Parties

Increasingly with industrial incidents of this scale, one of the major investigatory challenges that must be quickly faced is to organize and conduct a thorough and timely investigation in an environment where multiple parties, both governmental and non-governmental, assert investigatory authority or a right of access. Although all of the parties claim to pursue a common goal of determining "the cause," competing and sometimes conflicting bases of legal authority, interests and expertise present special challenges for investigation management. The aftermath of the Kinston explosion provides a good illustration of the challenges presented and the importance of managing all aspects of the investigation process.

Numerous governmental agencies, both federal and state, responded to the Kinston incident. Federal agencies included: the Bureau of Alcohol, Tobacco, Firearms and Explosives (ATF), the U.S. Chemical Safety and Hazard Investigation Board (CSB), the Occupational Safety and Health Administration (OSHA), the Environmental Protection Agency (EPA) and the Federal Bureau of Investigation (FBI). State agencies included: the North Carolina Department of Labor's Occupational Safety and Health Division (NCOSHA), the North Carolina Department of Environment and Natural Resources (NCDENR), the State Bureau of Investigation (SBI), the State Police, the North Carolina Attorney General's Office, and the Division of Emergency Management. In addition, numerous local law enforcement, public safety, and emergency management agencies responded.

Local public safety and emergency management officials planned their response in part because of initial reports that an airplane from a nearby airport had hit the Kinston plant. It was later determined that such an event did not occur.
At the request of local officials, ATF initially assumed the lead role in the investigation to determine if the explosion resulted from a terrorist act or some other criminal activity. In the initial week following the explosion, ATF's presence was dominant, and ATF proved to be the primary governmental authority to be dealt with not only because of its vigorous assertion of authority and site control, but also because of the sheer size of its investigation taskforce – several dozen ATF agents were involved in the onsite investigation. Once ATF concluded that the cause of the explosion was accidental and not the result of terrorist or criminal activity, ATF ended its onsite investigation, and the CSB became the primary federal agency investigating the cause of the explosion. (Because North Carolina has a State OSHA program, federal OSHA served the limited role of providing North Carolina OSHA with technical support.) The CSB established an onsite trailer for its investigation team and maintained an onsite presence for approximately four months. The EPA also established an onsite trailer for several months to evaluate, monitor and coordinate environmental activities in coordination with NCDENR.

On the state side, NCOSHA played the lead investigatory role. NCOSHA established a mobile command center onsite, and its investigation team remained onsite for approximately four and one-half months.

Although these various agencies share substantial common governmental and public interests, they all have their own specific statutory missions and each possesses unique statutory authority to investigate various aspects of incidents such as the Kinston explosion. Therefore, West not only had to coordinate many basic investigation activities with multiple governmental entities, it sometimes also had to deal with investigatory jurisdictional and similar issues among the various state and federal agencies concerning site access, interviews, document collection and production, the timing and sequence of investigation activities, and examination, identification, collection, preservation and analysis of evidence.

In addition to the need for West to cooperate and coordinate with multiple governmental agencies, further substantial investigatory challenges were presented by the large number and various types of third parties asserting a right to conduct investigations. West's insurers and their counsel undertook investigations and put close to 100 companies -- companies that were either involved in the construction, renovation, or maintenance of the facility, as well as companies that had supplied materials or equipment to the plant -- on notice of possible claims against them. As a result, dozens of these companies sent investigators to the site. In addition, counsel pursuing a class action against West based on alleged effects from the explosion made visits to the site, as did counsel representing workers who had been injured or killed. In addition to demanding access to the site, third parties also requested information including documents and employee interviews.

Managing the sheer volume of site access and information requests from all of these governmental agencies and third parties, and coordinating the requests so that West's own investigation could proceed in a timely and thorough manner, required that investigation and information management systems be put in place by West. Important aspects of these systems are described below.

Site Security and Access Control

In the immediate aftermath of the explosion, ATF and local law enforcement personnel restricted access to the entirety of the West property. Once ATF relinquished control of the scene, West, in cooperation with CSB, EPA and NCOSHA, adopted additional site security and access measures. Persons with a need to be on the site for investigation activities were identified on and cross-checked against a roster of authorized personnel. To narrow the property's restricted zone, fencing was erected around the damaged facility and the immediate surrounding area where trailers for the various investigation teams were established. Access into the fenced area was controlled by a guard who cross-checked entrants against the authorized roster. Access into the damaged building itself and its immediate surrounding areas was further restricted through an entry system administered by a designated site access coordinator. (See discussion on Site Safety and Access below.) Only authorized persons with a proven need for access who had completed a site safety briefing and, in light of the extensive damage to the building, who had executed a liability waiver and release (except for government officials), were permitted to enter. Entries were performed after a safety assessment that included a structural integrity assessment, air monitoring and a personal protective equipment (PPE) assessment. All entries were under escort by a designated West representative, and radio communications were maintained between the escort and a site access coordinator.

Protecting Trade Secrets and Other Confidential Information

West's manufacturing process involved substantial trade secret, business confidential and proprietary information. Government employees are obligated under federal and state criminal statutes to not disclose trade secret or confidential business information. These criminal statutes, however, do not apply to non-governmental third parties. Therefore, before West provided third parties with access to the facility, information obtained from documents, or to West employees, West required the third parties and their representatives to execute confidentiality agreements protecting against the disclosure of confidential information. Also, any document containing trade secret or business confidential information was specifically identified as such by West.

Document and Interview Requests

A formal system for handling document and interview requests was established. All requests for documents or interviews, whether by government agencies or by third parties, were required to be in writing and to be submitted through West's counsel at their onsite trailer. All document and interview requests were logged in. All documents provided in response to the requests were identified with a numbering system for each party and were specifically identified as containing trade secret or business confidential information. All documents provided were subject to the terms of the confidentiality agreements entered into by the third parties. During the investigation phase, a total of over 32,000 pages of documents was produced to government agencies and third parties. A schedule for interviews was established so that they could proceed in an orderly fashion and be coordinated with other investigation activities.

Access to Evidence

Measures were taken by West to ensure that all investigating parties had access to any area in the facility or item of interest to them, and to ensure that evidence was preserved. Steps taken to restrict access to the site and to preserve areas, materials and equipment have been described above. Any investigating party had the opportunity to request access to specific locations and to request that specific materials or equipment be preserved. When tests of equipment were requested or planned, written work plans and protocols were developed. Interested parties had an opportunity to review and comment on proposed plans and procedures, and to observe, document, and photograph or videotape the implementation of the protocol. After in situ examination, materials or equipment of interest to any investigating party were identified, tagged and logged into an evidence inventory. A master log of retained evidence items was maintained and made available to investigating parties. Access to the retained items was provided upon request. Any potentially destructive examination or testing of materials was done only after notice to interested parties and development of an agreed upon protocol.

Dealing with Multiple Requests for Materials in Limited Quantities

Some of the substances in the plant that were of interest to investigating parties were available, after the incident, in only limited quantities. Parties were requested to specifically identify to West any material where a split sample would be necessary. If sufficient material was available, split samples were provided. If sufficient material for split samples was not available, all parties interested in the material were requested to identify the tests and analyses they wanted to have performed on the samples. Testing protocols and a testing laboratory were then agreed upon, and the results and costs were shared among the interested parties.

Safety and Site Access

Safely accessing the major damage zones was a significant issue in managing investigation activities. Access was greatly restricted due to the damage from the explosion and ensuing fire. The "Tower" area that housed rubber compounding and the immediate surrounding rooms were completely covered with debris and were wet from fire water and rain. The only access was on foot and required walking on top of debris. Potential overhead hazards prohibited access to some areas until the hazards could be inspected and, if necessary, removed or secured. Stairs to the second floor were destroyed in the explosion, preventing access to the mixers and offices until ladders could be safely positioned and landings created at the top of ladders. The roof structure sagged or collapsed in the Raw Materials Warehouse, necessitating some demolition before some areas were accessible.

The approach taken to allow inspection, documentation and evidence collection was to inspect and document highly damaged zones from the perimeter, mitigate hazards in selected locations to allow access further into the damage zones, followed by more extensive documentation from the "safe" locations. The work was scheduled in stages so that investigators had the opportunity to inspect and document the scene from safe locations, then investigators would make subsequent entries for more extensive inspection once areas had been rendered safe. Once all areas were made accessible and investigators had the opportunity to complete their documentation, more extensive debris removal began in order to search under the debris for evidence. The site was divided into zones, and each zone was searched as debris was cleared. The process took several months before all zones were fully accessible.

Debris, damaged structural components, and other obstructions required to be moved for safety reasons were retained close to their original location or relocated to a lay-down area. After a short period, the ground around the building became so congested with removed materials that investigation operations were being impeded. A protocol was developed and agreed to by all involved parties concerning debris that was to be retained and debris that could be disposed of. The only materials to leave the site were items agreed to in the disposal protocol or retained in evidence under chain of custody.

Several retention areas and an evidence cage were created. Once equipment and product had been cleared from portions of the facility not involved in the explosion and fire, the available space was turned into an indoor material and evidence retention area. A chain link fence was erected to create an evidence cage. Open floor space was marked in a grid that corresponded to the building zones, and retained materials were moved to the corresponding grid as zones were cleared of debris. Pieces of equipment and large debris were placed in lay-down areas.

Site access for all activities was through an entry permit system with an escort in radio contact with the site access coordinator. Access to the site was allowed only after legal entry formalities were cleared and a safety briefing was attended. Due to the large number of parties involved, entries were conducted in groups with each group having an escort.

Absence of Residual Dust from the Dust Explosion

The West investigation team ultimately proved that the explosion was a dust explosion involving polyethylene dust. Not until several months after the incident, however, was conclusive forensic proof found. Unlike most dust explosions, there were no residual accumulations of combustible dust, even in protected locations like flanges of I-beams supporting the second floor. Moreover, housekeeping in the facility was extremely good with dedicated cleaning crews working on all shifts, every day. All witness interviews plus pre-incident photographs verified that work areas were kept very clean to meet stringent customer and FDA requirements for the pharmaceutical industry.

Because the area above the suspended ceiling in the ACS was not a work area, however, it was not viewed as requiring a similar cleaning regimen. Some deposits were known to be above the suspended ceiling as a result of the separation of a pneumatic conveying line for a non-combustible raw material and the migration of clay dust from outdoor storage silos due to negative building pressure. The plant was continually working to improve dust control and had replaced the original dust collection system about a year before the incident. The belief of plant personnel was that the material above the suspended ceiling was non-combustible. Post-incident tests of solids collected in the dust collectors supported this belief, as the mixture of dusts in the dust collectors were, in fact, non-combustible.

Some investigators drew the conclusion just days after the incident that the explosion was a dust explosion based on the *assumption* that the dust above the suspended ceiling was a combustible dust. In the ensuing months, however, they found no combustible dust above the suspended ceiling in the ACS. Not until West's investigation team found a coating of melted, resolidified polyethylene on structural beams supporting the second floor of the ACS, was actual presence of combustible dust above the ceiling positively identified. Polyethylene dust that had not been consumed in the explosion had melted and deposited on various surfaces as a solid coating with a light tan appearance. The coating was unremarkable, easily mistaken for a dirty surface caused by concrete dust, soot from the nearby fires, stains from fire water and rain, or other contamination sources.

Direction of Propagation

The directions of flame and blast wave propagation were investigated to understand damage patterns and provide insight into initiation of the explosion. The Kinston incident posed challenges because the location of highest damage was near the opposite end of the room from the initiation source, and most investigators focused their efforts in the zone of highest damage. As the discussion below points out, it is incorrect to assume that the location of highest damage is where the event initiated.

Observations at the site were used to assess the direction of the blast forces. It is important to note that blast forces on building surfaces are applied perpendicular to the surface of the component, which may not be in the direction of the blast propagation. Therefore, it is necessary to consider several damage indicators to determine the direction of the significant explosion or explosions. Indicators include structural deformations, debris throw and equipment movement.

Figure 1 shows lateral bracing and a floor beam east of Mixer 1 that was deformed eastward. It is interesting that none of the other mezzanine floor beams in the ACS and none of the mezzanine columns were deformed along this column line. The apparently localized damage to the bracing and floor beam led some investigators to early conclusions that the explosion was “seated” just to the west of the bracing. Once drawings were reviewed and plant personnel interviewed, it was determined that there was a concrete block wall along this column line was driven into the bracing by the blast. The presence of the wall increased the out-of-plane loads carried by the bracing, which was then transferred to the floor beam, causing the larger deformations in this area. Thus, the localized damage was caused by the structural configuration, not an adjacent localized explosion.

Figure 1. Out-of-Plane Displacement of Steel Bracing

Figure 2 shows an interior wall north of Mill 1 that is deformed to the north. In the area to the south and east of Mill 1, there was evidence that a significant amount of debris was moved during the emergency response operations and initial investigation by ATF. Therefore, the investigation team could not use the location of the debris in certain areas as evidence of the blast direction. However, items that were supported at higher elevations could be used as reasonable directional indicators.

Figure 2. Deflection of Interior Wall Northward

Figure 3 shows a second level floor slab that uplifted significantly and separated from the floor framing. The mixer floor slabs immediately adjacent to this area suffered no damage. It is important to determine the actual structural conditions before drawing any conclusions from this type of evidence. Most floor slabs are not designed for upward loads; there may be no reinforcement for uplift and there may be little or no attachment of the floor deck to the floor framing. A static load of only about one-half psi would have lifted the second floor slab. The floor slabs at the mixers were heavily reinforced due to the heavy equipment loads. Without proper analysis, one could inaccurately determine the explosion source and magnitude.

Figure 3. Uplift of Second Level Floor Slab

Building debris can also be used as an indicator of the blast direction and magnitude. An interior wall panel found in Mill 1 (see Figure 4) shows that blast forces in this area were directed northward. Locating or determining the origin of pertinent pieces of debris can sometimes be a challenge. Figure 5 shows the area west of the mills, where there was a significant amount of debris. The doorframe shown in this figure originated from a wall along the west side of the mixers.

Figure 4. Debris Thrown into Mill 1

Figure 5. Building Debris West of Mixer 1 (note doorframe)

Use of the information discussed above and many additional damage indicators allowed the West investigation team to develop a vector diagram (Figure 6) showing the direction of the blast forces throughout the building. As shown, the center of the explosion might seem to be in the area just south of Mill 1. An early witness report indicated that the event originated in this area as well. However, the investigation team was unable to identify critical evidence that would support the hypothesis that the initiating event took place in this area. Later witness reports indicated that the explosion originated north of Mill 2, at the opposite end of the room from Mill 1. In addition, all personnel working in the first floor mill area were knocked down or thrown to the south as shown in the personnel vector diagram in Figure 7 before major damage occurred. All of these witnesses reported an initial sighting of flame to the north of their positions. Detailed inspection of the equipment and ductwork in the area north of Mill 2 confirmed these accounts. Therefore, it was shown that the event initiated in the area north of Mill 2 and the explosion propagated to the south, where the highest blast pressure and most significant damage occurred south of Mill 1.

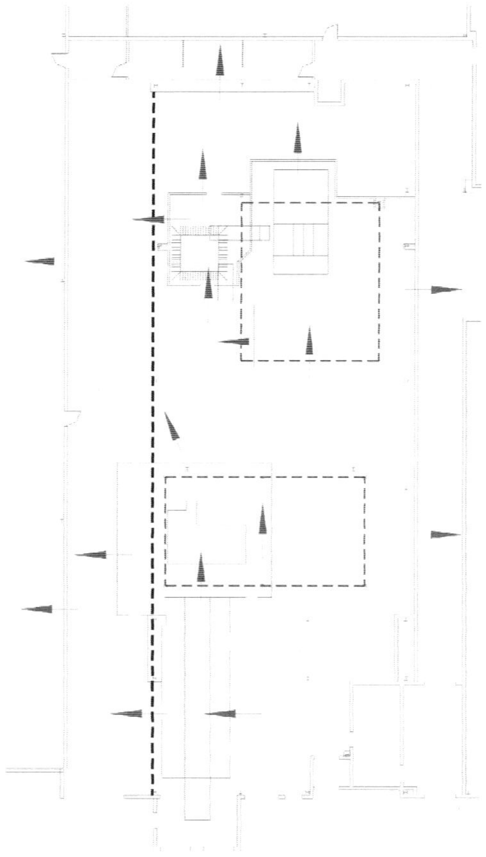

Figure 6. Vector Diagram Showing Direction of Deformation and Debris Throw Around ACS

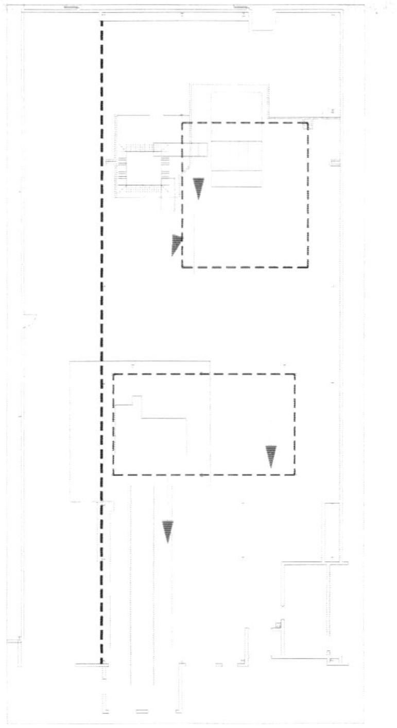

Figure 7. Vector Diagram Showing Movement of Personnel During the Explosion

Determining Process Status

Considerable effort was expended to determine the status of rubber compounding operations at the time of explosion. Rubber was produced in batches, and there were a number of changes each day in the rubber formulations being compounded. The investigation team wanted to determine the actual batches being processed in the mixers and the status of the mixers. Electronic process data was the target of initial efforts. Programmable logic controllers (PLC) controlled processes and an on-site computer server recorded process data. The computer servers were recovered as quickly as possible. Unfortunately, the computer room suffered blast and fire damage, and the hard drives were too badly fire damaged to recover any data. PLCs were recovered next. While the PLCs did not store process data, they contained process status information in memory. However, recovery and interrogation of the PLCs was delayed for approximately 4 weeks because they were only accessible by crane and other site and investigation constraints hampered the process. When the PLCs were finally able to be recovered and interrogated, it was found that their batteries had been depleted and any available memory had been lost. This underscores the urgency of recovery of any available process data in accident investigations. High priority should be given to rapid recovery and prompt interrogation of process control and data storage systems as delays may result in loss of available data.

Lacking control system data, the investigation team obtained information from a number of sources to determine process status. The information sources included:

- Witness accounts
- Batch scheduling documents
- Laboratory QA samples
- Process equipment and batch position
- Switch positions
- Samples of batches on/in each piece of process equipment
- Recovery of raw materials staged on conveyors for upcoming batches
- Recovery of product already processed

Gathering of the information took considerable time as much of the physical evidence was inaccessible due to safety concerns or being covered with debris. Witness accounts were particularly problematic. None of the injured workers' recollection of batch and equipment status fully matched the physical evidence, and in many cases differed significantly. Also, their recollection of timing of events was inconsistent or events were recalled in an incorrect order. The need for physical evidence to corroborate witness accounts is always desirable, but it proved to be essential here.

Determining Initiation of the Explosion

Initiation of a dust explosion requires that combustible dust be in suspension in air at a concentration above minimum explosive concentration (MEC) and be exposed to an ignition source with sufficient energy to exceed minimum ignition energy (MIE). A systematic approach was taken to search for initiation sources that satisfied these concurrent requirements. In addition, forensic evidence was sought to provide proof of initiation. For the sake of brevity, discussion is limited to the final result.

Initiation occurred inside ducts that provided cooling air to electric motors driving Mill 2. The ducts were at the north end of the room, consistent with the location discussed above based on vector diagrams and witness accounts. The physical evidence that identified the initiation source included outward bulging of the cooling air inlet header, presence of melted and resolidified combustible dust deposits inside the inlet header and smaller ducts leading to the motors, and melted and resolidified combustible dust deposits inside the motors including on the commutators and brushes. The ducts and motors are shown in Figure 8.

The headers were located above the suspended ceiling, the location where dusts had accumulated including some polyethylene dust. Careful inspection of the inlet air header revealed that it had split open in a number of locations from internal pressure and discharged material through these splits. Analysis of the motors determined that normal arcing at the brushes exceeded MIE.

Figure 8. Cooling Air Ducts for Mill 2 Motors with an Internal View of the Bulged and Split Inlet Header

The sequence of events was as follows. Fugitive combustible dusts accumulated in the cooling air inlet header and ducts. The dusts ignited at the brushes of a Mill 2 motor, and the flame propagated upstream in the riser to the header. The deflagration overpressured the inlet header, discharging pressure and burning material above the suspended ceiling, thereby initiating the dust explosion in the room.

Conclusions

The following conclusions concerning investigation methodology are drawn as a result of the challenging investigation at the West Pharmaceutical Services Kinston plant.

- Understand that numerous parties, both governmental and non-governmental, will assert a right to conduct a site investigation. Clear lines of communication should be established with government agencies asserting authority over the accident scene.
- Site security and access systems must be implemented in order to preserve physical evidence. These systems should include procedures for providing access to retained items by all interested parties, agreed upon testing protocols, and preservation and sharing of materials or analytical results where quantities are limited.
- Safety and site access pose major challenges at accident sites having a high degree of damage. Managing the accident site to allow access to investigators for inspection and documentation before, during, and after structures are secured and debris is moved may have to be accomplished in a staged approach by zones.

- A document control system, including collection, retention and production procedures should be implemented so that documents are preserved and made available, and to ensure that trade secret or other valuable proprietary information is protected.

- The location of greatest damage may not be where an explosion initiated. Moreover, the apparent direction of blast wave propagation may not be indicative of the direction of flame propagation in a dust explosion. Evidence of the early stages of the event is needed to identify the location of initiation.

- Interpretation of damage patterns requires knowledge of dynamic response and information concerning the original configuration. Some damage that appears to be the result of localized blast phenomena may simply be due to the structural configuration. Misinterpretation of such indicators can lead to incorrect conclusions about the location and nature of an explosion.

- Witness accounts, particularly from injured persons, may, and frequently do, contain inaccuracies. These witnesses may convey a high degree of confidence in their recollection, but the physical evidence is often at odds with their recollection. Other segments of their account may be accurate. While witness accounts must be considered, independent corroboration by other witnesses and, in particular, physical evidence is needed to confirm or discount the accounts.

Investigating Explosion Incidents – What Do You Look for and What Does It Tell You

Darrell D. Barker, P.E.
ABS Consulting

ABSRACT

Explosion incidents can have devastating effects on personnel and equipment leading to loss of life, severe injury, and business interruption. It is important to learn from these incidents to eliminate further occurrence. To understand the proximate cause or causes, it is almost always necessary to reconstruct the sequence of events leading to and through the explosion. Following an incident, we must rely on the evidence available to shape and validate or disprove our theories for cause and sequence. There are quite often conflicting eyewitness observations that must be squared with forensic evidence requiring a systematic approach to analysis so that we arrive at correct conclusions. Collection and preservation of this evidence is critical to the investigation process and must consider a number of potentially conflicting priorities including safety, restarting operations, and litigation issues.

This paper describes some of the key considerations for collecting data including what to look for and where to find it, systems for preserving integrity of data, legal considerations, and materials testing. An approach for analysis of explosion damage data and other forensic evidence is also presented with discussion and examples on how this information can be used to determine causes as well as rule in or rule out claims of damage. Particular attention is paid to quantification of explosion energy through damage analysis. This part of the incident analysis often plays a critical role in prevention or mitigation of future incidents.

INTRODUCTION

Processing hazardous materials brings with it an inherent risk of accidental release, often leading to severe consequences for life and property. This risk is often ignored or at least minimized by companies and individuals until an accident occurs and they are reminded of the devastation that can happen. The best approach to dealing with this risk is to minimize the potential for an accidental release. This can take two general forms: 1.) reduce the probability or 2.) reduce the consequence. The former is achieved through a through understanding of the process, an effective and proactive maintenance program, proper procedures, and most importantly, a workforce which views safety as a priority. The latter is addressed in a number of ways including reduction of inventories, separation of personnel from hazardous areas, and strengthening of buildings and equipment to minimize damage.

Even when these steps are taken, accidents can still occur. Petrochemical plants are complex facilities which present a tremendous challenge to eliminate all potential releases. For this reason, it is imperative that plant operators have in place emergency response plans to deal with potential accidents. Analysis of hazards is a key step in understanding potential consequences which is necessary to develop response plans and determine necessary equipment. An often forgotten piece in the response plan is preparation for incident investigation. This investigation is a required part of the process safety management (PSM) program but it should be more than just a compliance exercise. Accidents offer tremendous potential to learn more about the process and present opportunity for improvement, not only for the type of incident that occurred, but for similar process and hazards.

The Investigation Team

Incident investigations which are developed in a reactive mode following an accident are rarely effective and are destined for poor performance. The exception occurs when the appointed lead and team have prior experience and have learned the pitfalls firsthand. A better approach is to have trained incident investigation leaders and team members in several groups across the facility who can be called on to participate immediately following the incident.

Some considerations to keep in mind when appointing the internal investigation team include:

- Analyzing the aftermath of an accident requires a logical and systematic approach. Team members who can work within an established system while maintaining the ability to think beyond normal operations will be key assets to the investigation. Ideally, team members are selected in advance and observed to have these traits.
- The most knowledgeable personnel may be the focus of investigations by plaintiff's attorneys, state or federal safety officials and perhaps law enforcement. Having these personnel lead or participate may adversely impact the performance of the internal investigation team. Consultation with defense counsel is recommended to avoid this impact.
- Team members must often be relieved of all operations duties for at least some period of time which may range from a few hours to many months.
- Determining cause typically requires knowledge of systems and procedures involving many groups or departments. Personnel with a long history at the plant are usually much better at guiding the investigation than a newcomer.
- Team members and especially team leaders must be cognizant of the requirements for data and evidence preservation required by other investigation teams. Failure to do so can lead to charges of impeding and circumventing investigations and lead to a host of legal troubles.

One team member should be designated to manage safety aspects including attending all safety meetings, advising team members of hazardous materials areas, coordinating any necessary safety unit training, and insuring adequate personnel protective equipment is provided for the team.

Another team member should be designated as the lead for data collection and preservation including electronic information, physical property, chemical inventories, and personnel information. This person needs to be detail oriented and able to make logical decisions quickly.

First Things First

Once the investigation team is formed, an initial meeting is critical to lay out the focus of the investigation, assign roles to the members, review critical rules and procedures for data collection, and address safety. Team members should develop an initial task list which is continuously updated by the team leader. It is important for the team to get a first hand look at the site as early as possible to get initial impressions of the consequences and to scope safety and logistic issues to be addressed.

Large explosions produce tremendous damage and the challenge of finding critical evidence can be daunting. In general, debris of most interest is that originating at the source of the explosion. In many cases, these will be small fragments located far from the source which means the debris collection process must include more than the immediate area. Piping which has ruptured is of particular interest as well as fragments from reactor vessels. Starting form the center of the explosion, the collection should extend radially. The collection process should be systematic which utilizes a grid system overlaid on a site drawing to map debris pieces. A unique identifier must be assigned to each piece and denoted on the site plan. It is also helpful to coordinate photographs with the site plan.

A key decision to be made early is where to store data and property. Some forethought is required here because the storage area may be required for months or perhaps years with minimal access for normal operations. This area must be secure and adequate to handle all of the items that may be collected. It must be controlled access with logging of personnel entering the area to satisfy legal and law enforcement personnel that data is preserved and secure. A protocol for entry and removal of property must be agreed upon by all parties which require access. Since each piece of data or property is likely unique, sharing must occur but is often a source contention with competing interests by the various parties. Tracking of data and property is absolutely necessary so that they can be located at any time. This system should at a minimum, consist of a database with evidence description, unique identifier, location from which it was recovered, date and time of removal and return to the storage area and signature of the party involved in the removal as well as the person physically controlling the access. Data and property should be adequately marked with the unique identifier in such as way that it does not impact analysis or testing of the items.

Photographic evidence is a critical piece of every investigation. It is recommended that one person be designated to control collection, preservation and dissemination of photographs including prints and negatives. Date and time stamps are key requirements for all photographs to establish chronology. Film cameras are advisable if the photographs are to be used by any of the legal team to avoid the potential for manipulation of photographs that can easily occur with digital media.

What's the Point

The objective of the incident investigation team is to determine first the proximate cause and secondarily the root cause. Determining proximate cause will focus on physical evidence and developing an understanding of the sequence of events including initiation of reactions, equipment performance and failures, material release and knock-on events. Often this will require many of the team members to quickly learn a process which is unfamiliar to them. Team members should be prepared to devote a good deal of time learning the process, normal and abnormal events, control systems and equipment performance. They will also have to look for clues without knowing precisely what they are looking for.

Finding the root cause is a more wide ranging process which goes beyond physical evidence and examines maintenance systems, procedures, training, and communication. Establishing root cause usually has a much greater potential to improve safety and reduce future incidents because it potentially affects many more operations and processes within a facility than those involved in the incident at hand.

An example of proximate and root cause is demonstrated from a recent incident investigation which involved rupture of a product storage tank in a refinery. The rupture occurred when hot asphalt was introduced into a tank with a heel containing a high percentage of water. Flashing water to steam caused a pressure buildup which could not be relieved by the small vents leading to a rupture of about 25 percent of the tank perimeter at the roof line. Examining only proximate cause would lead to a recommendation to insure product heel is removed prior to filling a tank. During root cause investigation, it was determined that the tank had been used to hold products other than asphalt which gave rise to the water content. Operators controlling product movement and storage had no guidance on tank selection other than capacity. There were no procedures for assessing consequences of storing different products or for ensuring tank contents were clear.

It was further discovered that the product transferred to the tank was well over the prescribed temperature due to faulty coolers which had apparently functioned inadequately for months. These temperatures were duly noted in log books but there was no mechanism for these to be reported and tracked. Notifications to maintenance fell between the cracks and no repairs to the coolers were made. Without correction of these root causes, further accidents were certain to happen, potentially involving more than the product storage tank. The root cause analysis in this case had much more potential for improving safety but would not have been determined without first understanding the direct causes and events of the accident.

Unfolding the Truth

The general opinion of most people who haven't been involved in accident investigations is that eyewitnesses offer the most reliable source of information to determine what happened. Unfortunately, this is not always the case. Because people are not expecting the incident to occur, they aren't paying close attention to much of the key information necessary to arrive at the sequence of events and potential causes. Their statements may directly conflict with hard data from control systems and other physical evidence. The reasons for this vary widely but the investigator must carefully sort through witness accounts and avoid being thrown off track by inaccurate information. A witness' sense of timing is often the most inaccurate such as the time between or the number of explosions. It is critical to obtain eyewitness accounts as soon as possible after the incident because their remembrances can change over time as they are exposed to the recollections of other witnesses and observations of the accident site.

One useful technique to sort out the eyewitness testimony is to prepare a site plan which tracks witness locations over time as shown in Figure 1. It is necessary to establish timing which is often based on the first sound or on the most significant sound. Once their location is established on the site plan, which includes buildings and equipment, it can be used to determine why certain witnesses didn't see or hear certain events due to intervening structures or perhaps noise of rotating machinery.

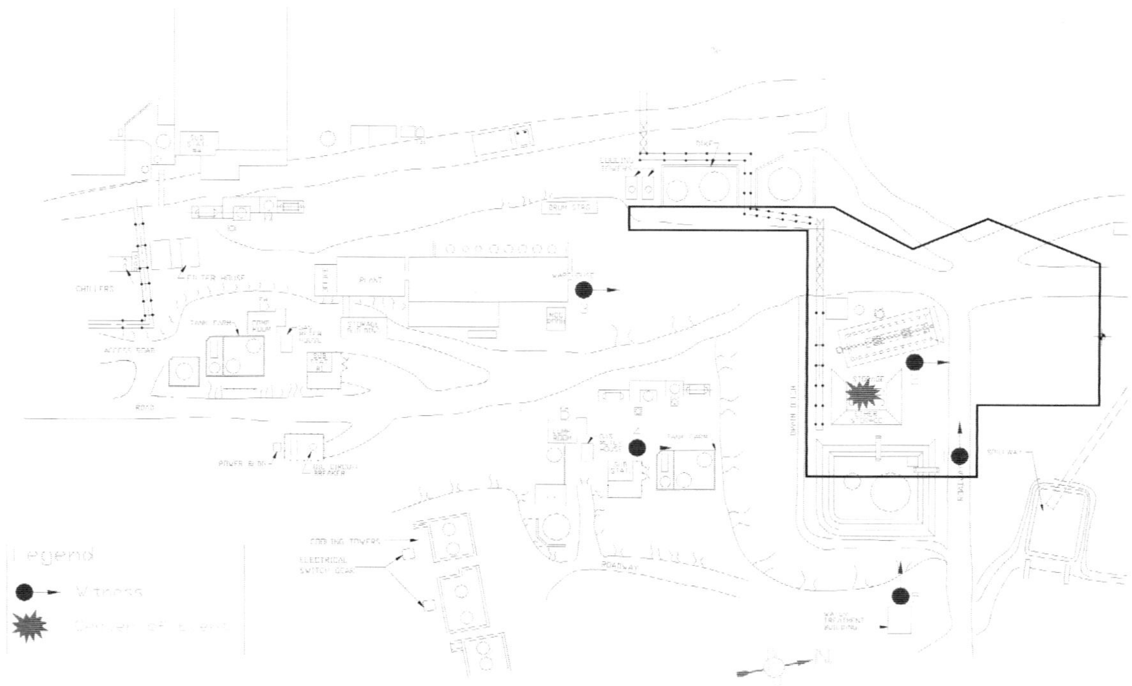

Figure 1 Site Plan with Witness Locations

Testing Issues

Since physical evidence is typically the most reliable information, it is important to collect all items which may yield clues. The investigator must make key decisions on which items should be retained and which can be discarded. It is rarely feasible to retain all evidence, especially for large explosion accidents. There is pressure to get the accident site cleared so that repair can begin to minimize downtime. It's a good idea for the investigation team to walk through the site with cleanup personnel and tag items to be kept. The investigation team needs to document the "as found" condition prior to pieces being moved.

The investigation team must also decide what type and extent of testing that should be conducted. Perhaps the most common are chemical analysis of residues on equipment, especially vessels and piping. This testing can be valuable in discerning the type of reactions that are occurring at the time of the accident. Test results must be viewed in conjunction with an understanding of the process to distinguish between normally occurring residues in the process and abnormal chemical compositions.

Mechanical testing of piping and reactor vessels can also offer clues as to the speed of reaction which can help distinguish between a slow buildup of pressure and a detonation reaction. Fragment size is normally a good indication of the intensity of the explosion. Small fragments are produced by detonation reactions while large fragments are indicative of an overpressurization from a slowly developing reaction. Establishing wall thickness is extremely important because it helps define the condition of materials and equipment prior to the incident. Many releases occur due to premature thinning caused by corrosion, often in inaccessible areas during normal operation. Thickness testing should be done on components beyond the obvious areas of release to establish the effectiveness of the maintenance and quality control systems to prevent accidental releases.

Explosion Energy Analysis

It is often necessary to establish the source energy of the explosion. This information can be used to help understand the types and quantities of materials involved and establish sequence of events in some cases. It is often used to help determine the validity of offsite damage claims. These claims are sometimes made by property owners seeking to receive compensation even though their property was not impacted by blast loads. Establishing the source energy allows the investigator to determine what the most probable and maximum credible blast loads were at the offsite property during the event and determine if the damage is blast related.

The usual procedure for the energy analysis is to collect structural damage information within plant boundaries. This will be the most reliable area to collect evidence because the condition of the buildings and structural information from drawings is often available. Permanent deformations are measured for each component along with span, support condition, loaded area, and material thickness. An example component is shown in Figure 2. The crease along the mid span line indicates the component stress at the critical location is just past yield. This information is used to establish a structural model for dynamic analysis. A range of blast loads are applied to the numerical model to determine what load is needed to produce the permanent deformation observed. In practice, a pressure-impulse (P-i) diagram is usually made to rapidly evaluate a range of loads and establish minimum blast load levels.

Figure 2 Metal Wall Panel at First Yield

Once the blast load level is established, source energy can be determined based on this load and the distance from the explosion to each component. The output from this process is a collection of source energy and distance pairs. Table 1 shows an example of the results of the explosion energy calculation. Ideally, each source energy would be the same since the loads are all due to the same explosion, but in practice there is a range of energies due to a number of factors including soft supports which reduce damage, variations in material properties, shielding or reflections of the blast load, and errors in deformation measurements. The collection of source energies will normally form a band centered on the most probable source energy as shown in Figure 3. Energies much higher or much lower than the average can normally be explained by the potential problems mentioned above. The average value would be used as part of the event sequence and proximate cause analysis. Upper bound source energies might be used for predicting blast loads offsite for evaluation damage claims.

Table 1 Explosion Energy Analysis Results

ID	Component	Material	Span (in)	Ru (psi)	Xm	Distance (ft)	Energy (in-lb)	Confidence Level
1	Beam	Steel	120	5.2	4.2	75	6.59E+09	90%
2	Wall Panel	Metal	48	2.1	1.7	150	1.98E+09	60%
3	Column	Steel	96	6	3.4	65	5.93E+09	50%
4	Girt	Steel	180	3.3	6.3	250	7.25E+09	70%
5	Roof Deck	Metal	60	1.2	2.1	875	4.61E+09	95%
6	Girt	Steel	180	3.5	6.3	200	6.59E+09	80%
7	Column	Steel	108	6.3	3.8	450	4.61E+09	50%
8	Wall Panel	Masonry	60	2.1	2.1	1125	5.93E+09	70%
9	Roof Deck	Metal	36	1.5	1.3	135	3.30E+09	40%
10	Roof Deck	Metal	48	1.7	1.7	279	6.92E+09	95%
11	Girt	Steel	144	2.1	5.0	865	7.91E+09	80%
12	Girt	Steel	168	2.5	5.9	400	7.58E+09	85%
13	Beam	Steel	175	4.5	6.1	540	8.24E+09	75%
14	Column	Concrete	120	7.5	4.2	350	1.25E+10	50%
15	Beam	Steel	132	3.5	4.6	150	5.27E+09	95%
16	Beam	Steel	124	2.5	4.3	275	7.25E+09	90%
17	Wall Panel	Masonry	108	1.5	3.8	300	5.93E+09	100%
18	Roof Deck	Metal	36	2	1.3	455	7.25E+09	95%
19	Girt	Steel	120	2.5	4.2	295	6.66E+09	85%
20	Wall Panel	Masonry	48	1.2	1.7	780	7.25E+09	80%
21	Column	Concrete	108	4.7	3.8	597	7.51E+09	85%
22	Wall Panel	Masonry	96	1.4	3.4	205	6.72E+09	70%
23	Wall Panel	Metal	96	2.7	3.4	127	6.66E+09	90%
24	Column	Steel	96	4.1	3.4	86	6.26E+09	75%
25	Girt	Steel	120	3.3	4.2	340	6.66E+09	90%

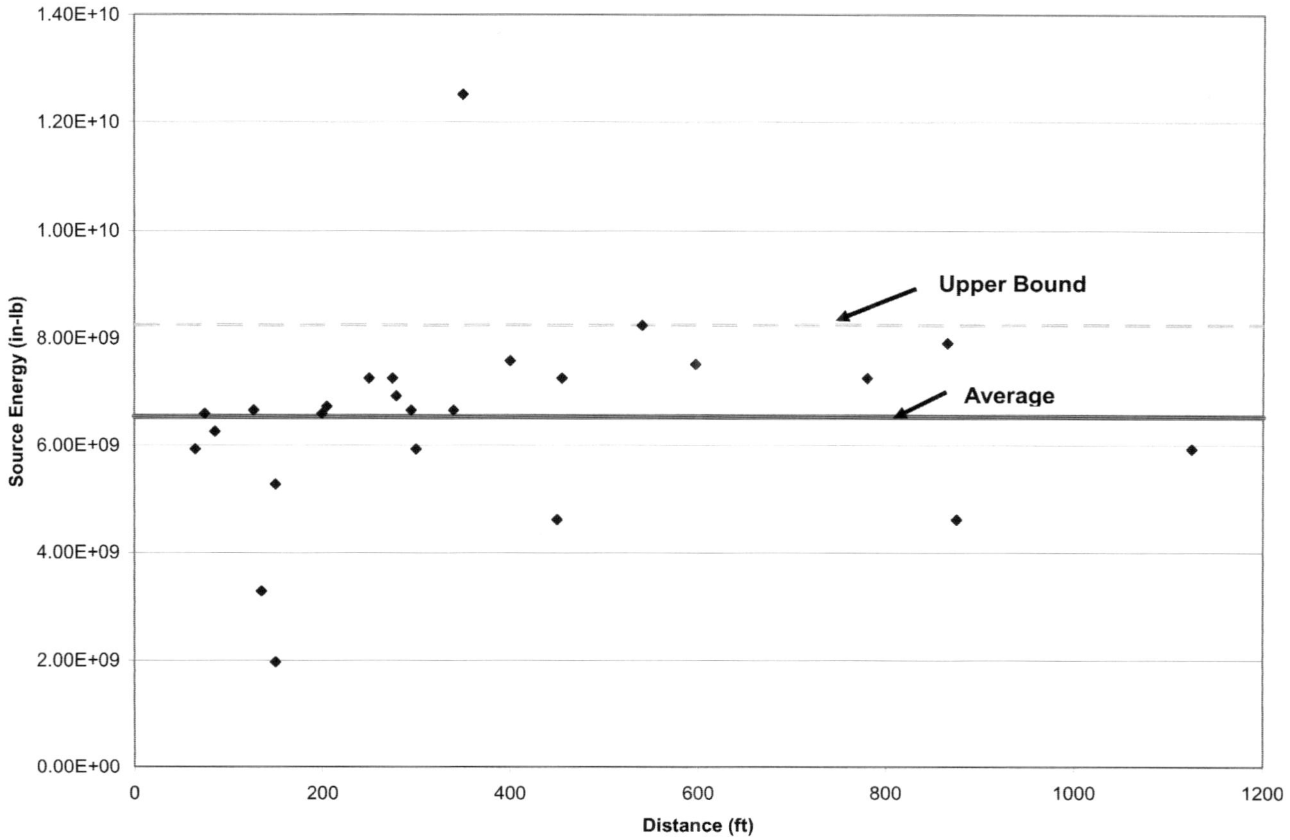

Figure 3 Computed Explosion Energy vs. Distance

A potential problem with this process is damage which has occurred prior to the incident. A lot of incidental damage occurs on a large plant site and at times it is difficult to distinguish between explosion related damage and that caused by vehicle impact or other ordinary causes. Plant personnel may be able to help distinguish pre-existing damage.

The most accurate calculation of source energy is produced by components whose permanent deformation is just past the elastic limit. The component will react much closer to the numerical model in the elastic and elastic-plastic region. Finding components with slight creases at support lines are among the most reliable.

A Last Word

Explosion accidents offer unique challenges to determine the cause and sequence of events. The evidence is a large puzzle with many pieces often missing. A systematic, logical approach to data collection and analysis is required to successfully establish what happened. Prior planning including training of investigation team members will greatly improve the success of the investigation.

Dust Explosion Case Study:
How a "Wrong" Material of Construction Caused a Dust Explosion

Dehong Kong, Ph.D., CSP
Flammability Group Manager
Senior Process Safety Engineer
Chilworth Technology, Inc.
250 Plainsboro Road, Building #7
Plainsboro, NJ 08536
Phone: 609 799 4449
Fax: 609 799 5559
Email: dkong@chilworth.com
Web: www.chilworth.com

Prepared for presentation at
Process Plant Safety Symposium
American Institute of Chemical Engineers
2005 Spring National Meeting
Atlanta, GA, April 10 – 14, 2005

ABSTRACT

A chemical reaction process began with feeding a combustible fine powder into a reactor through a charging hopper. The reactor and the charging hopper were fitted with a nitrogen inerting system to prevent dust explosions. Surprisingly, a dust explosion occurred when feeding the powder into the reactor from a FIBC.

Onsite investigations found that the design of the inerting system allowed a sufficient amount of air to enter the charging hopper. There were no mechanical, electrical and thermal sources of ignition inside the closed type reactor-charging hopper system prior to the explosion. Electrostatic discharges appeared to be the ignition source; however, the ignition source could not be identified without further analysis through pertinent tests. Therefore, various dust explosion and electrostatic tests were conducted to determine the minimum amount of electrostatic discharge energy for igniting the dust cloud and the likelihood of various types of electrostatic discharges capable of igniting the dust cloud under the given material, equipment, and process conditions. The tests results show that the powder suspended in air at the optimal dust concentration requires less than 3 mJ of spark energy to ignite. A glass-line pipe at the bottom of the charging hopper was found to be the most likely ignition source. Propagating brush discharges on the glass lining of the pipe were produced in the laboratory.

INTRODUCTION

The incident occurred in a reactor system of a newly built manufacturing plant. As shown in Figure 1, the reactor is made from carbon steel and lined with glass, and the charging hopper is made from stainless steel and unlined. The charging hopper is designed for dumping a powder chemical from Flexible Intermediate Bulk Containers (FIBCs). The discharge chute at the bottom of the charging hopper consists of a flexible stainless steel hose lined with polytetrafluroethelene (PTFE), a ball valve lined with PTFE, and a steel pipe lined with glass.

A combustible fine powder material is used in the reaction and was supplied in FIBCs. The reaction process begins from feeding the powder material into the reactor. When dumping the powder from a FIBC, the FIBC hung on an unloader is lifted over the top of the charging hopper. The FIBC is sealed to the charging hopper inlet pipe by inflating a ring-shaped balloon, which embraces the charging inlet pipe on the charging hopper. A dust collector is fitted to the charging hopper for the purposes of either removing the air displaced by the powder or reducing the dust concentration in the charging hopper or both.

A nitrogen purging system is installed to reduce the oxygen concentration in the reactor and the charging hopper in order to prevent explosible dust-air mixtures in the reactor system. The reactor is purged with nitrogen and then is the charging hopper. The dust collector is on when purging the charging hopper. A pressure conservation valve is installed in the charging hopper to prevent damages to the hopper from an overpressure. When the nitrogen purging is completed, an operator opens the iris valve in the charging hopper inlet. Through the glove window in the charging hopper, the operator cuts open the outlet of the inner liner of the FIBC using a knife held in a tool rest inside the charging hopper. The powder falls by gravity to the reactor.

At the time when the incident occurred, the operator on duty was watching the bottom of the charging hopper through an observation window above the glove window. Suddenly, he saw a flame rising from the bottom of the charging hopper. The FIBC was approximately half emptied at this moment. He felt a pressure from the direction of the flame. Some witnesses recalled that there were three booms, suggesting that secondary explosions occurred. The charging hopper suffered damages from an internal pressure. Some of the window glasses in the building were broken. Some corrugated metal walls of the building were bulged or dislodged. The FIBC was burned. Powder materials were spread on the floor and machine surfaces.

The damages to the charging hopper, building wall and window glasses, the dislodgement of the FIBC, and the spreading of the powder, showed a typical damage pattern of fuel-air explosions inside process vessels. The reactor and its glass lining remained to be intact.

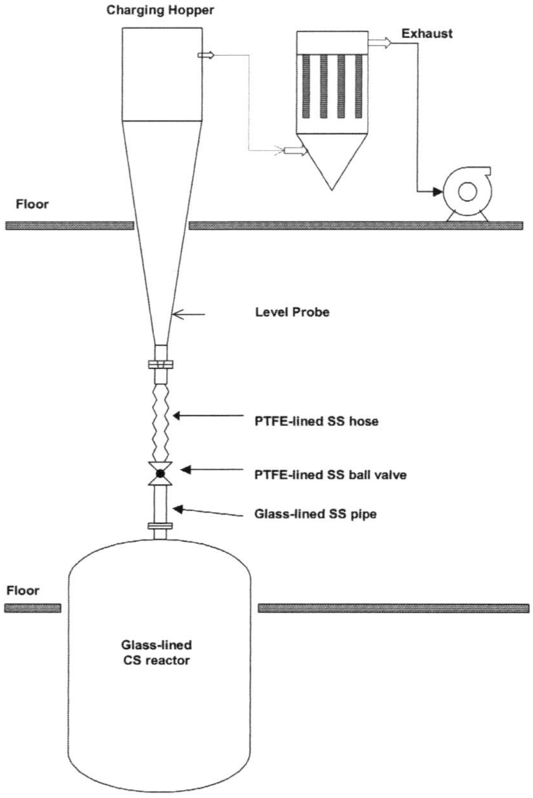

Figure 1 The reactor and powder charging hopper system.

ON-SITE INVESTIGATION

For safety and investigation purposes, the plant quarantined the area affected by the explosion after the site was secured. The investigation was conducted through a systematic collection of the evidence, measurements, and analysis of the evidence to identify the probable causes of the incident. The onsite investigation activities included:

- Inspection of the damages caused by the explosion to the equipment;
- Electrical measurements on the electrical equipment;
- Review of relevant documents about the process and equipment;
- Review of the available dust explosion hazards and electrostatic properties of the combustible powder; and
- Interviews of operation and engineering personnel.

The combustible powder was found to be the only combustible material that could be present in the reactor-charging hopper system at the time of the incident. Therefore, the explosion must be a dust explosion in the reactor-charging hopper system. For a dust explosion to occur, three elements need to be present at a location simultaneously:

- A combustible powder dispersed in air at an flammable/explosible dust concentration;
- An oxidant gas such as oxygen in air; and
- An ignition source of sufficient energy.

During the powder loading process, some portions of the powder can be dispersed in the headspace of the reactor and the charging hopper and the dust concentration can exceed the minimum explosible dust concentration (MEC). If a sufficient amount of oxygen is present, the dust-air mixture can be ignited by a sufficiently energetic ignition source.

The dust collector fan was pulling approximately 100 CFM, while the nitrogen purging flow was around 5.9 CFM. This mismatch would have caused a negative pressure in the charging hopper. The negative pressure would have initiated the pressure conservation valve in the charging hopper to open (set value is -2 inch water column). Consequently, a sufficient amount of air would have been drawn into the charging hopper, forming an ignitable dust-air mixture.

After the explosion, the plant personnel found some combustion products on the pile of powder in the reactor; however, their origin could not be identified. No damages to the glass lining of the reactor were detected using a spark detector. The operation documents indicated that the reactor was probably purged adequately and all the valves in the reactor were closed prior to powder loading. Therefore, it could be possible that the amount of oxygen in the reactor was insufficient to support a dust explosion and the combustion products found in the reactor were from the charging hopper.

As the source of fuel and air had been determined, the investigation was focused on identifying the source of ignition. Several common ignition sources were evaluated, including:

- Frictional sparks and/or frictional heating;
- Hot surfaces and self-heating;
- Electrical equipment; and
- Static electricity.

Since the powder feeding operation was done in a closed system, the ignition source would have been inside the reactor-charging hopper system. Moreover, the electrical equipment around the reactor system was properly rated. Therefore, the electrical equipment outside the reactor system could not be the source of ignition. Upon inspecting the equipment, it was found that there were no moving parts inside the reactor system except for the agitator.

However, the agitator drive motor, which is normally turned off during powder feeding, was not running at the time of the incident. The plant building had lightning protection system, and there was no lightning at the time of the explosion. The reactor system at the time of the incident was at ambient temperature. Therefore, the dust explosion could not have initiated by a mechanical ignition source, thermal ignition source, or lightning.

A powder level probe was mounted at the lower part of the charging hopper. It measures the powder level by detecting the change of Radio Frequency (RF) admittance due to the presence of a solid material. During the onsite investigation, a wiring problem was found in the powder level probe. A question was raised whether the wiring fault could have resulted in the dust explosion.

It was also noticed that the powder is probably an electrically insulating material, which might give rise to electrostatic discharges. In addition, the FIBC and the glass and PTFE linings can cause electrostatic discharges. Finally, the following elements were determined to be the potential ignition sources, which could give rise to electrical arcs or electrostatic discharges:

- The powder level probe which was improperly wired;
- The powder which is probably an insulating material;
- The FIBC which is fabricated from insulating materials; and
- The pipes between the charging hopper and reactor, which have insulating linings.

Notably, these potential ignition sources could not be confirmed or negated without further quantitative analysis.

FURTHER EVALUATIONS OF THE IGNITION SOURCES

It is stated in the Material Safety Data Sheet (MSDS) that the powder may form an explosible dust cloud under certain conditions. However, no explosion hazard data can be found from the MSDS. The dust explosion hazard properties of the powder need to be measured for the investigation.

Following the onsite investigation, the combustible powder, the level probe, the PTFE-lined stainless steel hose and the glass-lined pipe were tested to assess the likelihood of ignition by the potential ignition sources identified during the onsite phase of the investigation. The general methodology is as follows:

- Determine the minimum energy required to ignite the powder suspended in air at the most ignitable dust concentration;

- Determine the maximum energy that can be released by the individual potential ignition sources for the present dust explosion; and

- Assess the probability of a potential ignition source by taking into account of the present test results, process conditions, site inspection results, and witnesses.

A detailed account of the evaluations is given below:

Dust Explosion Properties

Based on the onsite investigations and reviews of available information for the fire and explosion hazards of the powder, the present dust explosion incident appeared to be caused by an electrostatic discharge. However, there were several possible electrostatic ignition sources in the reactor-charging hopper system. Each type of electrostatic discharge can release a certain amount of energy. To determine whether a potential ignition source could have initiated the dust explosion, the ignition sensitivity of the powder must be determined.

A dust-air mixture with a given dust concentration requires a certain amount of discharge energy to ignite. There exists an optimal dust concentration at which the dust-air mixture requires the least amount of energy to ignite. This minimum amount of energy for ignition is called the minimum ignition energy (MIE). To be capable of causing an ignition, the energy released from an electrostatic discharge must be greater than the MIE of the ignitable mixture.

The MIE of this powder was measured to be less than 3 mJ (**Table 1**). The MIE of a dust is affected by many factors, including the chemical properties, particle size, and moisture content. Commonly used powders usually have MIEs from 1 mJ to 1 J. For example, the published MIEs for aluminum powders are as high as 10 mJ and as low as less than 1 mJ. Clearly, the MIE of the powder involved in the incident is very low compared with the commonly used powders. Moreover, it is not much higher than the MIEs of vapors of common organic chemicals in air (usually between 0.1 mJ to 1.0 mJ [1]). As discussed below, the MIE of this powder is less than the potential discharge energies of some possible electrostatic discharges in this rector-charging hopper system.

Bulk powder can retain an electric charge for a period of time, called charge relaxation time, which is determined by the volume (electrical) resistivity of the powder and its dielectric constant. Electrostatic charging occurs on "high-resistivity" materials even when the material is in contact with ground. Such materials have volume resistivities above 10^{10} Ω·m. Examples include organic powders, synthetic polymers, and quartz. High-resistivity powders can produce electrostatic discharges such as corona, brush, bulking brush (cone), and propagating brush discharges. The chemical powder was found to be an electrical insulator. At an ambient humidity (53% R.H.), its volume resistivity is 7.2×10^{11} ohm·m and the charge relaxation time is 2.5 hours. When the air humidity is decreased to 21% R.H., its volume resistivity is greater than 10^{14} Ω·m and its charge relaxation time is increased to 5.5 hours.

The charge densities (charge to mass ratio) of the powder after flowing through a pipe of metal, plastic and glass were measured under relatively high air velocities (20 m/sec). The charge densities obtained under such conditions represent the "worst-case" charging level and can be used to show the chargeability of powders. At an ambient humidity (51% R.H.), the charge density of the powder was on the order of 10^{-6} Coulombs/kg (C/kg). When the humidity was reduced to 18% (R.H.), the charge densities were increased approximately 10 times.

Notably, the powder is an electrical insulator, on which a large amount of charge can be generated in powder handling processes. Moreover, the powder is very sensitive to electrostatic discharges.

Table 1 Ignition sensitivity and electrostatic property tests results

Parameters	Results
Minimum Ignition Energy (mJ)	<3
Volume Resistivity (Ω.m)	>10^{14} at R.H. = 21% (low R.H.); 7.2×10^{11} at R.H. = 53% (ambient R.H.).
Charge Relaxation (Decay) Time (Hour)	2.5 at R.H.=53% (ambient R.H.); 5.5 at R.H. = 21% (low R.H.).
Charge Density (Charge to Mass Ratio) (Coulombs/Kg)	At R.H. = 51% (ambient R.H.): -1.0×10^{-6} for stainless steel; -2.4×10^{-6} for plastic; -2.4×10^{-6} for glass. At R.H.= 18% (low R.H.): -1.6×10^{-5} for stainless steel; -1.6×10^{-5} for plastic; -8.3×10^{-6} for glass.

Level Probe

The powder level probe that is mounted at the bottom of the charging hopper consists of a sensing element (stainless steel rod) and a metal shielding element. The metal rod and shield are isolated from each other and from the steel casing (ground) by PEEK (polyether ether ketone) insulation. The resistance between the sensing element and the casing was measured to be 9.3 GΩ (9.3×10^9 Ω). The probe is inserted though a 2″-ID nozzle into the charging hopper and is mounted to the nozzle with a flange in the probe and an identical flange in the nozzle.

The level probe could produce two types of ignition hazards: electrical arcs from the electrical circuits and electrostatic discharges from the metal elements. The sensing element was connected to the electronic circuits through a spark protector. The spark protector protects the electronic circuits from having a high voltage on the sensing element. If the voltage on the sensing element exceeds a designed value, for example, due to electrostatic charging, a spark discharge in the spark protector will take place and the charge will be bled to the ground through the spark protector. Therefore, no high-level electrostatic charge can be retained on the sensing element.

The shield for the probe should be connected to the ground; however, after the dust explosion incident, it was found that the shield was ungrounded. In the laboratory test, the capacitance of the shield was measured to be 15.0 pF when the casing and the sensor element were not grounded; and 211 pF when the casing or the sensor element was grounded. Because the casing of the probe was grounded through the charging hopper the capacitance of 211 pF is a representative value. If the shield can be charged to 10 kV, it could retain 10.6 mJ, which is sufficient to ignite the cloud of the powder.

There is a cuff at the other end of the probe in which wires for the sensing element and the shield are enclosed. A laboratory test was conducted to measure the maximum potential that could be built on the metal elements of the probe. A corona probe attached to a high voltage source set at -30 kV was used to apply electrostatic charge to the shield and sparks were observed inside the cuff. Thereafter, it was found that the insulation between the sensing element and the shield was electrically broken down.

The gap between the electrical connector of the shield and the surface of the cuff was much smaller than the gap between the shield and the 2″-ID nozzle. Therefore, if the shield became electrostatically charged, it would give rise to a spark between the shield and the cuff rather than the gap between the shield and the nozzle. Since these discharges are confined inside the casing, they could not ignite the dust cloud in the charging hopper. Therefore, the shield element could not be the ignition source for this dust explosion.

FIBC

As the powder is dumped from the FIBC, electrostatic charge can be generated on the powder and the FIBC. The spout (inner liner) of the FIBC is not in contact with the charging hopper. If the charged spout comes into contact with a grounded conductor, such as the lid for the iris valve, an electrostatic discharge may occur. Usually, electrostatic discharges from insulating surfaces can release energy less than 4 mJ and these discharges are called brush discharges. If a brush discharge on the spout did occur, the energy of the discharge could have exceeded the MIE of the powder (< 3 mJ). However, brush discharges have not been identified as an ignition source of dust explosions in air in reported dust explosion accidents and laboratory tests [2].

In principle, a propagating brush discharge might occur when filling a large-size FIBC made from ordinary plastic materials with a breakdown voltage greater than 4000 V. However, the risk of having a PBD on the FIBC during the emptying is considered to be low. There is lack of firm evidence that PBDs have occurred on FIBCs [3]. Moreover, this hypothesis is not supported by the witness that the flame started from the bottom of the charging hopper suggesting that the ignition did not take place on the FIBC.

Powder Bulk in the Reactor

As shown by the laboratory test results, the powder will become electrostatically charged as it is dumped from the FIBC and falls through the charging hopper. As the powder accumulates in the reactor, an electric field is produced in and around the powder. If the amount of charge per unit mass of the powder is sufficiently high, the electric field strength may exceed the breakdown strength of air, causing electrostatic discharges on the powder surface. Such electrostatic discharges are called bulk brush (cone) discharges. The energy that can be released from a bulk brush discharge on a powder surface is determined by the particle size and the dimension of the powder bulk (vessel diameter). The larger the particles size and vessel diameter, the greater is the energy of a bulk brush discharge. Experience indicates that these discharges are not capable of igniting dusts having MIEs greater than 10 mJ [1].

The energy of bulk brush discharges can be estimated using the method of reference [2]. The diameter of the reactor is 8 feet (2.44 m). All the particles are smaller than 75 μm. If the median particle size is assumed to be 75 μm, the estimated energy of bulk brush discharges in this system is approximately 3 mJ. Since the median particle size is smaller than 75 μm, the discharge energy of bulk brush discharges would be less than 3 mJ. Therefore, the likelihood of ignition of the present dust explosion by bulk brush discharges could not be determined by the MIE of the powder and the estimated discharge energy of bulk brush discharges.

As discussed above, the reactor was probably purged adequately. Even if some amount of air were entrained into the reactor by the falling powder, the oxygen concentration in the reactor would have been lower than in air. The energy required to ignite the dust-air-nitrogen mixture would be greater than the MIE of the dust in air (< 3 mJ). Therefore, the likelihood of ignition by a bulk brush discharge in the reactor under this circumstance is considered low.

Pipes with Insulating Linings

When an insulating layer material is in direct contact with a conductive substrate, the electric field from the surface charge is mainly exerted through the insulating layer to the countercharge produced in the conductive substrate. Charge can accumulate on the insulating surface until electrical breakdown occurs through the layer. In other words, the surface charge density on the exposed layer surface is limited only by the electrical breakdown strength of the insulating material, which is higher than that of air. When a "short-circuit" of the two layers of charges of opposite polarities is achieved, or if a grounded metal object is brought into contact with the charged insulating layer, large surface potential gradients (electric field parallel to the surface) will be produced, which triggers massive lateral discharges to the point of the short-circuit or the point of the contact with the grounded metal object. This kind of electrostatic discharge is called propagating brush discharge (PBD). The energy released in a PBD has been stated as high as several joules.

At the bottom of the charging hopper, the powder passes through a PTFE-lined flexible stainless steel hose, a PTFE-lined ball valve, a glass-lined pipe and a glass-lined inlet. As the powder reaches the inlet of the reactor, the powder particles can gain a free-fall velocity of approximately 10 m/s. Therefore, the rate of charge generation on the powder and the insulating linings would be high. These conditions suggest that the pipe section at the bottom of the charging hopper poses a high risk of PBD.

The glass-lined and PTFE-lined pipes were tested to determine their propensities to give rise to PBDs. The breakdown voltage of the glass and PTFE linings were measured to be greater than 26,000 V. It has been known that insulating sheet materials, possessing a dielectric strength (or breakdown voltage) greater than 4,000 V, can support PBDs. Therefore, both linings have the capability to generate PBDs.

The pipes were tested further to see if PBDs can be produced on the linings. The steel pipe was connected to electrical ground. A metal brush connected to a high voltage power source was inserted into the pipe to spray charge onto the lining. After a sufficient charging time, the corona brush was removed from the pipe and the power supply to the brush was turned off. A grounded spherical electrode was then brought near to the charged lining surface to induce discharges from the insulating lining. The PBD tests were conducted under both ambient (R.H. = 59% - 60%) and low humidity (R.H. = 26% - 27%) conditions. For the low humidity testing, the pipes were conditioned at less than 20% R.H. for 12 hours. As shown in **Table 2**, PBDs were observed on the glass-lined pipe at the low humidity condition. The test results suggest that the glass-lined pipe was the most likely ignition source for the subject incident. Moreover, the location of the glass-lined pipe is consistent with the witness.

Table 2 Glass-Lined Pipe and PTFE-Lined Flexible Hose

Parameters	**Breakdown Voltage (kV)**	**PBD**
PTFE-Lined Flexible Hose	>30 kV at R.H. between 26% and 60%.	PBD was not observed at both R.H.= 59 –69% and 25-26%
Glass-lined pipe	>26.5 kV at a R.H. of 59-60%;	PBD occurred at a R.H. = 26-27% but did not occur at

	>30 kV at a R.H. of 26-27%.	59-60%

* R.H. Relative humidity

CONCLUSIONS

The dust explosion was most likely initiated by a propagating brush discharge (PBD) that could have occurred on the glass-lined pipe at the bottom of the charging hopper. The presence of enough oxygen in the "nitrogen-purged" reactor-charging hopper system was due to that the dust collector fan induces an airflow far greater than the nitrogen supply flow. This mismatch between the fan and nitrogen supply produced a negative pressure in the charging hopper. The negative pressure resulted in opening of the pressure conservation valve, which allowed enough air to enter the charging hopper.

The PTFE lined pipe may also give rise to PBDs; however, PBDs were not observed on the PTFE lining in the tests conducted. The powder collected in the reactor could be able to produce bulk brush discharges on the surface of the powder. The energy of such bulk brush discharges was estimated to be 3 mJ or less based the method of reference [2]. However, the least amount of energy required to ignite the powder suspended in the nitrogen-purged reactor would be much greater than the measured MIE in air (< 3 mJ). Therefore, the likelihood of ignition by a bulk brush discharge is considered low in this case.

LESSONS LEARNED

The present investigation was completed effectively within a short period of time due to the following:

- The incident scene was preserved as it was after the fire fighting.
- Full access to sources about the equipment, process, and operations were provided for the investigation.
- Specialists in relevant disciplines (engineering and fire/explosions) participated in the investigations from the early stage.
- For those potential ignition sources that could not be ruled out based on onsite investigations and/or review of process documents, conclusions were reached through various pertinent tests, which included:

  - ✓ Dust explosion hazard characteristics of the powder involved in the incident, such as the Minimum Ignition Energy (MIE); and
  - ✓ Electrical properties that determine the electrostatic ignition hazards, such as the volume resistivity, the capacitance of electrical device, the charge per unit mass that

can be generated on a material, and the propensity to give rise to propagating brush discharges.

Whenever a process involves combustible materials, the fire and explosion hazards associated with the process should be analyzed thoroughly and carefully with all pertinent expertise, including those on fire and explosions. The relevant fire and explosion properties of materials, such as the dust explosion hazards, under the actual process conditions, should be fully quantified before the manufacturing process is started. The MSDSs may not provide the data required for achieving such a goal [4]. The ignition hazards of plant items that can be exposed to combustible materials should be carefully evaluated.

Based on the hazard characteristics of all plant items, the basis of safety for the plant operations can be defined. The basis of safety for plant operations can be achieved through sound engineering and safety practices. The materials for constructing plant items and equipment should not introduce fire and explosion hazards. For example, unnecessary use of insulating materials, such as the glass and PTFE linings for the powder feeding pipe section in this case, should be avoided. Electrical grounding of plant items and electrical equipment should be done thoroughly. In the present case, the shield element in the level probe was not grounded and could give rise to energetic sparks that can ignite the powder suspended in air under certain conditions.

REFERENCES

1. NFPA 77, “Recommended Practice on Static Electricity”, 2000 Edition.
2. M. Glor, B. Maurer, & R. Rogers, “Recent Developments in the Assessment of Electrostatic Hazards Associated with Powder Handling”, Loss Prevention and Safety Promotion in the Process Industries, Vol. 1, 1995 Elsevier Science.
3. Britton, L.G., Static Hazards Using Flexible Intermediate Bulk Containers for Powder Handling, Process Safety Progress, 12, No. 4, October 1993.
4. D. Kong, “Flashpoints Are Affected by Process Pressure”, Chemical Engineering, Volume 111, No. 13, December 2004, pp. 50 - 53.

Paper TH002-1

APPLICATION OF INHERENTLY SAFER PROCESS CONCEPTS IN DUPONT

C. Curtis Clements
Principal Consultant, Process Safety and Risk Management
DuPont Engineering Technology
E. I. DuPont de Nemours and Company, Inc.

Prepared for Presentation at
American Institute of Chemical Engineers
2005 Spring National Meeting
Process Plant Safety Symposium
Atlanta, Georgia
April 11-13, 2005

February, 2005

UNPUBLISHED

Abstract

The DuPont Company has a long history and tradition of success in the area of process safety, including the application of inherently safer design principles long before they were recognized as such. Over the past 15 years, however, the company has embraced the Inherently Safer Process (ISP) concepts, as articulated by pioneers in the field such as Trevor Kletz, and has attempted to proactively apply them as a means of improving process safety and reducing the risk of accidental, catastrophic incidents. This paper will provide a brief history of ISP in DuPont, up to and including the methods used today for getting "inherently safer thinking" into both the company's new projects and existing processes. It will also include a presentation of some case studies, that illustrate the successful application of ISP via these methods, which resulted in significant safety improvements while also providing enhanced business value.

In The Beginning

The application of inherently safer design principles in DuPont goes back to the company's earliest days, with the construction of the first black powder mills (see Fig. 1) on the banks of the Brandywine River, just outside of Wilmington, Delaware.

Figure 1 – Powder mills on the Brandywine River – circa 1830

While the inherently safer process design principles of minimization, substitution, moderation and simplification that we work with today had yet to be articulated, the design of those early powder mills shows that some of the concepts were certainly employed at least back that far. Specifically, the mills were designed to moderate, or limit the effects of a black powder explosion. The mills had three massive walls, made of granite blocks, three courses thick, designed to channel the energy of a blast. The open fourth side, which faced out across the river, and the light gauge roofing, were intended to allow the fire ball and over-pressure to blow out and be channeled in a safe direction. The mills were located along the Brandywine River in places where the opposite bank was very steep. This both precluded the possibility of building anything in the blast zone and provided additional secondary containment for the blast energy and debris. The spacing between the mills was selected to prevent an explosion and fire in one mill from propagating to the adjacent facilities.

The design and operation of those early mills were also intended to limit the probability of ignition sources being present. The tracks, upon which iron-wheeled carts were used to move materials between the mills and storage buildings, were made of wood. Non-sparking tools were used to work with the raw materials, intermediates and finished products. Today, we might classify these as applications of the ISP principle of simplification.

Early Lessons

By and large, these concepts served the DuPont Company well for many years. Yes, explosions did occasionally occur. And as a consequence, a number of brave powdermen – including several members of the DuPont family, lost their lives, taking that last journey that they euphemistically referred to as "going across the river." Nevertheless, the concept worked for over 80 years, in that no explosion destroyed more than the mill in which it originated. It worked, that is, until one day in 1889, when a massive explosion, the results of which are shown in this picture (see Fig. 2), shattered the quiet in the Brandywine River valley.

Figure 2 – Destroyed powder mills – circa 1889

The explosion was so violent, that it shook the ground and rocked buildings throughout the Wilmington area, causing residents to run outside, terrified that they were experiencing an earthquake. The reality, however, was much worse, as the plume of smoke rising above the horizon gave testimony to the magnitude of the tragedy that had befallen the community, one in which almost every family had relatives working in the powder yards. What was different about this event was, that for the first time, a single explosion had destroyed two powder mills – the two largest mills in the powder yard. What the resulting investigation revealed was, that over time, the size of the powder batches had been increased in successive increments, until, at the time of the massive blast, the basic design and spacing of the mills was no longer adequate to contain the amount of energy available. The lesson learned, from an inherently safer perspective, was minimization – the need to limit the size of the batches to a level that could be safely contained.

20th Century Learnings

Now let's fast forward almost a full century, to 1984. In December of that year, another infamous process safety incident would further the application of ISP within DuPont. The tragic event in Bhopal, India provided the driving force. At the time, a major business in DuPont's agrichemicals portfolio was critically dependent upon methyl isocyanate, or MIC, as a key ingredient in the manufacture of a very profitable insecticide. At the time, DuPont was purchasing MIC from its West Virginia-based supplier and shipping it in railcars half way across the country to a plant located just outside one of the most populous metropolitan areas in the United States. Routinely, several million pounds of MIC were maintained in storage on the site at any given time. The Bhopal tragedy immediately put this entire business at risk, as concerns about the safety of making, shipping and storing MIC threatened the supply chain for this product.

The answer was provided by the Company's research and development organization in the form of a totally new process, one that utilized the ISP principles of substitution and minimization. Specifically, an alternative chemistry was developed that allowed for use of a significantly less hazardous material than methyl isocyanate as a starting raw material, thus eliminating the risk of shipping and storing MIC. Alternative reactor technology was then utilized to convert the new raw material to MIC in situ. The MIC was then immediately reacted and consumed as part of the next reaction step, limiting the total amount of MIC in in-process inventory at any point in time to something on the order of 10 to 20 pounds. Thus, the need to ship and store MIC was completely eliminated and a roughly five order of magnitude reduction in hazardous material inventory was achieved. From a business impact perspective, this is still perhaps the single most significant application of ISP for the Company to date.

A few years later, in 1992, another incident opened the door once again for a further demonstration of the value of inherently safer design. This one, which occurred quite a bit closer to home as it were (i.e., about 10 miles from the corporate headquarters), involved a runaway decomposition reaction of ammonium dichromate, or ADC, the starting material for the production of chromium dioxide.

The first step in the manufacture of chromium dioxide involved the conversion of ADC, an orange crystalline material, to chromic oxide, a dull green powdery material, via a controlled decomposition reaction, as shown below.

$$(NH_4)_2Cr_2O_7 + \text{heat} \rightarrow Cr_2O_3 + 4H_20 + N_2 + \text{heat}$$

As listed at the time in NFPA 49, ammonium dichromate was assigned a reactivity hazard of 1. Initiation of the decomposition reaction requires a temperature of approximately 180°C. At about 225°C, the reaction generates sufficient heat to be self-sustaining, and at higher temperatures, a runaway decomposition becomes possible. With the reaction producing four moles of steam and one mole of nitrogen for each mole of ammonium dichromate, there is obviously the potential for building pressure in any confining vessel. The reaction rate is highly dependent upon the degree of confinement, the burning geometry, and the actual point in the reaction mass where the reaction initiates. Under "worst case" conditions, however, a rate of pressure rise of up to one million psi/second is possible.

The processing equipment for converting ADC to chromic oxide was relatively straightforward, as shown in the simplified sketch (see Fig. 3) below:

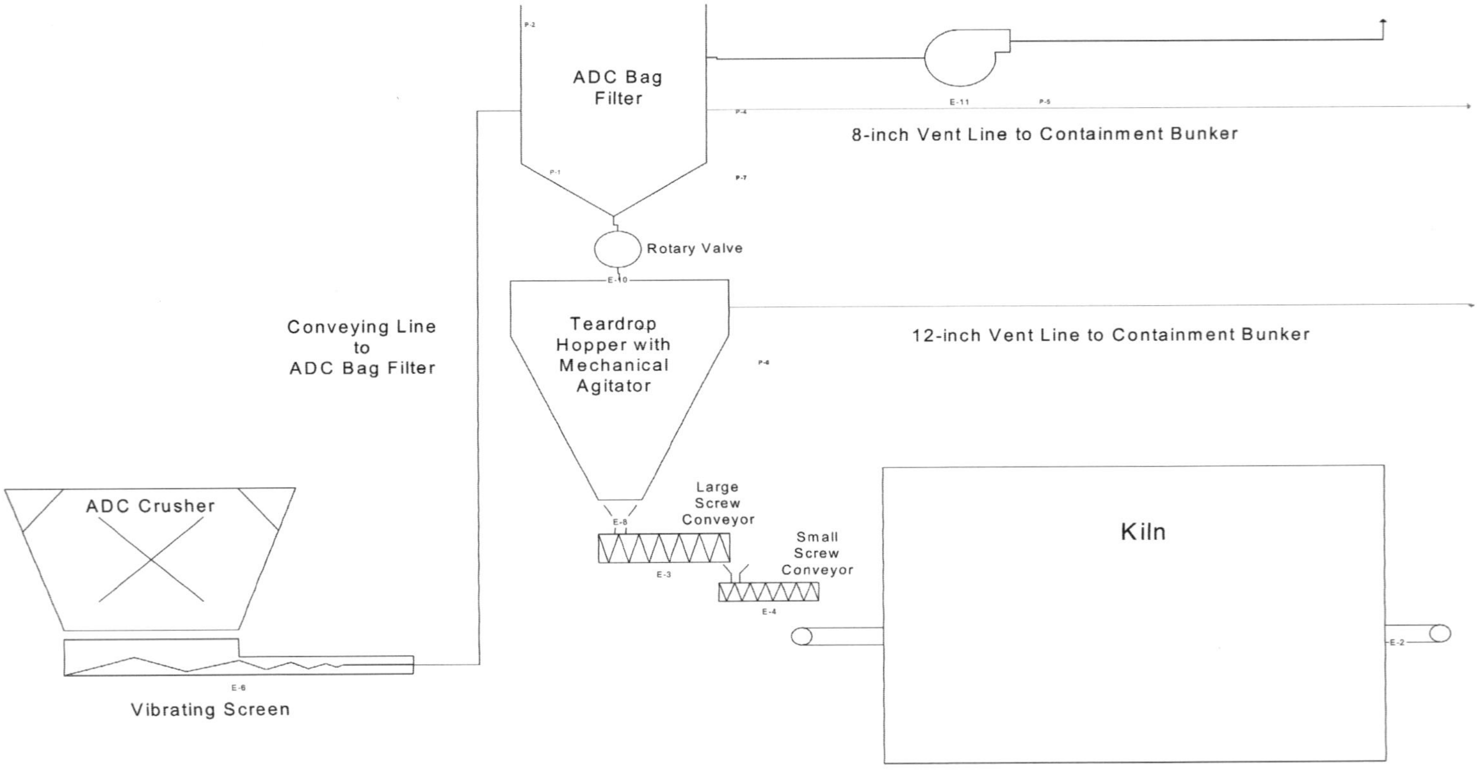

Figure 3 – Simplified Process Sketch – ADC Conversion Facility

Operators would dump from one to four 400-pound drums of ADC into a large open piece of equipment known as the ADC crusher. The material would be worked through the parallel shafts with interlacing fingers, which broke up any large clumps. At the bottom, a vibrating screen further helped to size-reduce the material and move it to the pick-up point of a pneumatic conveying system, which conveyed the ADC up to a bag filter. From the filter, the material gravity flowed through a rotary valve and into a large, agitated, teardrop-shaped vessel, known as the ADC hopper. This hopper could hold up to 800 pounds of ADC. With the slide gate at the bottom of the hopper open, the ADC dropped down into the larger of two, in-series, screw conveyors. The screw conveyors were used to control the rate at which the orange ADC was fed onto a conveyor belt, which transported it through the large, steam-heated kiln in which the decomposition reaction was carried out. At the back of the kiln, large drums collected the green chromic oxide as it came off the conveyor belt, ready to be sent on to the next step in the process. Recognizing the potential for pressure to be generated, should a decomposition occur in either the bag filter or the ADC hopper, both vessels were equipped with rupture disks. They discharged through the building wall, via 8-inch and 12-inch vent lines, respectively, and into a large, cinder-block containment bunker, designed to capture any chromic oxide produced while allowing the gas-phase decomposition products to safely vent to atmosphere.

This particular facility had been built in the late 1970s and had operated without major incident since start-up. On July 27, 1992, however, an ADC decomposition reaction initiated somewhere between the bag filter and the second screw conveyor and ultimately progressed to an explosive runaway. The resulting explosion over-pressured and ripped apart the lines connecting the process equipment, releasing a torrent of hot solids and scalding vapor into the building. The bag filter was burned up, electrical conduit was turned into spaghetti, the vent lines were "relocated," and the containment bunker was essentially demolished. A number of these effects can be seen in the following photographs (See Figs. 4, 5, and 6).

Figure 4 – Exterior of Process Building Showing Destroyed Containment Bunker

Figure 5 – Inside and Outside Views of Hole Created By "Relocation" of 12-inch Vent Line

Figure 6 – ADC Hopper, With Screw Conveyor and Remains of Bag Filter

Fortunately, there were no injuries as a result of this explosion. This was primarily due to the fact that no operators were present at the time, as the process had been put in "standby" mode sometime earlier in the shift and the operators had been detailed to work in other buildings. Had this event occurred during normal operation, it would very likely have produced several fatalities.

Why then, did it occur? At the time of the incident, per standard operating procedure, when the process was put in "standby," the ADC crusher and pneumatic conveying system were shut down and the slide gate which allowed flow from the ADC hopper to the screw conveyors was closed. To prevent potential bridging of the ADC, however, the rotary valve, the hopper agitator, and both screw conveyors, were allowed to continue to run. The incident investigation ultimately found that the barrel of the large screw conveyor had, over an extended period of time prior to the incident, apparently been subjected to a number of external impacts. These were presumably done by operators, in response to plugging problems in the conveyor. This activity finally resulted in deformation of the screw barrel, allowing the rotating element to come into contact with the inside of the barrel. During normal operation, the flow of material through the conveyor was apparently sufficient to remove the heat generated by the metal-to-metal contact. When placed in standby mode, this "cooling mechanism" was lost. Due to the extended length of time that the system was in standby, the barrel apparently began to heat up, and heat was transferred by convection and conduction back up to the immediately-overhead ADC hopper, which held, at the time of the explosion, approximately 350 pounds of ammonium dichromate. The physical evidence indicated that this confined reaction mass provided the major source of the explosion energy.

While the shift in market demand from electro-magnetic to electro-optical recording media ultimately resulted in the shutdown of this process, at the time, it was a very profitable venture. Understandably, there was an intense amount of management interest in addressing the incident investigation findings and returning it to a safe operating condition as promptly as possible. In response, the incident recovery team brought inherently safer process concepts to bear on the problem. Specifically, the team focused on using minimization, moderation and simplification to reduce the inherent hazard level of the process. Let's take a closer look at how these concepts were applied.

First, with the explosion potential of confined ADC now clearly understood, the most significant issue was how to minimize the amount of ADC held up in the process. At the time of the incident, operating practice had been to run the pneumatic transfer system until the ADC hopper was full, and then shut it down until the level in the hopper dropped. After the incident, the control scheme was modified to allow the pneumatic transfer system to run automatically in an intermittent, or pulse mode, to better match the processing rate of the kiln. This resulted in a reduction of in-process inventory from more than 350 pounds on average to less than 30 pounds, with no reduction in production capacity. Significantly less material meant lower loading density, less reaction mass and lower internal pressure generation in the event of a decomposition.

With significantly less material in the system, the operability problems associated with ADC bridging went away and control of flow to the kiln could be managed with sufficient precision using only the smaller screw conveyor. This allowed elimination of both the large screw feeder and the agitator in the ADC hopper. These examples of process simplification eliminated two potential sources of mechanical friction, reducing the opportunity for initiation of a runaway decomposition. In addition, the remaining mechanical equipment was reinforced and shielded to prevent internal damage from external mechanical contact. This reduced the likelihood of the creation of internal hot spots due to externally-induced damage.

Under the heading of moderation, the capacity of the over-pressure relief system was upgraded, the vent piping supports strengthened and the containment bunker re-built and reinforced to handle both higher static and dynamic pressures. This resulted in an overall system robust enough to withstand the greatest over-pressure that could be generated internally, running with lower internal inventory.

Figure 7, below, provides a simplified sketch of the "inherently safer" process.

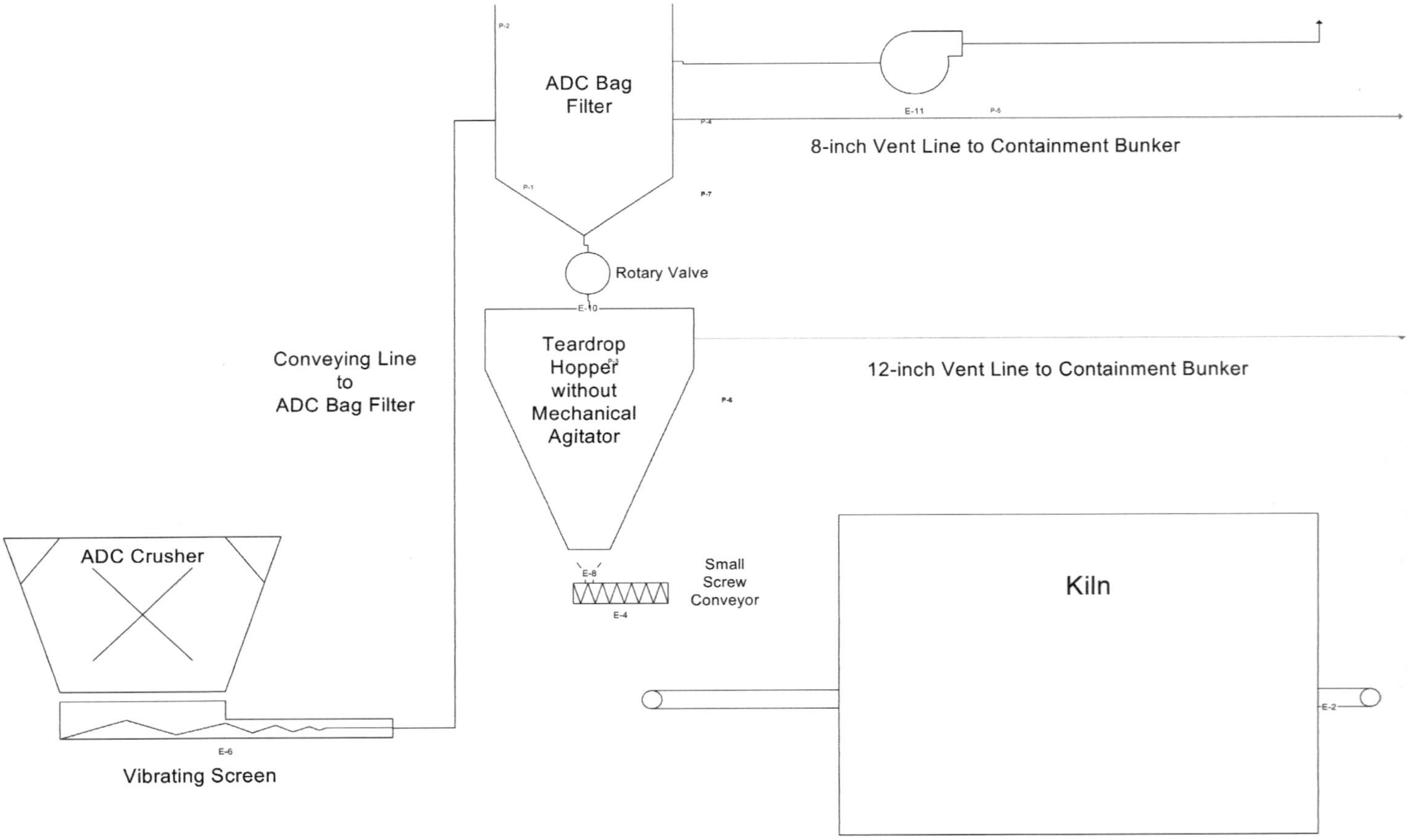

Figure 7 – Simplified Process Sketch – The "Inherently Safer" ADC Conversion Facility

Ultimately, the work done in response to this incident provided a practical demonstration within the company that the application of inherently safer principles was not limited to the development of totally new, "revolutionary" process designs. The concepts could be effectively applied in an "evolutionary" sense, through retrofitting existing facilities to either eliminate or significantly reduce the inherent potential for catastrophic accidents. Finding the best way to make this happen proactively, rather than reactively in the wake of process incidents, would take a few more years, and unfortunately, a few more incidents.

Answering the Wake-Up Call

During August and September, 1996, the company experienced four serious process incidents. The nature of the events, their potential severity and their proximity in time clearly caught the attention of senior management. As a result, the CEO chartered an ad hoc, multidisciplined, multi-functional corporate task team and gave it the assignment of determining how best to effect substantial and sustainable improvement in the Company's process safety performance. This team, known initially as the "zero process incident task force," ultimately came to be known by the acronym ZIP, which stood for Zero Incidents. . . Period! The ZIP team identified several key areas of emphasis as being critical to improving process safety performance and driving toward the goal of zero incidents, including:

1. Integration of process safety management with basic business processes;
2. Application of the principles and essential features of process safety management to lower hazard operations, as well as higher hazard processes;
3. Improved operational discipline in all manufacturing operations; and
4. Application of inherently safer process concepts to existing processes, and to new processes – as early in the development stage as practicable.

Additional teams were subsequently chartered, with the mission of putting these concepts into practice. In the case of ISP, there were two key goals to be accomplished: first, raise the level of awareness and knowledge of ISP in key areas of the organization, and second, find the best way to institutionalize ISP thinking into the project and process development work processes.

Meeting the first goal would involve an expansive training effort. Toward that end, four distinct ISP seminars were developed, each targeted for a different audience, as described below:

- ISP Orientation, a 45-minute overview session, provided an introduction to, and a basic awareness of, ISP principles for a general, non-technical audience.

- ISP For Leadership, a one-hour session intended to provide the basic fundamentals of ISP to individuals in business and functional leadership roles, to both build value for the concepts and to improve their ability to help their work teams successfully apply them.

- ISP For R&D, Project and Operations Team Members, a four-hour seminar, incorporating a combination of lecture and interactive workshops, intended to provide a detailed, in-depth review of ISP concepts, principles, methods and tools for their application, with a focus on functional issues, for individuals who work in areas dealing with process safety hazards on a day-to-day basis.

- ISP For Process Safety Practioners, a four-hour seminar, incorporating a combination of lecture and interactive workshops, intended to provide a detailed, in-depth review of ISP concepts, principles, methods and tools for their application, with a focus on the facilitation of ISP application in their work. This seminar is intended for process safety professionals, who will, in turn, provide ISP training to others.

The content of the training curriculum for each seminar was based upon materials purchased from, and used with the permission of, the IChemE and upon materials from internal company resources. In addition, copies of the CCPS "gold book," Inherently Safer Chemical Processes, A Life Cycle Approach, were purchased and distributed to each attendee in one of the four-hour seminars. To date, copies have been handed out to roughly 4000 process and project engineers, research scientists, and other technologists across the company. During a three-month period in 1999, every principal researcher, process development engineer and lab technician at the company's central research and development campus – over 1000 people – went through the in-depth program.

Mainstreaming ISP

This training effort helped to raise awareness levels and the overall level of technical functional competence relative to ISP. The next challenge was to institutionalize application of the ISP principles through existing work processes. The primary vehicle for doing this was the addition of specific requirements relative to ISP in selected company policies and practices, as detailed in the relevant safety and engineering standards and guidelines, including those covering Process Safety Management, Process Hazards Analysis, Management of Highly Toxic Materials, and Project Implementation. For example:

The Corporate Safety, Health and Environmental Standard covering process safety management states:

> *"New facilities involving project-level investment (typically calling for off-site authorization) must have . . . [a] Screening process hazard review (PHR)—as early as possible during basic data development, to identify potential acute hazards and concerns and to develop broad recommended changes in scope to significantly reduce these hazards (including consideration of the use of inherently safer technology) . . ."*

and,

> *"For all baseline PHAs, the PHA team shall give consideration to incorporating inherently safer process technology in the process under review."*

The Corporate Safety, Health and Environmental Standard covering the policies and practices for conducting process hazards analyses (PHAs) contains the following guidelines and requirements:

> "Consistent with the guidance in [the relevant sections of] the DuPont Engineering Guide to Project Implementation, research and technology organizations are responsible for identifying, evaluating, and using inherently safer process technology—from the development of basic chemistry to process design. Inherently safer technology should be applied where it reduces process hazards and makes good sense."

> "For processes including hazardous substances, [research and technology] reviews should be held and documented by the appropriate research organizations. The focus of these reviews should be to challenge the basis for the proposed chemistry and process design and consider and evaluate inherently safer process options."

> *"For all baseline PHAs, the PHA team shall consider incorporating inherently safer process technology in the process under review."*

"[The PHA report] shall document the team's consideration of inherently safer process technology and any resultant recommendations."

And finally, the Corporate Safety, Health and Environmental Standard covering the management of highly toxic materials states:

"Strong consideration must be given to the use of inherently safer technology during the development of process technology and design scopes of work and during modifications."

Compliance with the requirements of these company standards is routinely checked as part of our internal process safety management audits. It is important to emphasize, however, that the intent of these requirements is to force a conscious thought process aimed at the proactive identification and evaluation of opportunities to apply ISP principles, not to mandate their implementation. Actual decisions around implementation are, of course, dependent upon both technical and financial feasibility, based upon appropriate risk and cost-benefit analyses.

Typically, the identification of ISP opportunities is accomplished using a team approach, as part of a process hazards analysis. In most applications, a "What If/Checklist" technique is utilized. The team, having first identified the hazards inherent in the process under review, applies a brainstorming technique to generate ideas for the application of ISP concepts. This is followed by use of an "Inherently Safer Process Considerations" checklist, to further stimulate the thought process of the team. Alternatively, a HAZOP-type approach may be used. The hazardous materials and specific hazardous aspects of the process are paired with guidewords, such as less and no, to identify potential ISP opportunities. For new facilities, this is done as part of the research reviews and the hazards analyses conducted throughout the life cycle of the project, from those conducted at the basic data stage to those conducted during the detailed design. New PHA leaders are now given specific training in inherently safer process concepts and in the application of these techniques to identify potential ISP opportunities.

For existing facilities, during each cyclical revalidation of a baseline PHA, the team is encouraged to revisit and consider anew the application of inherently safer concepts, in hopes of identifying opportunities for further advances. In some cases, this effort results in the identification of inherently safer opportunities that can be retrofitted into the existing process. In others, it results in the identification of "possibilities" for incorporation in future generations of the process technology. As a minimum, it provides a vehicle by which PHA teams can clearly identify and document any existing "inherently safer aspects" of a process design, to ensure that they are not "lost" during future process change activities.

Putting Theory Into Practice

Since the requirements described above have gone into effect, there have been a number of applications of ISP, in both new and existing processes, ranging from small, incremental process safety improvements to major, step-change reductions in the inherent hazard potential of the processes. Some selected case studies, illustrating the depth and breadth of ISP application, are briefly discussed below. While the need to preserve proprietary trade secrets prohibits the amount of detail that can be provided in a public forum, they illustrate some of the potential results of a consistent, dedicated effort to identify and implement inherently safer process concepts.

Case Study 1 - Simplification

The initial design of a new process unit included a large, high-horsepower mechanical mixer, operating at high pressure, to mix a solid with a highly-flammable liquid. Operating experience with the mixer seals in similar equipment used in other process units, however, demonstrated a lack of reliability and the potential for a major, uncontrolled loss of containment event.

Consideration of inherently safer process possibilities led to the design of an alternative mixer – one with no internal moving parts and no mechanical seals. Following satisfactory demonstration of the new mixer in a pilot facility, the mechanical mixer was subsequently eliminated from the design and the new technology was incorporated into the production unit for start-up. This significantly simplified the process, from both a safety and a maintenance perspective, and eliminated the single most-likely scenario for a major fire or explosion. The benefit of the new, inherently safer technology was subsequently leveraged through retrofit of other process units having similar issues.

Case Study 2 – Moderation and Simplification

Another process unit, involving a highly-flammable material being handled at high-pressure in a partially-open structure, posed the potential for a catastrophic explosion under certain loss of containment scenarios. Specifically, there was the potential for the resulting vapor cloud to migrate into zones having a high degree of congestion and confinement. An ignition of the cloud would generate over-pressures of sufficient magnitude and duration to potentially collapse a nearby process building.

Through ISP brainstorming, the concept of using a barrier curtain, to prevent potential clouds from reaching portions of the process where an ignition would generate catastrophic over-pressures, was developed. It was also proposed that certain other pieces of the process equipment could be rearranged and relocated to be outside of the curtain. Other equipment could subsequently be eliminated and removed, further simplifying the process and reducing the level of congestion in the area. The ISP concepts were implemented following demonstration, through computational fluid dynamic and explosion modeling, of a dramatic reduction in damage potential with the curtain in place. Peak over-pressure projections showed a reduction of over 70%, enough to limit the effects on the building of concern to a negligible level.

Case Study 3 – Minimization and Simplification

For many years, an existing process had been run using a series of large, stirred-tank continuous reactors. The reaction, involving highly reactive and flammable organics, at temperatures above their normal boiling points, posed significant process hazards. These included the potential for fires and explosions resulting from reactor overpressure, major runaway reactions and other loss of containment scenarios.

An improved reactor process was conceptualized and developed, initially as part of an effort to reduce manufacturing costs. A process hazards analysis of the process was conducted, including consideration of the potential inherent safety advantages. The concept involved using one new, two-stage continuous reactor. The new reactor was capable of replacing all of the existing reactors, greatly simplifying the process and significantly reducing the number of opportunities for catastrophic events. It also resulted in a major reduction of in-process inventory, minimizing the amount of hazardous material contained in the process at any point in time.

Case Study 4 - Substitution

A new polymer extrusion process required use of a heat transfer fluid. During the PHA, it was determined that the intended fluid, a combustible liquid, would be operating at a temperature at or near its flash point.

To avoid the increased potential for fire or explosion in the event of a loss of containment, an alternative heating medium was selected, one using a non-flammable mixture of propylene glycol and water. As an additional benefit, it was determined that the new heat transfer fluid could be used in both the "hot" and "tempered" heat transfer systems for the process, allowing simplification of the overall heat transfer system.

Case Study 5 - Moderation

An existing process is run at pressures in excess of 25,000 psig and at temperatures above 200°C, under free radical conditions. At these conditions, there is the potential for extremely energetic, runaway decomposition reactions to occur.

A research and development effort led to a new process technology that runs at a pressure less than 1000 psig and at temperatures under 120°C. At these new conditions, the potential for runaway decompositions was effectively eliminated.

Case Study 6 - Moderation

A chemical plant had a large storage tank containing a regulated toxic and flammable material. The tank was in a diked area, but consequence analysis modeling indicated a potential for significant off-site impact in the event of a large release. High evaporation rates from the large pool surface area inside the containment dike were the major factor in the potential off-site impact.

Although monitoring devices were in place to detect any leakage from the tank, the site wanted, if possible, to eliminate or significantly reduce the potential for toxic vapors to have an impact on the surrounding community. As a result of the ISP analysis done during a cyclical process hazards evaluation, the recommendation was made to use a "dike inside the dike" approach. A revised dike was installed, closer in around the tank, in order to reduce the surface area available for evaporation by almost 60%. The new dike wall height was kept the same as the original dike wall, and the new wall design included a spillway section that would allow any liquid that exceeded the spillway height to overflow to a newly constructed covered containment pit. Additional consequence modeling for the new design shows that even under worst case scenario conditions, there is essentially no potential for off-site impact.

Other Examples – Minimization and Moderation

DuPont operates several processes that utilize highly-toxic materials as key ingredients. Large-scale releases of the chemicals in this category have the potential, under certain conditions, to result in Bhopal-type events.

Recognition of this potential has driven the application of inherently safer alternatives, focused primarily on reducing stored and in-process inventories, moderating the conditions of use, and limiting the potential impact of loss-of-containment scenarios. In some instances, improvements in process efficiency and supply-chain management have enabled significant reductions in on-plant inventories. In others, feeding the process directly from appropriately secured rail cars and using secondary containment have allowed the elimination of much larger storage tanks. Examples of moderation include the use of refrigerated storage, double-walled storage tanks, and flow-limiting orifices, just to mention a few.

A Look Behind and A Look Ahead

The case studies and examples cited above are just a few of the instances where inherently safer process concepts have been utilized to improve the safety levels and reduce the risk of plant operations. The key to recognizing and implementing the opportunities to use ISP has been the institutionalized application of focused, analytical thought processes at key stages in the process development, project engineering and plant operations work processes. To further encourage the development and application of inherently safer process technology, an effort has been made to recognize and reward significant accomplishments through granting Engineering and Safety Excellence Awards at the corporate level.

The journey toward the development of inherently safer processes within DuPont has been underway for over two hundred years now. It has been in the last ten years, however, that efforts to institutionalize the search for inherently safer alternatives, through existing process development and process hazards analysis work systems, has begun to pay some real dividends. And, while we have made significant progress, we really have only just scratched the surface of what is possible. So, in closing, let me take you back to the early 1800s one more time, when one of the DuPont Company's first safety slogans was "Make it safer, make it better, make it safer." Today, through our efforts to incorporate ISP into our facilities, we have revived that slogan, but in a slightly modified form. Today we say "Make it safer, make it better, make it inherently safer."

References

"Applying Inherent Safety to Mitigate Offsite Impact of a Toxic Liquid Release," Douglas J. Ferguson, Center For Chemical Process Safety of the American Institute of Chemical Engineers, 19th Annual International Conference – Emergency Planning: Preparedness, Prevention, Response – June, 2004

Inherently Safer Chemical Processes – A Life Cycle Approach, Robert E. Bollinger, et al, Center For Chemical Process Safety of the American Institute of Chemical Engineers, New York, NY, 1996

"Inherently Safer Process Design," Safety Health and Environment Department, Institution of Chemical Engineers, Rugby, UK, 1995

"Inherently Safer Plant Equipment Layout to Minimize the Hazards from Vapor Cloud Explosions, David D. Herrmann, Mary Kay O'Connor Process Safety Center 2000 Annual Symposium

"ISP Participants Guide," E. I. du Pont de Nemours and Company, Wilmington, DE, 1999

Paper (TH002-2)

Inherently Safe Process Plants: Practical Methods and Examples

David Kehn
Mustang Engineering, LP

Prepared for Presentation at
American Institute of Chemical Engineers
2005 Spring National Meeting
Process Plant Safety Symposium
Atlanta, Georgia
April 11-13, 2005

ABSTRACT:

Although the principles of inherent safety are well known, practical means of achieving inherently safer processes are not. This presentation will describe practical methods to achieve inherently safer processes in process plants. Several examples will be presented to illustrate the timely application of these methods in both onshore and offshore engineering and construction projects.

Table of Contents

ABSTRACT:

INTRODUCTION:
Table 1: Inherent Safety Hierarchy
Figure 1: Inherent Safety Time Line

DEFINITIONS:

ACRONYMS:

PROJECT PHASES AND GOALS:
Figure 3: Project Phases and Milestones
Select Phase:
Select Phase Project Goals:
Select Phase HSE Goals:
FEED/Define Phase:
FEED/Define Phase Project Goals:
FEED/Define Phase HSE Goals:
DETAIL DESIGN STAGE:
Detail Design & Execute Phase Project Goals:
Detail Design & Execute Phase HSE Goals

INHERENT SAFETY IMPLEMENTATION METHODS
Project Team Expectations and Deliverables:
Developing a Hazard Evaluation Procedures Plan:
Table 2: Typical HEPs by Project Phase
Site Selection with HSE Input:
Preliminary Hazard Review:
Essential/Safety System Survivability Assessment:
Figure 4: Essential Safety System Survivability Assessment
Figure 5: As Found
Figure 6: Inherently Safer
Essential/Safety System Performance Objectives:
Figure 7: Performance Fire Protection
Example Performance Objective – Water-based Fire Protection Systems:

POSSIBLE INHERENT SAFETY ACHIEVEMENT:
Table 3: Possible Inherent Safety Achievement
Table 3 (cont.): Possible Inherent Safety Achievement

INTRODUCTION:

While the principles of inherent safety and its hierarchy is well understood (Table 1) its application into process plant projects is inconsistent and rarely complete. Most projects will not maximize inherent safety for the most basic of reasons – lack of formal HSE/Risk Management strategy aligned with the project phases. This paper will:

1. provide background on project phases and their respective project and HSE goals

2. overview select inherent safety implementation methods to achieve inherently safer processes in process plants with examples

3. summarize possible inherent safety achievement

Table 1: Inherent Safety Hierarchy

Eliminate:	removing all potential of a hazard such as no flammable or combustibles and therefore no fire or explosion potential
Prevent:	depriving the hazard from occurring, that is, keeping the fluids within the pipe and therefore no fire or explosion
Detect / Control:	sensing a hazard as it occurs and implementing executive actions to reduce / restrain the severity of the event such as facility shutdown to reduce the release quantity or amount
Mitigate:	minimize or make less severe the consequences or results of a hazard scenario such as the thermal effects of a fire by deluge water spray systems, fireproofing, or robust structural design
Response/Recovery:	contingency measures in place, usually non-fixed and personnel intensive, to mitigate a hazard scenario such as firefighting, rescue, and evacuation or shelter-in-place including restoration of operations and affected areas

Process plant projects can maximize Inherent Safety design when:

1. the HSE professionals are aligned with project phases and project/business goals

2. the project team expectations include clearly defined HSE project deliverables and the procedures/processes to accomplish resolution of identified hazards

The ability to implement inherent safety is represented graphically in Figures 1 and 2. As the project progresses in its lifecycle, measures to reduce the hazard are increasingly less inherently safe by design and more dependent upon procedures and people. Additionally, the costs of implementing these changes increase significantly as the project progresses. The lowest cost to implement and the greatest benefit are achieved early in the project lifecycle.

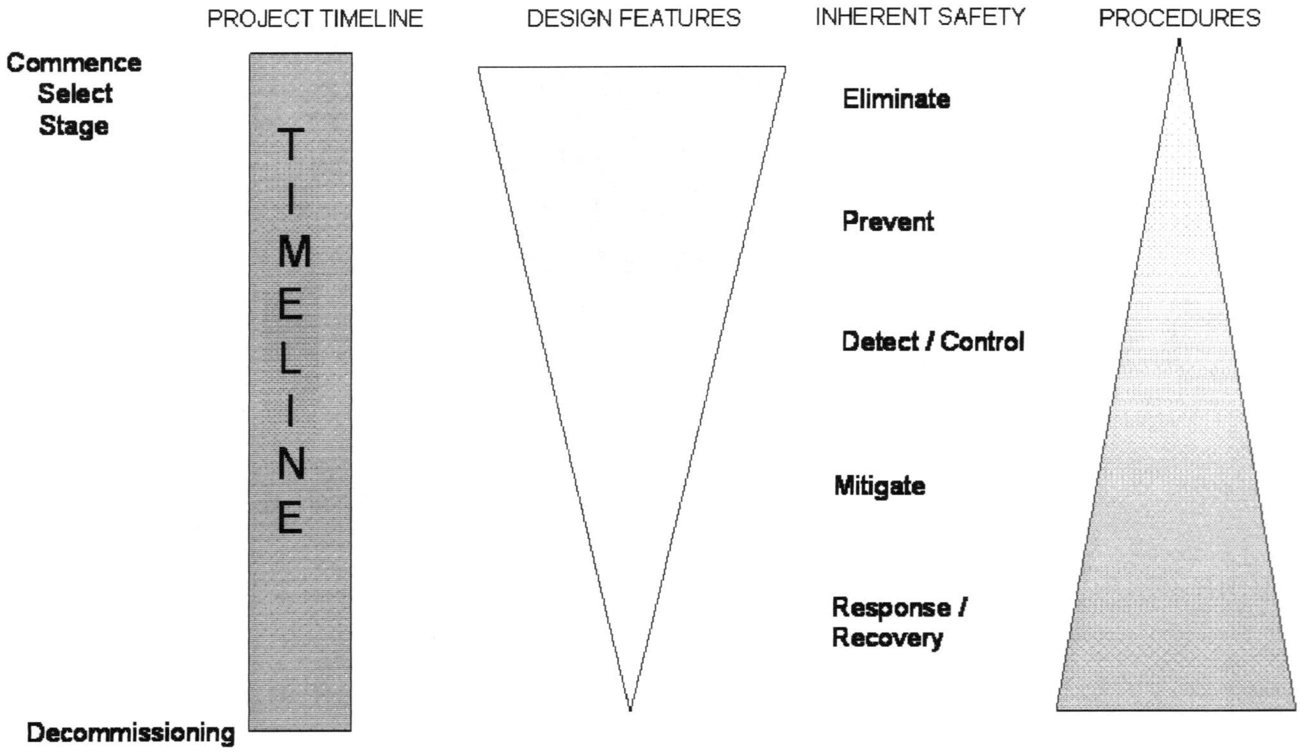

Figure 1: Inherent Safety Time Line

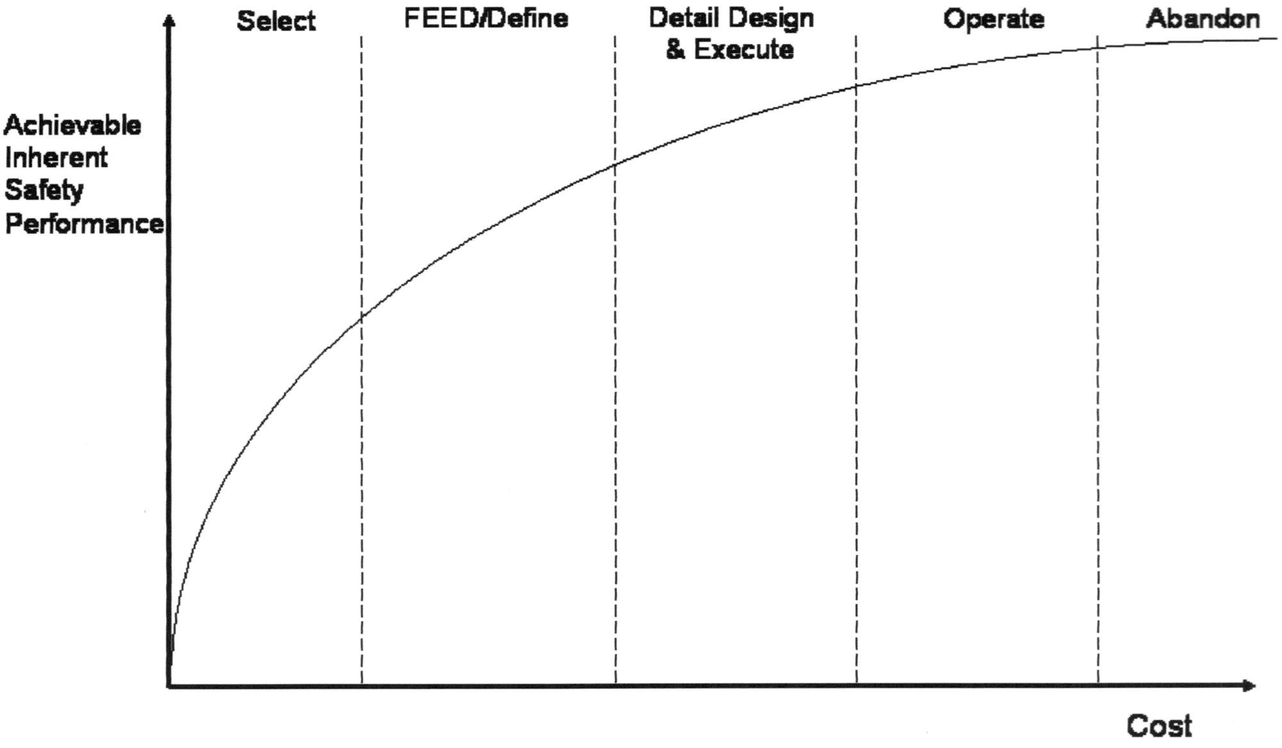

Figure 2: Influencing Inherent Safety over Project Lifecycle

DEFINITIONS:

Inherent:	involved in the constitution or essential character of something: belonging by nature or habit (Merriam-Webster 10th Ed.)
Safety:	the condition of being safe from undergoing or causing hurt, injury, or loss (Merriam-Webster 10th Ed.)
Hazard:	An inherent physical or chemical characteristic that has the potential for causing harm to people, property, or the environment (AIChE-CCPS, Guidelines for Auditing PSM Systems, 1993)
Hazard Evaluation Procedures:	The analysis of hazardous situations associated with a process or activity, using qualitative techniques to identify weaknesses in design and operation. (AIChE-CCPS, Guidelines for Auditing PSM Systems, 1993)
Risk	The combination of the expected frequency (events/year) and consequence (effects/event) of a single accident or a group of accidents. (AIChE-CCPS, Guidelines for Auditing PSM Systems, 1993)
Risk Management:	The systematic application of management policies, procedures, and practices to the tasks of analyzing, assessing, and controlling risk in order to protect employees, the general public, the environment, and company assets. (AIChE-CCPS, Guidelines for Auditing PSM Systems, 1993)

ACRONYMS:

AIChE	American Institute of Chemical Engineers
API	American Petroleum Institute
CCPS	Center for Chemical Process Safety
DAR	Deliverable Action Requirement
Ed	Edition
ESSA	Essential/Safety Systems Survivability Assessment
FEED	Front End Engineering Development
HAZID	Hazard Identification Study
HAZOP	Hazard and Operability Study

HEP	Hazard Evaluation Procedures
HSE	Health, Safety, and Environment
MOC	Management of Change
NFPA	National Fire Protection Association
PSM	Process Safety Management
RP	Recommended Practice
SIL	Safety Integrity Level

PROJECT PHASES AND GOALS:

Projects progress from conceptual ideas until a completed and operating process plant. Many companies identify phases or stages within this project execution continuum. Along with these identified phases companies will usually identify gates or criteria to be met before a project can proceed (that is, a stage gate process). These criteria are typically based upon maximizing a company's financial performance and project cost certainty but occasionally include HSE criteria. Figure 3 illustrates typically defined project phases and their milestones. This paper will concern itself with the design intensive phases of Select, FEED/Define, and Detail Design & Execute.

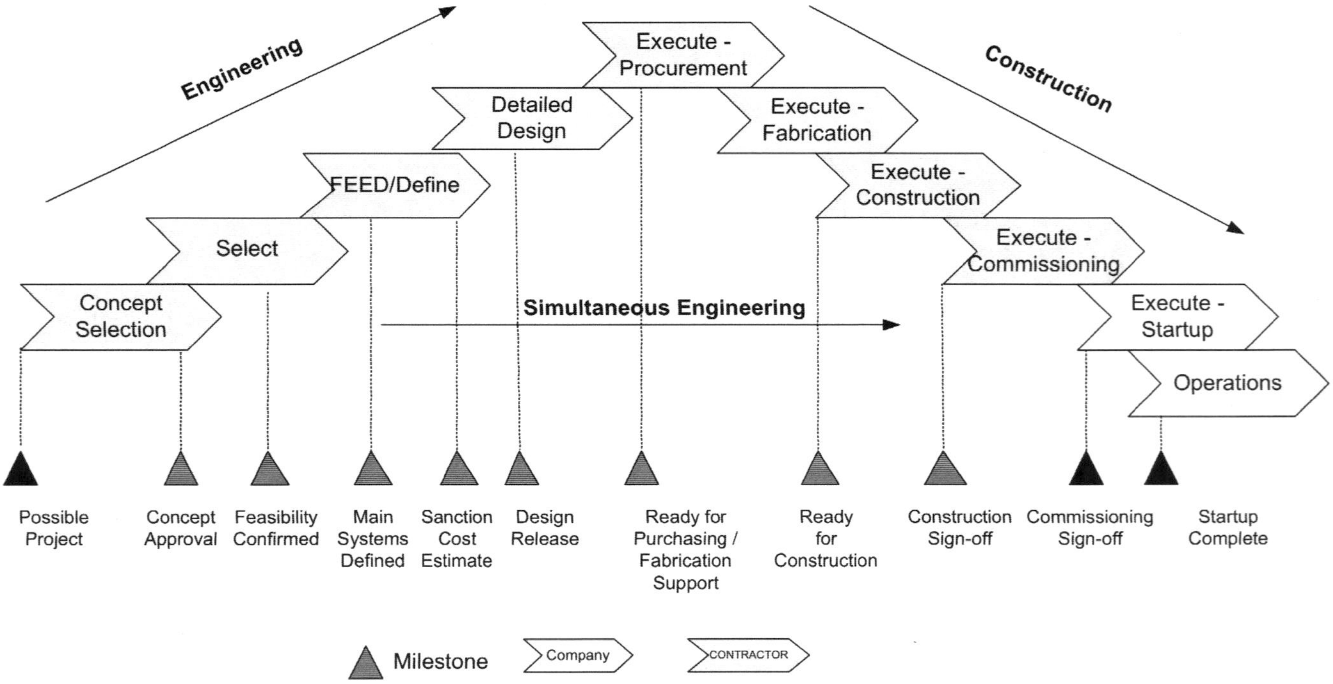

Figure 3: Project Phases and Milestones

Select Phase:

Select Phase Project Goals:

Typical high-level project goals for the Select Phase are to identify, evaluate, and rank technically feasible options for their major risks, both business and HSE, and selecting one development strategy for the FEED/Define Phase. Costs and schedule will be developed for the execution of the FEED/Define Phase and sometimes for the completed project to determine if the project is a business unit priority.

Select Phase HSE Goals:

The goals of the HSE process during the Select Phase are to:

- provide safety engineering input at the earliest phases of the development
- ensure major risks associated with the concepts being considered for the development are evaluated including questioning fundamental decisions
- confirm regulatory compliance including environmental permit plans and field surveys
- generate a comprehensive listing of major risk issues to be addressed in the FEED/Define Phase
- complete preliminary HSE philosophy (including risk management and risk acceptance strategy) with goals and expectations for the selected development option and essential safety systems
- complete preliminary Hazard Evaluation Procedure (HEP) Plan for FEED/Define including cost estimate and schedule.

FEED/Define Phase:

FEED/Define Phase Project Goals:

Typical high-level project goals for the FEED/Define Phase are to further define the selected development strategy to provide sufficient detail for project funding/sanction. The FEED/Define Phase may include developing and sending a bid package, evaluating bids and award of long-lead items such as compressors, turbines, and generators.

FEED/Define Phase HSE Goals:

The goals of the HSE process during the FEED/Define Phase are to:

- define all key hazards and ensure these are addressed in the design process and deliverables

- ensure engineering decisions are simple and robust from a safety viewpoint
- optimize configuration and layout of the concept from the safety viewpoint
- ensure that the cost estimate and schedule reflect all planned HEPs and HSE hazard concerns
- ensure operations are involved in safety related decisions
- maximize opportunities to improve inherent safety

DETAIL DESIGN STAGE:

Detail Design & Execute Phase Project Goals:

Typical high-level project goals for the Detail Design & Execute Phase, are to complete all FEED/Define Phase developed plans in accordance with the funding/sanction approval as well as HSE obligations until operations team acceptance. Execute is the construction, fabrication, commissioning, and start-up consistent with Detail Design plans, schedule, and budget.

Detail Design & Execute Phase HSE Goals

The goals of the HSE process during Detail Design & Execute Phase are:

- ensure HSE hazards are continually identified and managed until final resolution
- minimize impact of HSE on project costs and schedule
- confirm the facility is ready for start-up

INHERENT SAFETY IMPLEMENTATION METHODS

Two (2) of the more significant inherent safety implementation methods are aligning the project team expectations and identifying all the project HSE deliverables, especially the HEP Plan. Not typical hazard evaluation techniques of site selection with HSE input, preliminary hazard review, essential/safety system survivability assessment, and essential/safety system performance objectives are discussed.

Project Team Expectations and Deliverables:

Ideally, there is a Corporate or Corporate Engineering procedure that addresses the:

- HSE expectations for the project team including HSE and Operations personnel
- each persons roles, responsibilities and competencies

- the formal risk management process including tracking and resolution of recommendations or hazard action items

- the expected HEP Plan

The HEP Plan items should be reflected in the project contract documents, scope of work, and the list of deliverables, actions, and requirements (DAR). The HSE DARs should be defined consistent with other project DARs including the resources, responsibility, and deliverer for each budgeted and scheduled HSE DAR.

Developing a Hazard Evaluation Procedures Plan:

Experience has shown that to meet the commitments of inherent safety effectively and within project cost and schedule constraints, the application of a robust HSE/Risk management strategy is required. A key part of this strategy is the application of hazard evaluation procedures (HEP) to aid in the identification, understanding, and evaluation of hazards. There are a vast number and range of HEP techniques that can be utilized to support this process. Each company and project should develop an HEP Plan defining the various hazard assessments and studies to be conducted during the project.

The HEP Plan should aim to provide both a more inherently safe plant and lower cost design through focusing effort in the early project design phases. A proactive process adopted where opportunities to reduce risk are identified in time to implement them, rather than a reactive one, which depends on retrospective analyses of the completed design will result in inherently safer plants. It is important that the HEP Plan evolves in advance of the project program to ensure appropriate and timely influence can be exerted on the design process at each project phase.

Developing the HEP Plan is usually an iterative process within each project phase. As with projects, the HEP Plan will have increasing definition and accuracy as the project progresses from the Select to Detail Design & Execute Phases. The HEP Plan for each phase should support and reflect the overall project goals for that phase.

A preliminary HEP Plan should be completed early in the project in conjunction with the development of the phase's Project Execution Plan. The HEP Plan should be issued after completion of the planning HEPs such as HAZIDs, review of safety concerns from similar projects, and performance objectives for safety related systems, as appropriate. Hazard concerns from completed HEPs may lead to additional HEPs to understand or accept the risk of the hazard. The HEP Plan should be revised as necessary to reflect changes to project milestones, availability of deliverables, and other changes.

In developing the HEP Plan, consider:

- HEPs should lead engineering throughout the project – that is, be a gate or condition for progressing the design. This ensures that the design is influenced at the right time so that changes can be made with minimal (or zero) cost and schedule impact.

- HEP should be proactive. In this way, the analysis can positively influence the design process by suggesting alternative solutions. Hazard and Operability Studies (HAZOPs) are confirmation studies evaluating only what has been done and rarely challenging the fundamental decisions (e.g., hot oil or pressurized water for heat transfer) necessary to maximize inherent safety.

Table 2 shows some typical HEPs by Project Phase.

Table 2: Typical HEPs by Project Phase

Select Phase	FEED/Define Phase	Detail Design & Execute
Hazard Identification (HAZID) Global Project and focused if applicable	HAZID early in FEED/Define	HAZOP: - Early detail design (implement design freeze) - Vendor packages - Final design (implement MOC) - Validation/Pre-Startup Safety Review
Comparative Concept Risk Analysis between concepts and/or existing plants	Layout Reviews for constructability and maintainability	Human Factors Assessment
Site Selection HSE Input, Coarse Worst Case Scenario, and Facility Siting Analysis	Safety Integrity Level (SIL) Classification and Architecture Determination	Essential Safety System Survivability Assessment
	Preliminary Facility Siting/Consequence Analysis including flare and vent impacts	
	Significant Noise Source Study	
	Preliminary Hazard Review	

Typical HEPs are used consistent with Center for Chemical Process Safety descriptions and practices except the Preliminary Hazard Review and the Essential Safety System Survivability Assessment. Site Selection with HSE input is also not common.

Site Selection with HSE Input:

Inherent Safety begins with Site Selection. Usually completed during the Select Phase, site selection does not consistently include HSE input other than environmental due diligence. The evaluation factors rated for site selection typically are limited to economic factors such as infrastructure, workforce, environmental permitting complexity, and taxes. Just a few HSE evaluation factors to considered include assessment of:

- worst-case scenarios from the site and transportation to nearby communities
- impact of neighboring facilities worst-case scenarios on the proposed facility
- geological site activity and potential impact on design and loss of containment scenarios

Usually, site selection factors have defined criteria sheets that describe a rating (e.g., Unacceptable to Excellent, 1 to 5). Factors are averaged by category and explanations and/or costs are usually provided for Unacceptable (5) and Poor (4) ratings.

Evaluation of sites then includes the overall average, averages by category, costs above and below a baseline, and explanations of negative HSE issues or constraints that may override the averages and costs.

Preliminary Hazard Review:

The Preliminary Hazard Review is conducted near the end of FEED/Define to ensure the funding/sanction cost estimate and execution plans include all significant issues/hazards and provisions for their safeguards. This is usually completed as a structured what-if for the areas of:

1. Site Conditions, General Layout & Equipment Arrangement (a few topics are:
   - Weather Design Basis
   - Geotechnical & Underground Interferences
   - Flare
   - PSM/RMP Facility Siting Reports
   - Roads & Traffic Patterns & Clearances, etc.)

2. Buildings and Structures (a few topics are:
   - Buildings and their Functions
   - Building Occupancy Criteria
   - Building sizing basis and sensitivity, etc.)

3. Process Systems (a few topics are:
   - Process Materials and their Hazards
   - Operating Phases (startup (normal, post-shutdown, post-maintenance, etc.)))

4. Utilities and Services

5. Fire Protection (not already covered)

6. Environmental (not already covered)

7. Safety (not already covered)

8. Construction Plans and Construction Safety

9. Commissioning and Startup

Essential/Safety System Survivability Assessment:

The Essential/Safety System Survivability Assessment (ESSA) is a frequently completed offshore HEP that has good potential for onshore plants as well. The goals of this study are to provide independent design confirmation that the:

- essential (i.e., safety) systems and their critical functions, usually as performance objectives, have been identified

- their design and configuration are appropriate to the credible major hazard scenarios

An independent, knowledgeable HSE engineer interfacing with appropriate discipline engineers, reviews design documentation to complete this study. The methodology (Figure 4) is to review all project systems to identify those essential safety systems necessary to prevent, detect, control or mitigate, respond or recovery to a major hazard scenario. Critical functions for each safety essential system are then developed. The capability of each safety essential system fulfilling its essential functions is assessed during credible major hazard scenarios. That is, the safety essential system should meet the applicable design codes and not fail other than safe if vulnerable to applicable hazard scenarios. If the safety essential system is not fail-safe, then is it vulnerable, sufficiently redundant, or sufficiently hardened for the duration the system is expected to function.

Conclusions are recommendations for improvement and/or identification of the credible major hazard scenarios where the essential system fails to fulfill its critical functions.

An Essential/Safety System Survivability Assessment is a confirmation study and should only be started when most relevant project documentation is sufficiently set or firm. Typically it is completed in the Detail Design & Execute Phase prior to "Issue For Construction" documents.

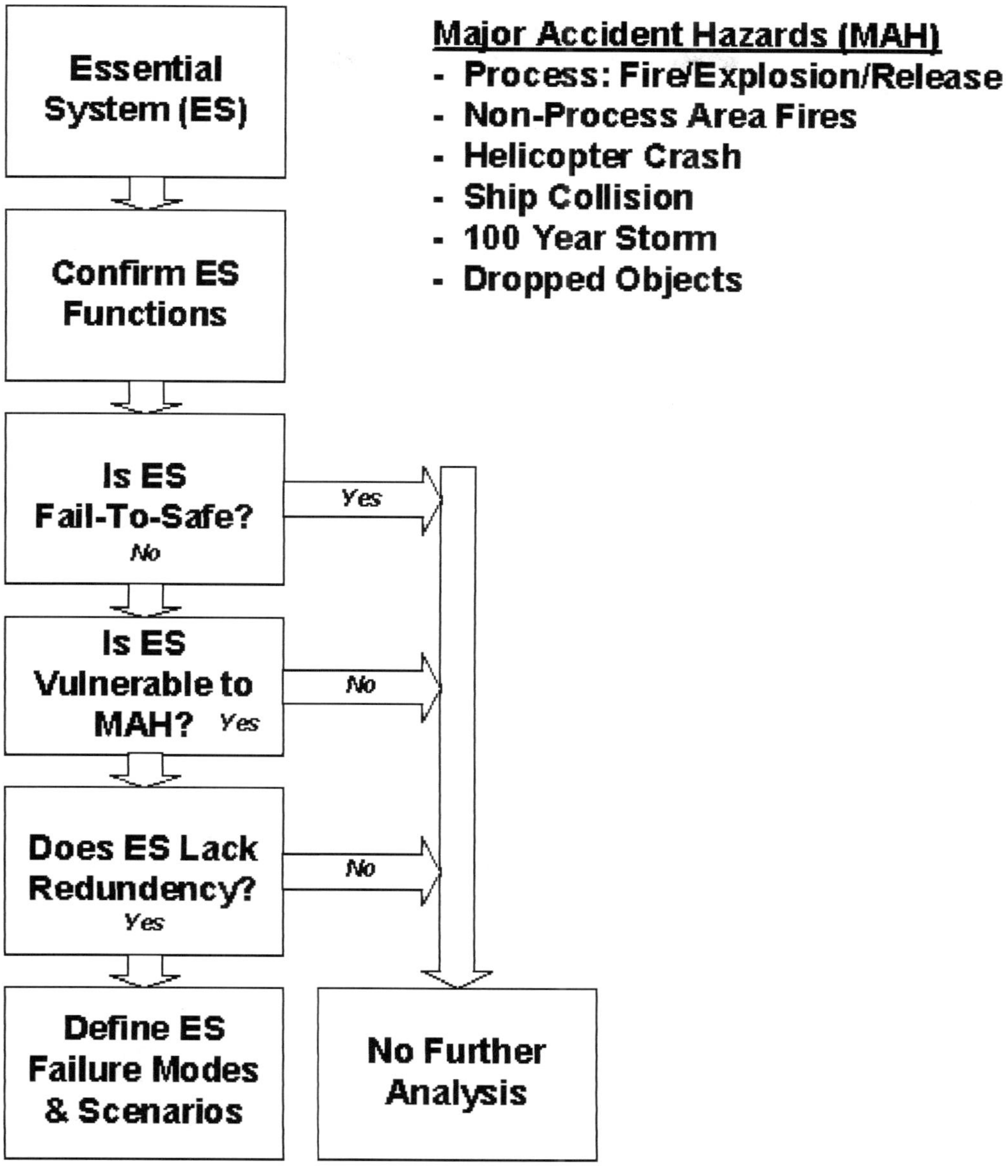

Figure 4: Essential Safety System Survivability Assessment

A fire protection/Safety Instrumented System example finding is shown in Figure 5 – As Found. An inherently safer design is shown in Figure 6 – Inherently Safer.

Which is inherently safer?

"As Found", Figure 5, actuation solenoid does fail-to-open tripping the deluge valve but the detection, power, and actuation cable are redundant and harden for 15 minutes. But a fire scenario could fail the power supply causing multiple deluge valves to open / trip probably overtaxing the fire water supply. This design would probably be recognized by instrument / control engineers as inherently safer.

"Inherently Safer", Figure 6, actuation solenoid fails-to-close not allowing the deluge valve to trip on power failure. But the deluge valve control is also hydraulically / pneumatically connected to the thermal sensing elements (pilot sprinklers or thermal plugs) making the electronic detection and control fully redundant. This design would generally be recognized as inherently safer by fire protection engineers.

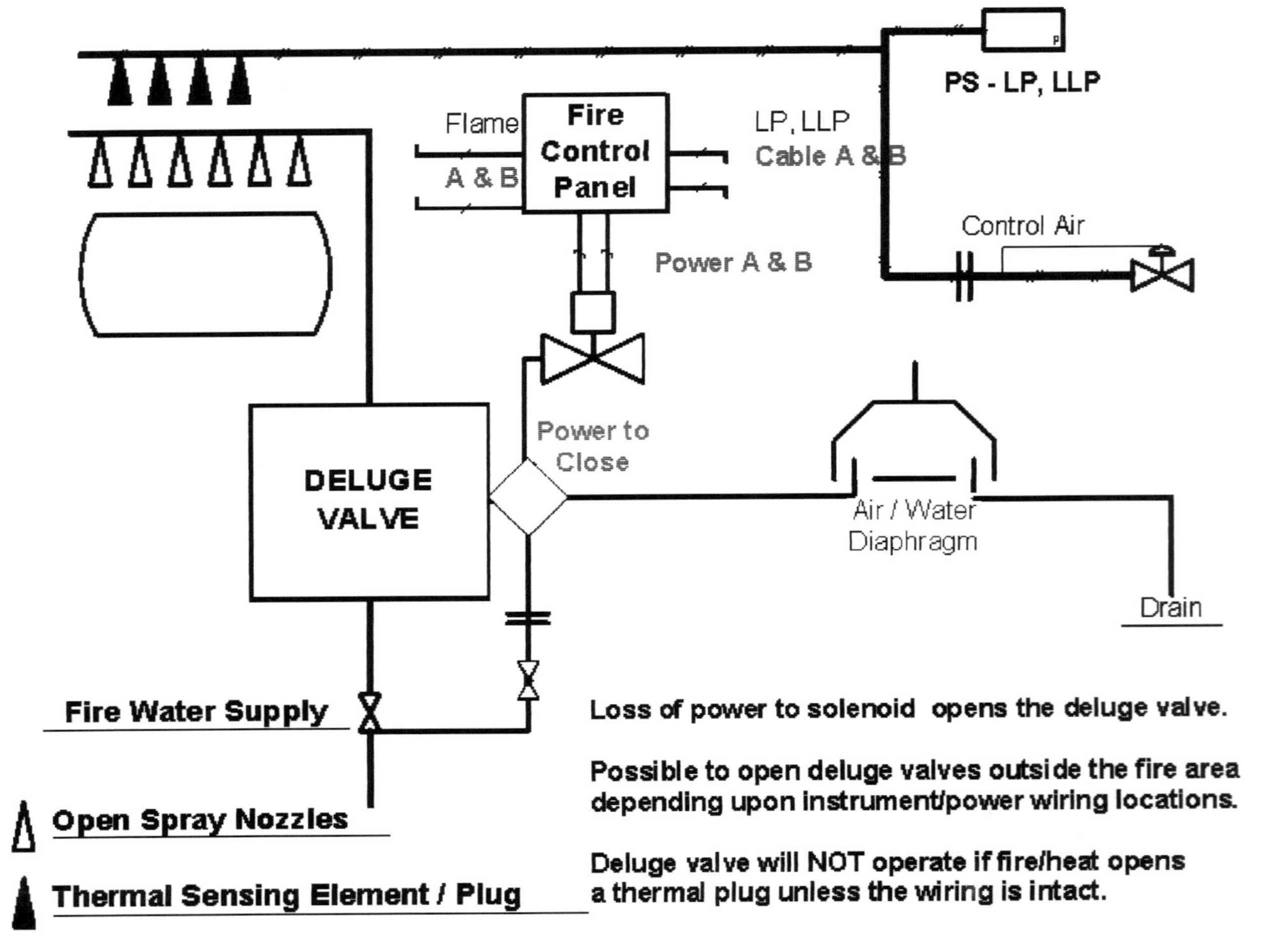

Figure 5: As Found

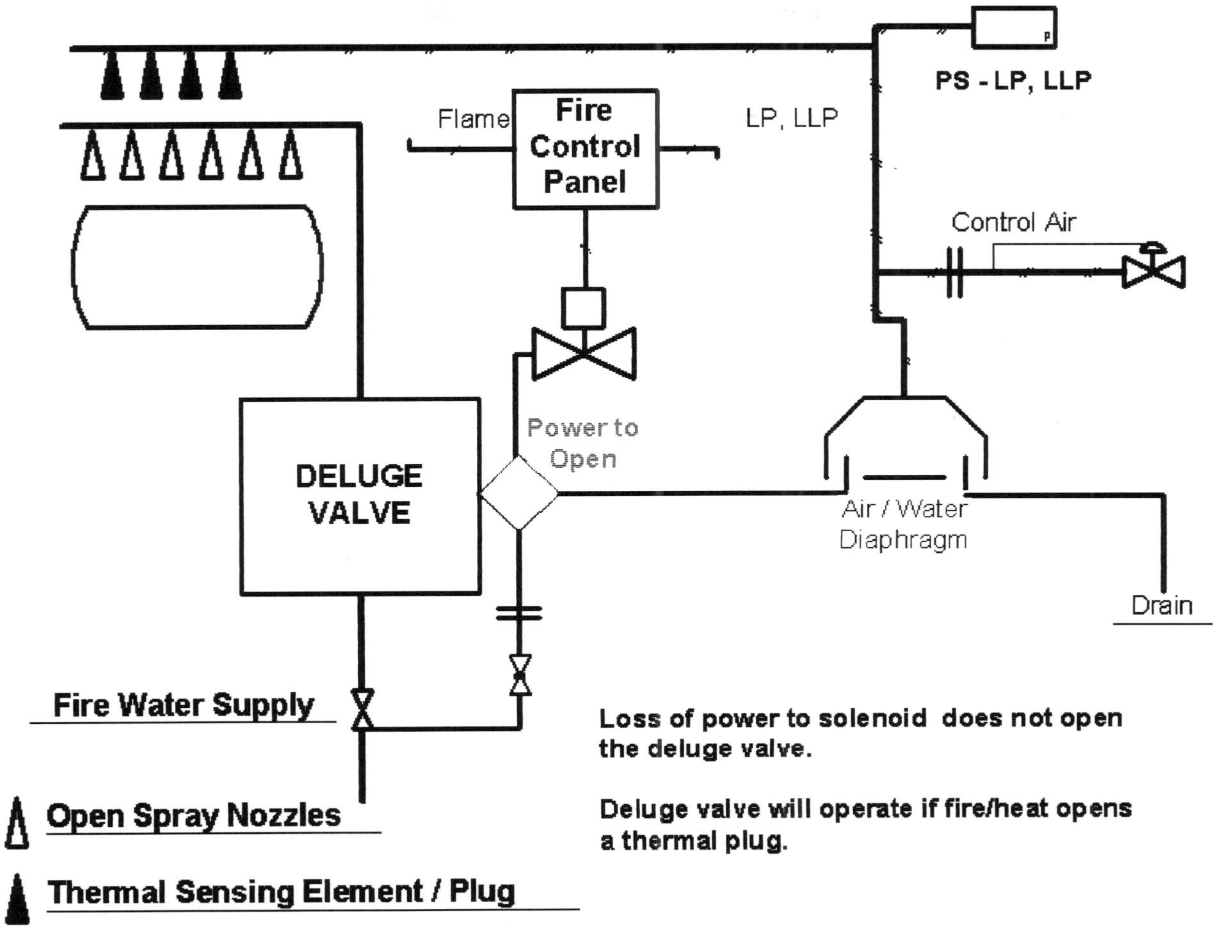

Figure 6: Inherently Safer

Essential/Safety System Performance Objectives:

To support the confirmation hazard evaluations and risk acceptance it is helpful to identify essential/safety systems and their performance objectives or goal. That is, what is expected to be accomplished? The performance objectives of a flare system, consistent with API RP 521 "Guide for Pressure Relief and Depressurizing Systems" for example could be to:

- safely dispose of hydrocarbon materials resulting from a process upset or emergency so that HSE consequences are either prevented or mitigated
- capable of operating during emergency situations with or without support utilities and with no operator intervention

A macro view of performance objectives applied to fire protection and the relationship between systems that limit the fire size and duration and those that limit the consequences is shown in Figure 7. A balance in damage/outcome is needed between fire, size, and duration and fire protection features for an acceptable design consistent with the risk acceptance procedure and fire protection philosophy.

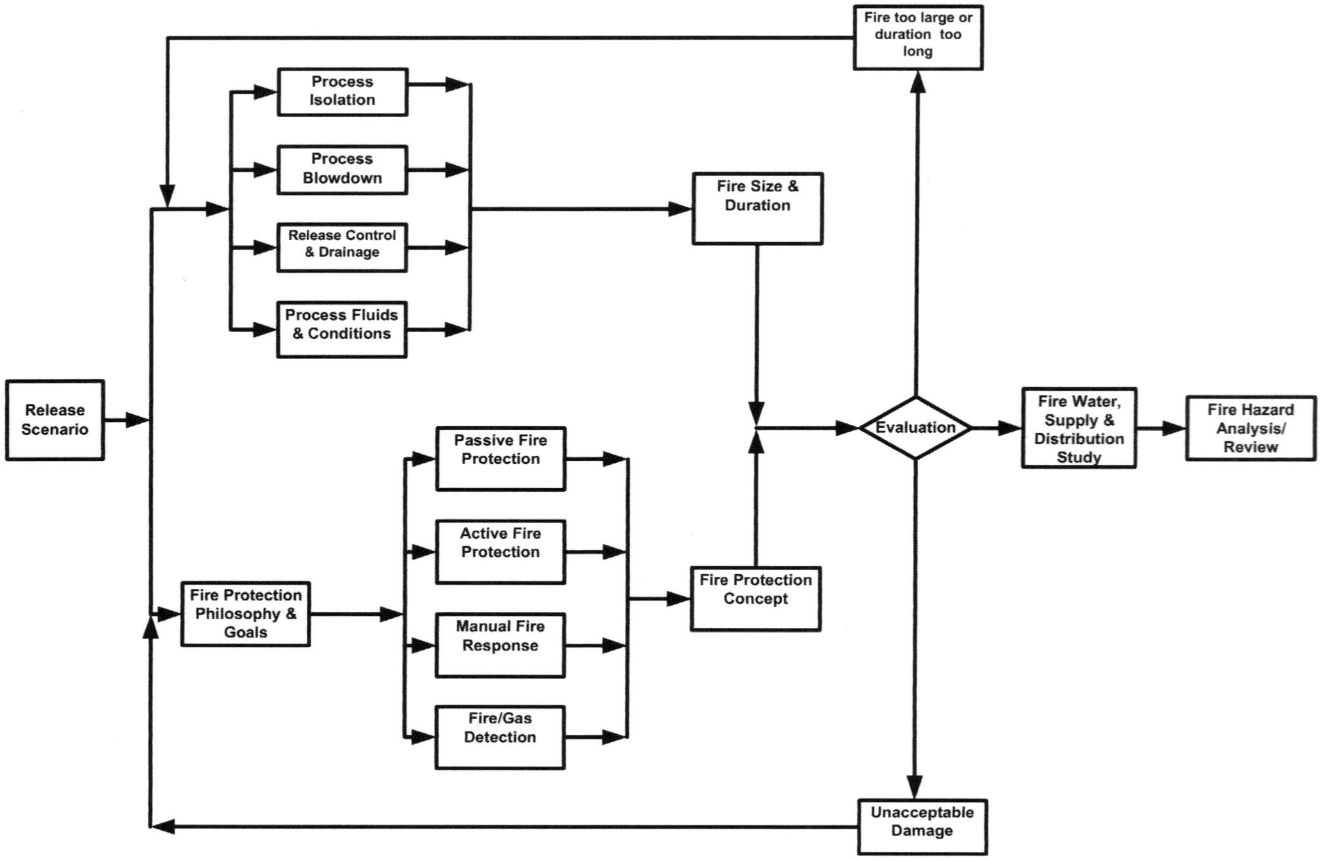

Figure 7: Performance Fire Protection

Fire, size and duration limiting features include process isolation, process blowdown, release control and drainage, and process fluids and conditions. A difficult to implement fire protection option is to reduce the operating pressures and the piping / vessel sizes. Lower pressures and lower volumes can reduce fire size and duration but are often at odds with the necessary production rate to be profitable.

Example Performance Objective – Water-based Fire Protection Systems:

For water-based fire protection systems performance objectives would involve considering the possible objectives by design credible fire scenarios include:

<u>Fire Extinguishment</u> – The objective of firefighting is to extinguish all fires. However, it is isolation of the fuel that is the method of extinguishment appropriate to process gas fires or pressurized liquid fires. Extinguishment is appropriate for ordinary combustible materials such as found in shops and warehouses. Foam-water systems (NFPA 11 "Standard for Low-,

Medium-, and High-Expansion Foam" and NFPA 16 "Standard for the Installation of Foam-Water Sprinkler and Foam-Water Spray Systems") can be used to extinguish liquid pool fires once the flow of fuel is stopped.

Fire Control – Application of water to the fire and its immediate surrounding area to absorb the heat produced. Typically used when the fuel release cannot be stopped and there are no important equipment or other exposures requiring protection. Automatic sprinklers (NFPA 13) and deluge sprinkler systems (NFPA 13 "Standard for the Installation of Sprinkler Systems", NFPA 15 "Standard for the Installation of Water-Spray Fixed Systems for Fire Protection", and API RP 2030 "Application of Fixed Water Spray Systems for Fire Protection in the Petroleum Industry") generally assume fire control.

Fire Exposure Protection – Application of water to equipment and other surfaces to minimize the damaging effects of heat. Typically used when the fuel release cannot be stopped and there are important equipment or other exposures requiring protection. Water spray systems (NFPA 15 & API RP 2030) generally assume radiant heat fire exposure protection.

Release Control – Application of water spray for the dissipation or as a barrier between a release source and an ignition source. Typically used when a gaseous fuel release cannot be stopped and ignition prevention is desired.

Application rates vary depending on the type of fuel, the type of incident anticipated and the intended objectiv of the water spray system. API RP 2030 suggests application rates that range from 0.1 gpm/sq.ft. to 0.25 gpm/sq. ft. and above. NFPA 15 (Water Spray) also recommends various application rates. The protection of:

- cables utilizes the minimum application rate of 0.15 gpm/sq.ft.
- fire source protection the minimum application rate is 0.5 gpm/sq.ft.
- Area coverage utilizes a minimum application rate of 0.25 gpm/sq.ft.

These are minimum application rates based upon radiant heat exposure from pool fires – not flame impingement. Hundreds of gallons per minute are necessary for cooling of flame impingement. The size of the fire and the quantity of protection needed for the particular protected area should be assessed to determine if these minimums are appropriate for the fire exposure.

In some cases significantly higher flows are required to provide the protection desired. Single large high pressure equipment such as valves may demand flows of 1,000 to 4,000 gpm and more to provide the desired results. The purpose of these special water spray systems is to control the rate of burning and provide for exposure protection to the equipment, piping and the exposed structure. The nozzles are designed to push a release and fire from the area of the equipment off to the side. If there is no ignition, the action of these water spray systems is to push the vapors and liquids away from the equipment and disperse the vapors by ingesting air into the release. The nozzles are selected, aimed, and located to provide strong ventilation to the area, fire control, and exposure protection.

POSSIBLE INHERENT SAFETY ACHIEVEMENT:

Inherently Safer plants can be achieved in process plant designs. This authors experience is that the design and construction of the inherently safer plants is cost neutral and operating costs can be greatly reduced. Some projects will cost more (maybe 1 – 3%) in evaluating design options and HEPs because of additional personnel and consultant time. Other projects have cost less because fewer and simpler essential/safety systems were needed. Operating costs are greatly reduced when the process plant design is robust reducing waste or rework product and, hopefully, fewer mistakes and accidents.

Table 3: Possible Inherent Safety Achievement

Eliminate:	By end of Select Phase: 1. Maximum implementation of elimination of hazards is usually complete. Usually this is process design driven and involves inventories, process fluids and conditions. An excellent eliminate example is pressurized water ethylene oxide (EO) reactor designs that eliminate hot oil circulation systems that are typically 10,000 to 40,000 gpm. Likewise, the production of ethylene oxide glycols without concentrating ethylene oxide. Another eliminate example is a fluorination process that consumes more than 100 rail cars of anhydrous hydrogen fluoride (HF) each year. Alternates investigated were fluorination with aqueous HF acid and production of HF gas from fluorspar without liquefaction.
Prevent:	By end of Select Phase: 1. site selection inherent safety can be accomplished 2. completion of the Hazard Evaluation Procedure Plan for the project (See also Table 2) By end of FEED/Define Phase: 1. minimum inventories are usually established as well as essential/safety systems. Continuing with the fluorination process, the two (2) rail car unloading was located to an enclosed building with a fume scrubber to eliminate possible releases. This relocation into a building with a scrubber reduced the societal risk by two (2) orders of magnitude.

Table 3 (cont.): Possible Inherent Safety Achievement

Prevent (cont.):	By end of FEED/Define Phase or Early Detail Design: 1. use of flanges can be minimized to reduce potential leak points 2. frequent failure items such as eliminating expansion joints and sight bullet glasses in vessels must be identified before purchasing vessels and equipment By end of Detail Design: 1. Human Factors Assessments have been completed throughout the design beginning with work station evaluation and continuing into placement and specification of valves, sample stations and other frequently operated or maintained equipment.
Detect / Control: **Mitigate:**	By end of FEED/Define Phase: 1. All key HSE hazards have been identified 2. All Essential/Safety systems have been identified and their System Performance Objectives are understood 3. Cost estimate, schedule, and Scope of Work including engineering contractor deliverables, actions, and requirements reflect all safety and hazard concerns 4. Operations personnel have been involved By end of Detail Design: 1. essential/safety system survivability assessment completed to assure that all essential systems are fail-safe, redundant, and/or hardened to expected hazards to perform their function for the duration of the possible hazard
Response/Recovery:	By end of Select Phase or Early FEED/Define Phase: 1. emergency response/recovery basis in terms of scenarios, expectations, capabilities, and available resources is confirmed.

Inherently Safer Technology: ***An Evolutionary Thing***

Tim Overton
The Dow Chemical Company
2301 N. Brazosport, Blvd.
Freeport, Texas 77541
Phone: 979-238-1358
Email: toverton@dow.com

George M. King
The Dow Chemical Company
2301 N. Brazosport, Blvd
Freeport, Texas 77541
Phone: 979-238-1329
Email: gmking@dow.com

ABSTRACT

As a long-time advocate of inherently safer technologies, Dow has a rich history of designing chemical processes that meet intrinsic standards for safety. Inherently safer technologies are not, however, a panacea for safe operations. Inherent safety must be complementary with other risk reduction strategies and is just one resource in our risk management arsenal. Effective use of this and other strategies help Dow uphold the stringent safety standards set by our company and the chemical industry, and contribute to the industry's reputation as being one of the safest in the world.

Why Inherent Safety?

Trevor Kletz, in his 1991 book *"Plant Design for Safety"* said,

> *"The essence of the inherently safer approach to plant design is the avoidance of hazards rather than their control by added-on protective equipment."* [1]

Dow embraces the use of inherently safer technologies, first and foremost because it is the right thing to do. Our Environmental, Health and Safety policy states, *"At Dow, protecting people and the environment is part of everything we do and every decision we make."*

Using inherently safer technology to reduce hazards in our process operations is just one of the ways we live out this policy for employees and the communities around the world where our plants are located.

The use of inherently safer technologies also is good for business. The chemical industry is a highly competitive, often commodity-based business. A one or two cent margin on a ton of product can have a significant impact on profitability, therefore any time we can implement new technology that saves money, our businesses are very interested.

Inherently safer technology can result in lower capital cost in new plant design, and typically produces lower operating costs, greater reliability and quicker start-up times. Plants with inherently safer technologies tend to be more simple in design, easier and more friendly to operate, and more tolerant of errors.

A Strategy that Starts Early

Dow supports the position advocated by Trevor Kletz[1] and much of the chemical industry that the most effective time to influence the inherent safety of a chemical process is during the initial design, not after the facility is built and in operation.

Dow's strategies begin at the earliest stages of plant development with the application of inherently safer technologies that take advantage of five key approaches.

1. **Intensification**, which calls for using smaller quantities of hazardous substances within a process. For example:

   - MDI plants at Dow's LaPorte, Texas, manufacturing site began a campaign in 1986 to reduce phosgene inventory through improved process control and in situ production. The table shows that implementing a series of major capital projects reduced phosgene inventory by 50 percent over 10 years while increasing MDI production – just one example of the impact of an evolutionary strategy for inherently safer technologies.
   - Dow's Union Carbide subsidiary redesigned the acrolein derivatives unit in Taft,

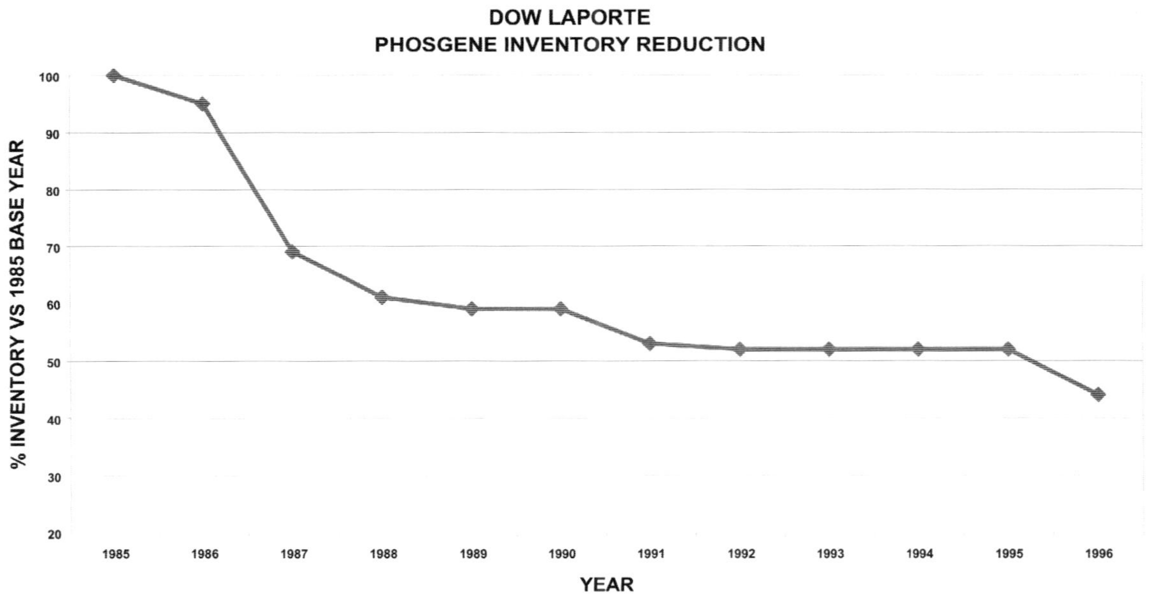

Louisiana during the mid-1980s to dramatically reduce acrolein inventory. In addition, by partnering with a key customer, another operating unit was added to convert acrolein to a less hazardous derivative at the Dow site prior to shipment to the customer site, thus reducing transportation risks.

   - Dow also uses continuous reaction processes instead of batch reaction processes to produce certain types of polyethylene and polypropylene plastic resins. This reduces the inventory of flammable material in these continuous reaction processes.

- Direct Chlorination using 95% Chlorine direct from Cl2 cells rather than purified Cl2 substantially reduces the amount of Cl2 storage.

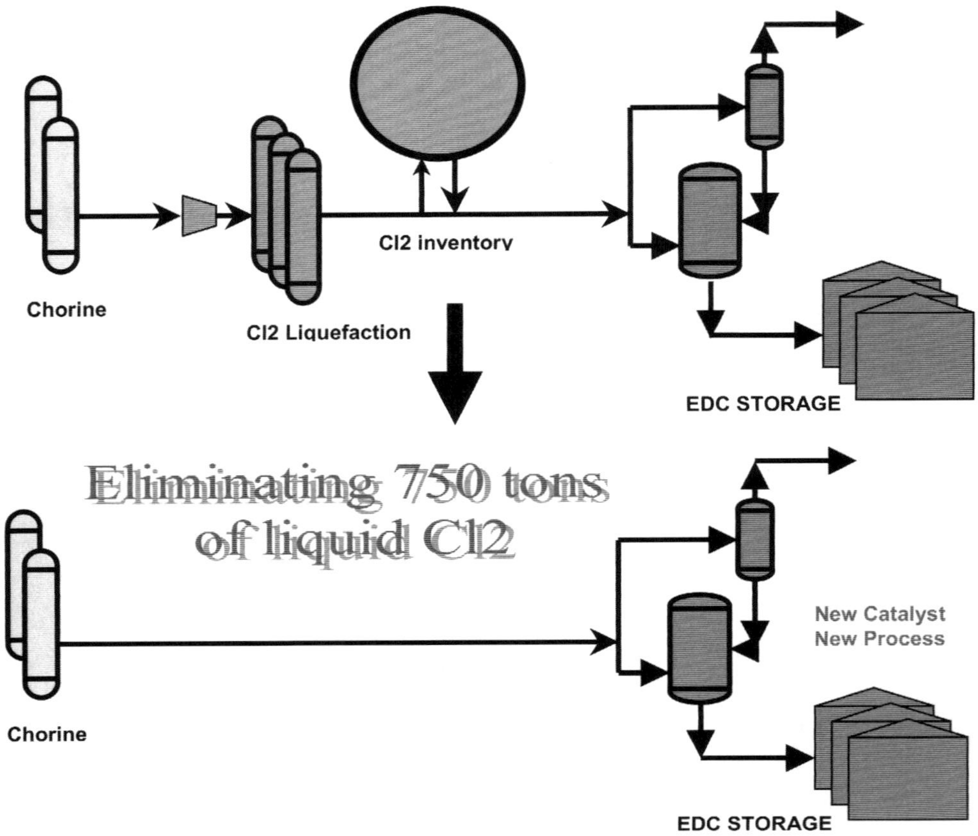

- Dow's licensed EO METEOR™ technology significantly reduces the portion of the process with concentrated EO.

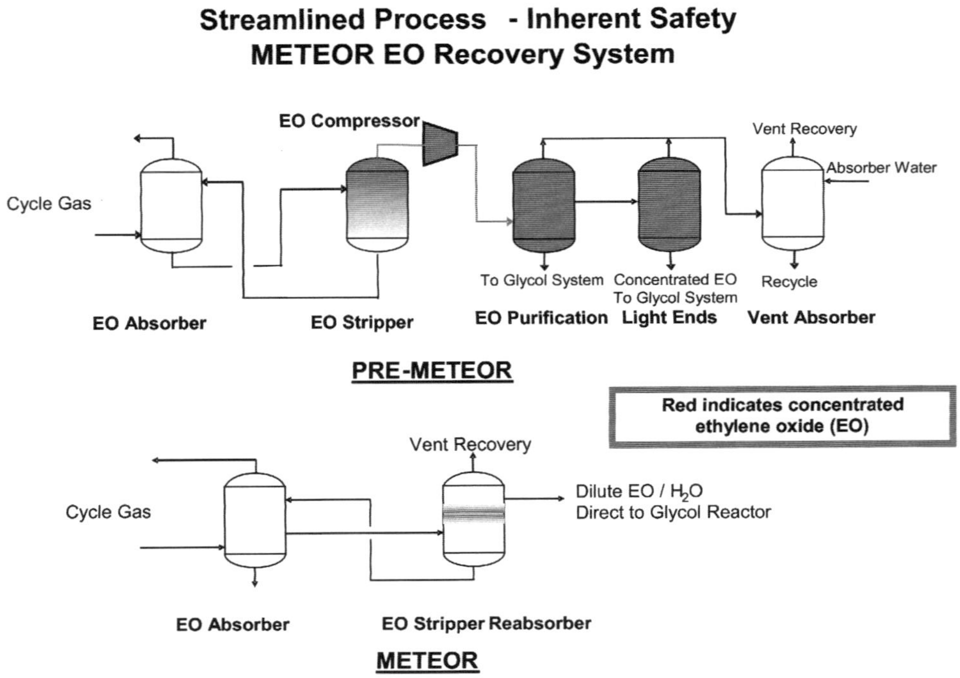

2. **Substitution**, a strategy to replace one material with a less hazardous substance if the chemistry of the process will allow. For example:

   - Dow had been using benzene as an azeotroping agent in an oxide derivatives unit, but replaced the benzene with a less hazardous chemical.

   - Dow replaced chlorine gas with sodium hypochlorite (bleach) as a water purification chemical at selected non-chlorine producing sites, and replaced Varsol in a process with another, less hazardous solvent.

3. **Attenuation**, which calls for using less hazardous conditions, such as lower temperature or pressure or a less hazardous form of material, if the chemistry allows. For example:

   - Dow's newest process for producing diphenyl oxide uses a catalyst to operate at temperatures and pressures that are significantly lower than previous processes that it replaced.

   - Storing highly toxic or flammable materials (e.g. chlorine, LNG) as a refrigerated liquid at atmospheric pressure rather than at atmospheric temperature under pressure reduces the potential consequences of an accidental spill.

   - Using an aqueous hydrogen chloride (HCl) solution rather anhydrous HCl in selected processes also reduces the potential consequences of a spill.

   - At one of our facilities near a populated area replaced the use of aqueous 36% HCl with 17% HCl. Although this did double the number of deliveries (i.e., increased transportation risks, increased risk associated with connection/disconnecting fill lines during deliveries), given that the partial pressure has reduced by a factor of 1000 (i.e., reduced toxic hazard in the event of a spill) it was viewed as an overall reduction in inherent risk.

4. **Limitation of effects** calls for designing a plant or process to minimize the impact of a release of material or energy. For example:

   - Multiple processes within Dow use small diameter piping or restricting orifices to limit the flow rate in the event of an accidental release or operational error, making it almost impossible for a reaction to occur too quickly.

   - In some cases, Dow uses process equipment designed to contain the maximum foreseeable pressures from the process, so pressure relief is not needed.

- Using high wall dikes and other types of secondary containment to severely limit the surface area available for evaporation of toxic liquid spills minimizes the impact, as shown in the illustration below.

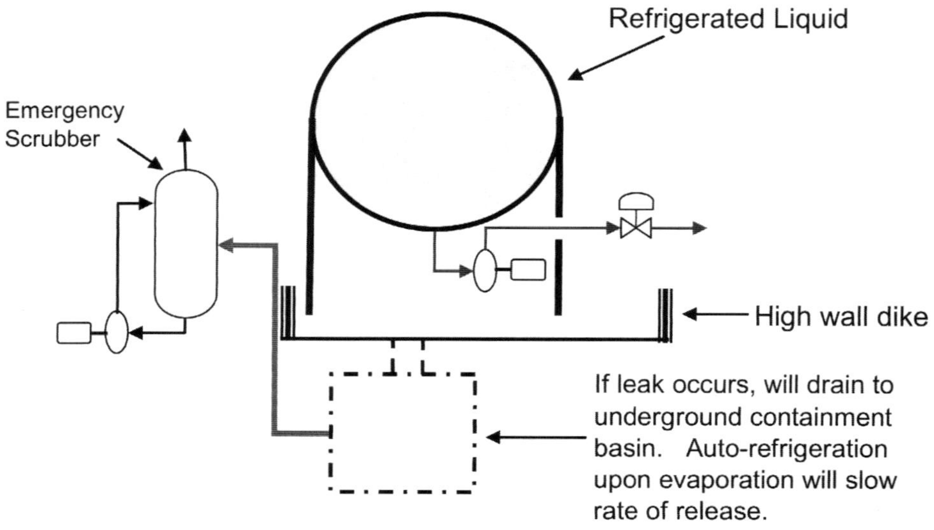

5. **Simplification and error tolerance** is designing to eliminate or tolerate operating errors by making the plant more user-friendly and reliable. For example:

    - Redesigned Latex reactor cleaning equipment eliminates the potential for incorrect installation that could result in unwanted reactivity or cross contamination.

    - Dow eliminated the use of hoses in several hazardous services in favor of hard piped connections.

Dow and other leading companies in the chemical industry consider all of these approaches and apply them where appropriate. In general, new plants of today have more of these features built into them than older plants of the past.

<u>Other Safe Operations Strategies</u>

Inherently safer technology, when properly applied, generally can become a competitive advantage. Having said that, inherent safety is not the solution to every challenge, but rather is an evolutionary process that takes several strategies into account. With inherently safer operations as a first step, from there Dow employs:

- Passive safeguards, which involve highly reliable systems that have a small likelihood of failure.
- Active controls and mitigation, including Safety Instrumented Systems, which tend to be less reliable than passive systems yet still highly effective.
- Procedures to assure that operating and maintenance practices are being carried out in a consistently safe fashion.

- A formal Operating Discipline Management System to ensure that all layers of protection necessary to improve chemical process safety are carried out for the life of our plants.

Implemented alone, none of these strategies can achieve the desired result. Together, they make a notable impression on process safety.

Lessons Learned
With decades of experience in applying inherently safer technology, Dow has learned several lessons along the way.

1. **Inherently safer technology is an evolutionary process.** It's never too late to consider inherent safety[2]. Dow uses an Inherent Safety Checklist as part of our Reactive Chemical / Process Hazard Analysis which is conducted on a 3-5 year cycle throughout the life of a chemical process. This is an opportunity to identify new technology breakthroughs that can be economically applied for greater inherent safety. Occasionally, significant risk reduction is achieved this way.

   With existing plants, it is not often economically feasible to apply new inherently safer technology. If the proposed change involves different chemicals or significantly different operating conditions, then it is likely that a major capital expenditure will be required to replace existing equipment.

2. **Inherently safer technology is best implemented early in a project when a chemical process is first designed.** New chemical processes will eventually replace the processes of today and when these new facilities are designed, fundamental changes in chemistry and materials can be made cost effectively. Process design depends on the physical and chemical properties of the materials being handled; changes can have far reaching implications in the equipment design and are best applied early in the engineering process.

   An Inherent Safety Checklist is now required as part of Dow's Global Project Methodology, our engineering work process. This checklist facilitates Inherently Safer Technology Reviews at two stages early enough in the project life to cost effectively make fundamental changes if needed.

3. **Inherently safer technology can be misapplied and result in higher risk if the "big picture" is not properly considered.** This mistaken approach merely shifts the risk from one area to another. Consider these examples:

   - Minimizing the size of a raw material storage tank will reduce potential impact in the process area, but if the tank is too small to unload an entire tank car of material at once, potential risk will increase. Why? During loading or unloading, when equipment is being temporarily connected and disconnected, the likelihood of a spill goes up. A smaller tank size would require more product transfers and therefore increase the risk. Smaller storage capacity on-site also means more frequent shipments, which increases transportation risks.

- Substituting bleach for chlorine in water treatment reduces risk at the site of the water treatment plant but increases the amount of chlorine required at the bleach manufacturing site. Is overall risk reduced? The irony of this high profile inherently safer step is that chlorine is still the active ingredient in the bleach. Someone, somewhere has to use chlorine to produce the bleach in the first place so the water treatment unit can handle the safer form of chlorine. If the bleach manufacturer who now has to use more chlorine is located in an area that is highly populated, chlorine must be shipped through that community. This scenario merely shifts, not eliminates, the risk. Also, because bleach requires more product to get the equivalent amount of active ingredient, shipments are actually increased. This results in additional transportation risks from the increased shipments, not to mention increased energy consumption and associated air emissions.

4. **Mandating Inherently Safer Technology is illogical. It would be unsound to attempt to regulate the use of inherently safer technology.** There are many facets to the effective implementation of inherently safer technology. Government regulation tends to dictate a "one-size fits all" approach. The complexity of process plants essentially prevents any prescriptive rules that would be widely applicable.[3]

5. **Inherently safer technologies are not a panacea for safe operations.** Inherent safety is only one tool that the chemical industry uses to keep its employees and surrounding communities safe. Inherently safer technology must be employed along with other risk reduction strategies. But IST is just one resource in our risk management arsenal.

   Because of the diverse chemical processes used to produce the thousands of products the chemical industry provides to the world, a customized approach to inherent safety is more appropriate. As seen in previous examples, a view of the entire value chain is needed to assure that risk is truly being reduced by proposed changes. Focusing too narrowly may result in ineffective misapplication of the inherently safer principles.

   Further, it takes expert knowledge of a chemical process and the entire value chain to properly evaluate the risk reduction potential of a proposed change. The people with the best knowledge of the chemical processes to make appropriate IST decisions work within the chemical industry. As demonstrated in this paper, the experts within industry are capable and willing to identify the applications that will bring true risk reduction.

Conclusion

Dow has used inherently safer technology for years in an effort to continually provide safer plant designs for the manufacture of a variety of chemical products. We strive to build inherent safety into our plants from the beginning and look for opportunities to further reduce process risks during the life of the process. We have learned that it is important to look at the "big picture" when applying inherently safer principles to assure that true risk reduction has occurred.

We are committed to protecting people and the environment as part of everything we do and every decision we make. The application of inherently safer technology helps us live up to that commitment.

References

1 Kletz, Trevor, *"Plant Design for Safety"*, Hemisphere Publishing Corporation, 1991

2 Carrithers, G. W., Dowell, A, M, and Hendershot, Dennis C., *"It's Never Too Late for Inherent Safety", International Conference and Workshop on Process Safety Management and Inherently Safer Processes*, October 8-11, 1996, Orlando, FL, 227-241, New York: American Institute of Chemical Engineers.

3 Mannan, M. Sam, *"Challenges In Implementing Inherent Safety Principles in New and Existing Chemical Processes"*, Mary Kay O'Connor Process Safety Center, 2002.

Paper Number 4

Sub-Atmospheric Pressure Storage and Delivery For Gases

Karl Olander and Peter Van Buskirk
ATMI, Inc.
7 Commerce Drive
Danbury, CT 06810

Prepared for Presentation at
American Institute of Chemical Engineers
2005 Spring National Meeting
Process Plant Safety Symposium
Atlanta, Georgia
April 11-13, 2005

December 10, 2004

UNPUBLISHED

INTRODUCTION

The use of compressed gases and their associated hazards have remained largely unchanged over the last 25 plus years. This paper examines how innovative changes to store and deliver gases can reduce the risk inherently posed by materials under pressure.

The consequence of a release from a pressurized cylinder and, to some extent, the probability of the event itself, is largely a function of pressure. Two distinctly different methods to achieve low [sub-atmospheric] pressure operation are described herein and contrasted with the operation of standard high-pressure cylinders. The technologies are based on [1] the use of sorbents or [2] an embedded set pressure regulator. Both approaches achieve low-pressure discharge from the cylinder: one altering the state of the gas while the other achieves similar performance using purely mechanical means.

Both technical approaches have been implemented as products and are in use within the semiconductor industry. One objective of this paper is to expand general knowledge of how they function and find uses in a wider range of industries and applications where similar risk reduction, process reliability and efficiency benefits can be achieved.

ADSORBED PHASE GAS SOURCES

A significant innovation in gas packaging, "sub-atmospheric pressure gas sources", became available, in the mid-nineties.[1] A "gas source" is, in reality, a high surface area sorbent, 1200 meters squared per gram, holding a gas physisorbed within the pore structure. Sorbents typically adsorb gas at 30-60% by weight to a final pressure of ca. 700 torr. A typical isotherm is shown in Figure 1.

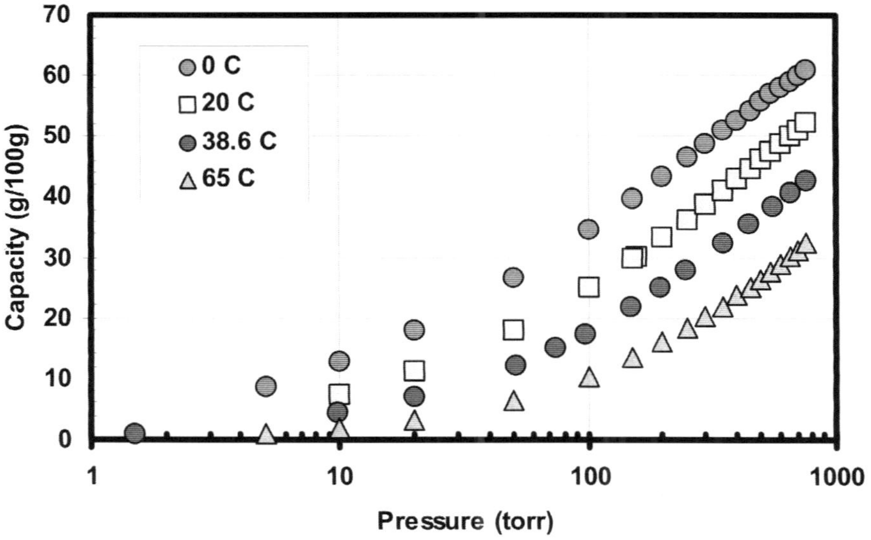

Figure 1. Arsine Adsorption Isotherm As A Function Of Temperature.

Shipped at a pressure slightly below one atmosphere, gas sources are intrinsically safe have the characteristics of [toxic] solids. Unless perturbed by temperature, gases stay adsorbed and release rates from open cylinders are negligible. The combination of a gas stored on a sorbent removes pressure as a risk factor and creates an inherently safe gas storage option.

Since introduction, about 50,000 "adsorbed phase" packages have been shipped and the technology has become a preferred methodology to store and deliver highly toxic dopants used in the semiconductor industry. Recently NFPA 318 was ammended to differentiate the use of sub-atmospheric pressure sources.

In most cases the motive force to discharge the gas is provided by a vacuum source, typically a pump, discharging to a process. In the semiconductor manufacturing many processes are run under high vacuum.

Sub-atmospheric Pressure Gas Sources Applications

The DOT has approved transportation of 28 hazardous materials in "adsorbed phase" packages to a maximum internal volume of 450 liters [2] Of the materials listed in Table 1, hydrides, corrosives and flammable materials will likely remain primary candidates for adsorbed phase packaging. Strong oxidixers like F2, ClF3 and NF3 are incompatible with adsorbents in use today.

Table 1: Frequently Used Specialty Gases.

Material	Hazards		IDLH [ppm]	LEL [%]
Ammonia	Corrosive		300	15.5
Arsine	Toxic	Flammable	3	4.0
Boron Trifluoride	Corrosive	Toxic	25	N/A
Chlorine	Toxic	Oxidizer	10	N/A
Carbon Monoxide	Toxic	Flammable	1200	12.5
Dichlorosilane	Flammable	Corrosive	50	4.1
Fluorine Mixes	Corrosive	Toxic	25	N/A
Germane	Toxic	Flammable	---	---
Hydrogen Bromide	Corrosive	Toxic	30	N/A
Hydrogen Chloride	Corrosive	Toxic	50	N/A
Hydrogen Fluoride	Corrosive	Toxic	30	N/A
Hydrogen Sulfide	Toxic	Flammable	100	4.3
Nitrogen Trifluoride	Oxidizer	Toxic	1000	N/A
Phosgene	Toxic		2	
Phosphine	Toxic	Flammable	50	1.6
Silane	Flammable		---	1.4
Silicon Tetrafluoride	Toxic	Corrosive	100	N/A
Trichlorosilane	Flammable	Corrosive	100	N/A

The initial use of adsorbed phase gas sources was coupled with processes run at sufficiently low pressures to withdraw and deliver the stored inventory. The use of a mechanical pump opens up a broader spectrum of applications. Using a bellows pump, for example, adsorbed gases are withdrawn and delivered at rates of over 10 SLPM; while the cylinder remains sub-atmospheric pressure. In this arrangement only a small fraction of the total inventory is ever above atmospheric pressure.

Adsorbed phase gas sources limit the worst case scenario because the amount of gas which can be released is inherently small. At 70°F the amount of "free or unadsorbed gas" in a cylinder, at 650 torr, is ca. 0.5% of the total inventory. Unlike pressurized cylinders, which can release 100% of their contents, losses from an adsorbed phase package are governed by diffusion. Studies with adsorbed phase gas sources have shown that "worst case" release rates are reduced approximately 10,000 fold vs. high pressure cylinders with restrictive flow orifices [RFO]. Gas releases are intermittent and small, measured in milligrams per minute.

In addition to risk reduction, system maintenance costs are also reduced. Service pressure has been shown to directly influence the lifetime of components used in gas delivery manifolds. A tied diaphragm cylinder valve can last 5-10 times longer operating at sub-atmospheric pressure.[3] It is postulated that the integrity of valves and regulators will be similarly extended.

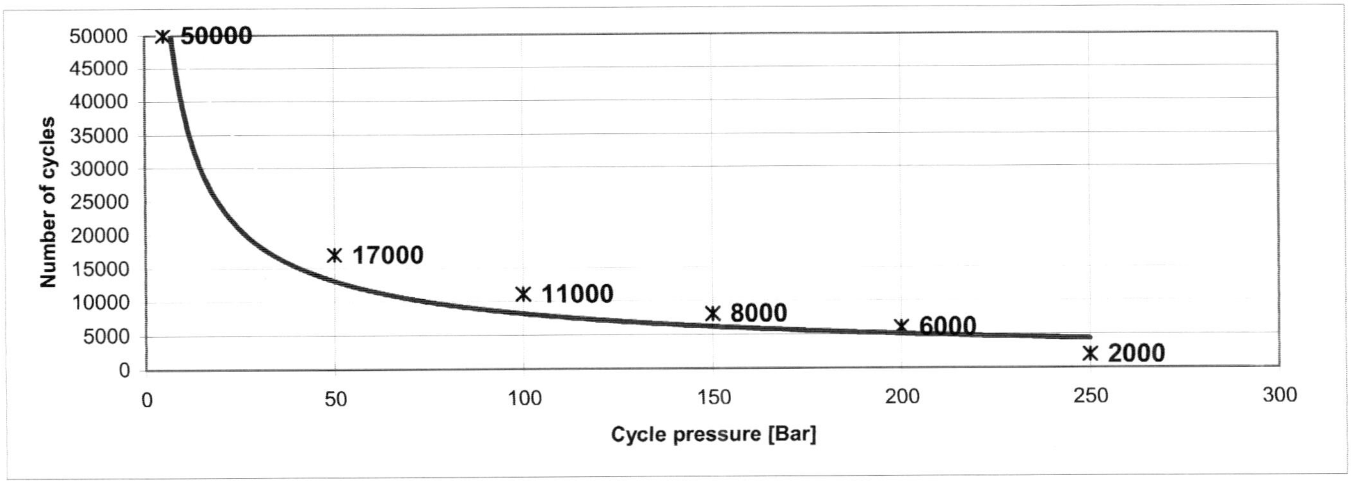

Figure 2. Cylinder Valve Lifetime Versus Operating Pressure

PRE-REGULATION OF PRESSURIZED CYLINDERS

The need to extend sub-atmospheric pressure gas delivery to a wider range of gases and achieve higher fill densities led to the development of a purely mechanical delivery protocol. The vacuum actuated cylinder [VAC] incorporates a pair of set pressure regulator inside the cylinder. Pre-regulating the delivery pressure effectively decouples the gas inventory pressure from downstream operations and mimics the behavior of the adsorbed phase gas source.

The Failure Mode and Effects Analysis for a conventional high pressure cylinder can be generally divided into two categories [1] cylinder related failures and [2] failures with the valve and downstream connections and components. Applying the condition that flow not occur unless the downstream manifold be at a prescribed sub-atmospheric pressure [e.g. 500 torr], removes most of the potential failures from consideration. There remains the possibility of loss of containment or rupture of a pressure relief device [where applicable], both catastrophic events. These however are extremem;y low probability events. Catastrophic cylinder failures are estimated at once per 10,000 years . Neck leaks are more common but release rates are typically < 1 sccm and are usually detected during manufacture or distribution.

Sub-atmospheric Pressure Operation

The Vacuum Actuated Cylinder [VAC® source] achieves sub-atmospheric pressure delivery, by placing set pressure regulators inside the cylinder before the valve.[4] With the VAC package, discharge pressure is fixed over the life of the cylinder--sub-atmospheric pressure [e.g. 500 torr] being common. Super-atmospheric pressure operation is also possible. The VAC system can be used with compressed and liquified gases as well as gas blends.

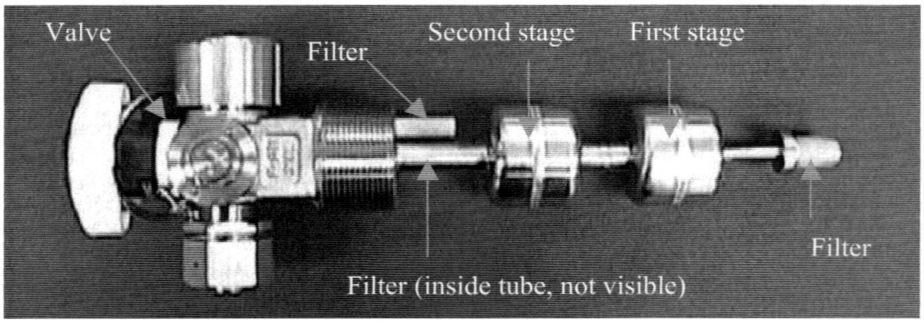

Figure 3: Cylinder Valve With Set Pressure Regulators Attached.

Figure 4: VAC Cylinder Cutaway Showing Regulator Installation.

When set up for sub-atmospheric pressure delivery, the regulator is normally closed. Placing the regulator inside the cylinder affords protection from physical damage and makes it tamper proof. Filters are placed upstream and downstream of the regulator to prevent particles from interfering with it's operaiton. Placement of the regulators is shown in Figures 3 and 4.

As gas flows through the regulator, a bellows unit reacts to pressure changes by expanding or contracting to move a poppet in the regulator orifice and module the discharge pressure. For sub-atmospheric pressure operation dual stage regulation is used to provide greater flow precision and offset supply pressure effects.

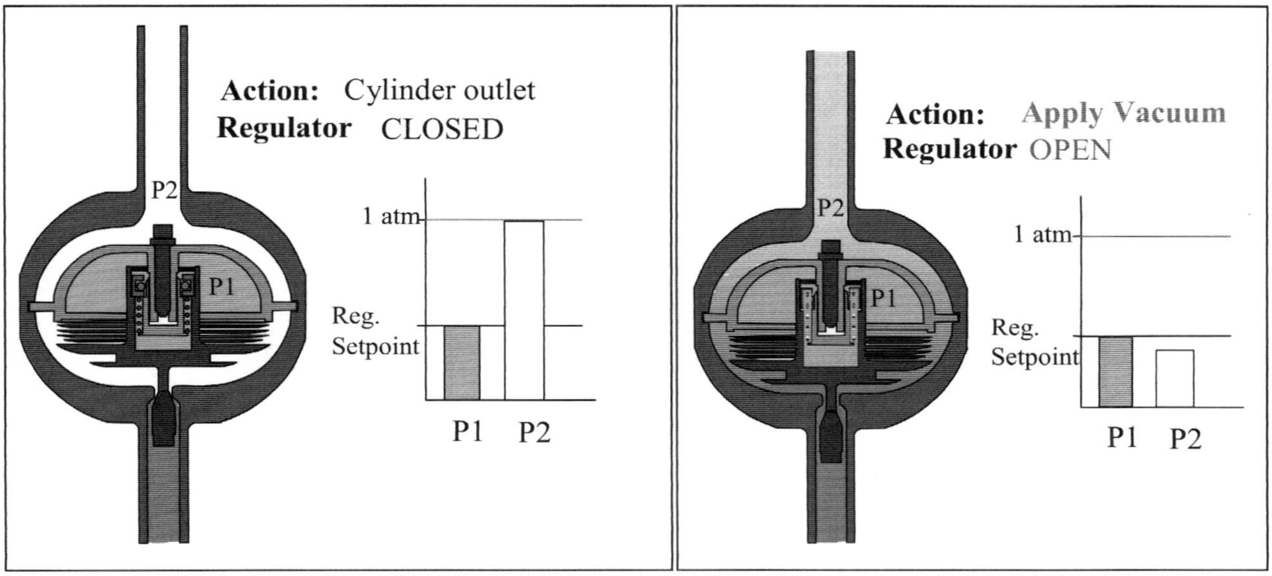

Figure 5. Regulator Operation

Lifetime estimates for the regulator are robust—less than a one percent failure rate is predicted after five years service. A regulator failure would result in a no flow situation and require replacement of the cylinder.

Pressure transducers located in the delivery manifold provide another safeguard. Should the delivery pressure exceed a set threshold, a pneumatic operator is actuated on the cylinder valve to isolate the cylinder and insure low pressure operation is maintained.

Using the VAC package in a sub-atmospheric pressure mode may require a vacuum assist for processes run at super-atmospheric pressures. This is depicted in Figure 6.

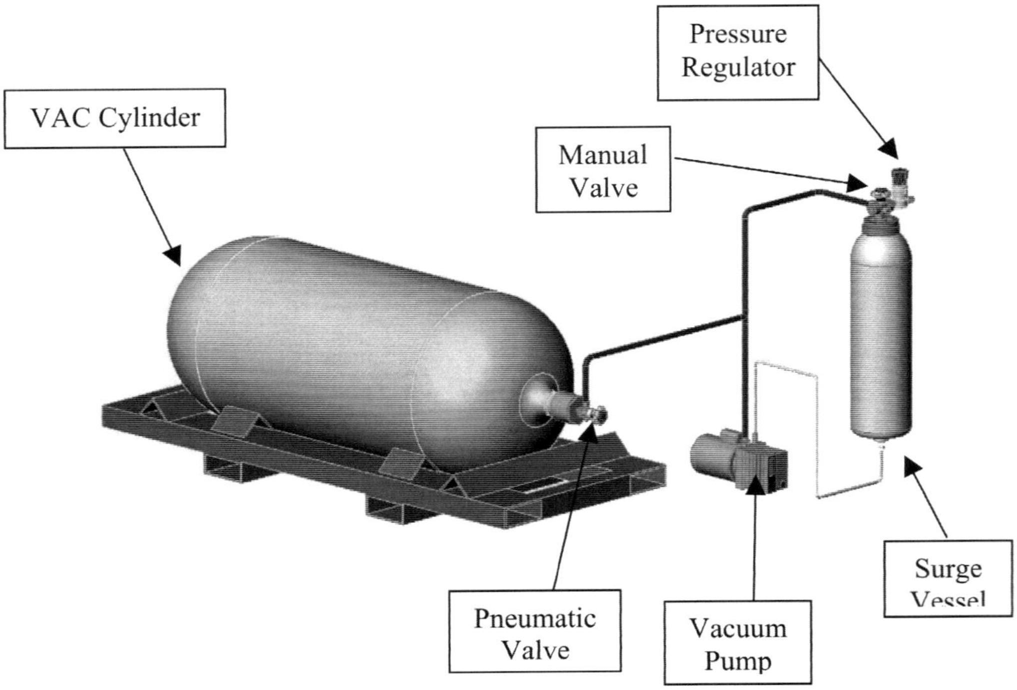

Figure 6. VAC Cylinder and Pump Dispense System.

Safety and Reliability Benefits of Low Pressure Operation

On the basis of Fault Tree Analysis studies, a VAC cylinder set to operate at nominally 600 torr provides approximately an 800 fold relative risk reduction factor compared to a standard high-pressure cylinder. Operating at 50 to 100 psig affords a relative risk reduction of 10-20 fold vs. high-pressure cylinders. [5]

Operating a gas delivery system under reduced pressure is expected to extend lifetimes and performance of valves and regulators. Under reduced pressure, connections should be easier to make leak tight. Valves seal more effectively and are less likely to leak and/or generate particles and purge/vent cycles will be more effective with less material to remove. Reducing the frequency of occurrence of component failures reduces the probability of a release. For example, components in corrosive gas service are replaced frequently; regulator failures being a major source of downtime and repair costs.

Efficiency And Capacity vs. Risk

For compressed gases, risk is proportional to the pressure of the cylinder contents. Increasing the cylinder pressure improves utilization of the contents and reduces change-out frequency, however, it also increases the maximum release rate and releaseable quantity. It is common today to operate at lower pressures [capacity] to reduce the risk. RFO's are essential to limit discharge rates but often limit efficient utilization of the contents by inhibiting flow at lower pressures. [6]

Cylinders can be pre-regulated to operate at low positive pressures, e.g. 50 psig. This occurs when the prospect of an inboard leak is undesirable and/or nominal pressure is required for gas distribution. In this case, an RFO is selected, consistent with process flow requirements. Using silane as an example, reducing the discharge pressure affords a 4-15 fold reduction in worst case release rate compared to a standard 800 psig cylinder with an 0.010" RFO. This is summarized in Table 2.

Silane cylinder fill densities are often a function risk tolerance. As compared to high-pressure cylinders, VAC permits higher fill densities while reducing the risk. In the silane cases, as much as 10% of the product may be returned as an unused heel because of flow restrictions arising from low fill pressures and small RFOs. Dropping the discharge pressure in conjunction with a larger orifice allows most of the contents to be delivered.

Table 2. Silane M aximum Flow Rates at 800 psig and 50 psig Through Various Orifices.

Cylinder	Pressure	Capacity	RFO	Release Rate
Standard	800 psig	5 kg	0.010"	30.5 slpm
VAC	50 psig	15 kg	0.010"	2 slpm
@1600	50 psig	15 kg	0.014"	4 slpm
psig	50 psig	15 kg	0.020"	8 slpm

Closely matching cylinder output rates to actual demand is accomplished by operating at low[er] pressure and selecting the appropriate restrictive flow orifice. This approach insures that leaks, should they occur, are as small as possible.

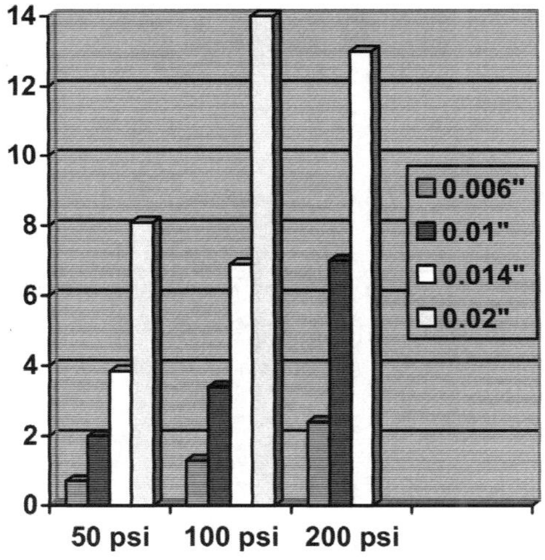

Cylinder Delivery Pressure

Figure 7. Silane Flow Rate [SLPM} as a Function of Pressure and RFO Size.

Ventilation standards for compressed gases are based on the expected [non-catastrophic] maximum release rates using standard RFO sizes and pressures. For the 30.5 slpm silane release in Table 2, the recommended ventilation rate is 360 CFM. The recommended exhaust rate using a VAC cylinder pre-regulated to 50 psig ranges between 21-86 CFM based on the RFO employed. [7] Pre-regulating the discharge pressure, albeit with larger RFOs, affords large reductions in exhaust rates from gas enclosures. Scrubbed exhaust and make-up air are estimated to cost between $3-4/CFM/year. [8]

The hazard posed by process materials, as well as, installation decisions within a facility are all influenced by the safety of the gas package. Continued innovation in how pressurized materials are stored and delivered promises to permit users latitude in choices of materials used and inventory quantities—based on risk management criteria.

COMPARISON OF GAS PACKAGING TECHNOLOGIES

The operation and performance of the two novel methods to reduce the risks posed by gases under pressure are summarized below. Innovative gas packaging has the potential to significantly lower risk, reduce operating costs and improve gas distribution reliability.

Table 3. Gas Delivery Methodology Comparison.

Category	Standard compressed gas cylinder	Adsorption technology	Mechanical technology
Storage mechanism	Pressure + containment per DOT regulations	Gas adsorption on a porous substrate	Pressure + containment per DOT regulations
Internal pressure	High	Low—650 torr nominal	High
External pressure	High, regulated downstream of the cylinder	Sub-atmospheric	Sub-atmospheric [regulated inside the cylinder
Accidental release probability	Highest	Lowest	Low
Consequence of an "event"	High	Small	Low
Delivery mode	Discharge Rate driven by pressure and orifice size.	Vacuum required (external pump) or apply heat to shift isotherm	Vacuum required (external pump) or process operating below set pressure threshold
Flow rate	High	Moderate	High
Chemistry limitations	None	Some—works with many gases.	None
Gas Mixtures	Yes	No	Yes
Purity	High	High purity essential	High

Storage density	High	Low to Moderate	High
Restrictive flow orifice	Yes	Optional	Yes

Adsorbed phase gas sources effectively remove pressure from consideration as gases are stored and delivered under sub-atmospheric pressure conditions. The technology has been widely accepted where consumption and flow rates are low to moderate.

Pre-regulation of cylinder pressure promises to become a cornerstone in gas handling safety for the future. Operating at significantly lower pressures reduces relative risk by 10 to 800 fold. Decoupling cylinder inventory pressure from delivery pressure, permits higher fill densities and leads to fewer cylinder changeouts and/or larger packages, both integral to achieving higher productivity. Increasing the lifetime and performance of components in gas delivery systems is an additional benefit.

The common theme in the two approaches presented here is to contain the pressure from the hazardous material within the storage cylinder. The latter typically have burst pressures exceeding 5000 psig and are effective and proven as containment vessels. Problems, if they occur, happen at the interface of the cylinder and the process. Effectively reducing or eliminating pressure at this juncture drives the risk reduction.

ACKNOWLEDGEMENTS

The authors.acknowledge help received from Jerry Sameth and Kevin Finn, Matheson-TriGas; Louis Restrepo and Pier Cararra, Omicron Safety and Risk Management; Bob Newton; Ceodeux, Inc.; Ron Buys, City of Austin Fire Department; Scott Stookey, City of Phoenix Fire Department; Franco Tamanini and Roger Benson, FM Global; Kent Carpenter, Jose Arno, Jim Dietz, Matt Donatucci, Russell Frye, Mike Wodjenski and Luping Wang; ATMI, Inc.

REFERENCES

[1] L. Wang, et al., "A New Era in Gas Handling Safety", Semiconductor Fabtech, 7th Edition.
[2] DOT Exemption E-12221.
[3] Figure 2. courtesy of Ceodeux, Inc. Lifecycle tests conducted in nitrogen.
[4] K. Olander and L. Wang, "Low Pressure Gas Sources", Gases and Technology, November 2002.
[5] Fault Tree Analysis conducted by Omicron Safety and Risk Reduction Technologies, Inc., Albuquerque, N.M. Report available from ATMI, Inc.
[6] James Hart, et al., "Incorporating More Gas Control with Cylinders", Solid State Technology, 11/20/03.
[7] CGA 2000 P-32 Silane Guidelines and NFPA 318 Standard for the Protection of Cleanrooms.
[8] P Naughton and k. Olander, "Energy/Exhaust Redduction Opportunities", A2C2 Magazine, October 2003.

Implementing Inherently Safer Design in an Existing Plant

Dennis C. Hendershot *

Rohm and Haas Company
Engineering Division
3100 State Road
Croydon PA 19021
DHendershot@rohmhaas.com

Jonathan A. Sussman

Chemical and Biomolecular
Engineering Department
University of Pennsylvania
Philadelphia, PA 19104-6393

Gerald E. Winkler
G. Lee Dill
Rohm and Haas Company
10 South Electric Street
West Alexandria OH 45381

* Corresponding author

Prepared for Presentation at the
American Institute of Chemical Engineers
2005 Spring National Meeting
7th Biennial Process Plant Safety Symposium
Session TH002 – Inherently Safer Process Design and Operations
Atlanta, GA
April 10-14, 2005

ABSTRACT

While the best time for implementation of inherently safer design alternatives is during early process development and plant design, there are often opportunities to modify an existing plant to significantly enhance inherent safety. We will describe a modification to a bromine raw material handling facility, replacing an existing bulk storage tank with cylinders. Incident consequence analysis and quantitative risk analysis were used to understand the relative inherent safety characteristics of the design options.

I. Introduction

The emphasis on the importance of early consideration of inherently safer design (ISD) alternatives can lead engineers and managers responsible for the operation of existing facilities to conclude that ISD is not relevant to them. They have an existing plant with established technology which is not easily or affordably changed. But, ISD is relevant to existing plants and processes – there are often opportunities to apply ISD principles to make an existing plant inherently safer, and many examples have been reported[1]. Furthermore, tools such as release consequence modeling and quantitative risk analysis (QRA) can be beneficial to understanding the potential benefits of ISD options, helping to gain management support for implementing ISD modifications to existing facilities.

Modifications to a bromine storage and process feed system illustrate the benefits of applying ISD to an existing plant. After many years of operation, the existing bromine storage tank required significant maintenance and repair. This prompted a review of other bromine supply options, including supply of liquid bromine from cylinders – an application of the ISD principle "minimize". Consequence modeling and QRA were used to understand the safety benefits of the proposed change to the bromine supply operation.

II. Facility Description

The plant used liquid bromine to produce a series of brominated polymers used in a variety of specialty products. The bromine was received in 15,000 pound tank truck shipments. Usage was fairly low, and bromine deliveries were required once every couple of months. When the bromine truck arrived, it was parked in an outdoor unloading area with spill containment. The truck was pressurized to about 25 psig using dry nitrogen, and the bromine was discharged to a 750 gallon storage tank. This storage tank was located in a concrete containment building which fully enclosed the storage tank, a small bromine weigh tank used to feed the bromine consuming processes, and most of the valves and piping associated with bromine handling. The containment building was kept at negative pressure relative to the outside atmosphere, and to the contiguous buildings. Ventilation air from the bromine containment building was discharged to a caustic scrubber, designed to remove bromine vapors in case of a leak or spill, and the discharge from the scrubber went to an elevated stack. Figure 1 is a schematic drawing of the existing bromine unloading and storage facility.

III. Properties of Bromine

Bromine (Br_2) is a heavy, reddish-brown liquid with an extremely irritating odor. Bromine exposure, either by liquid contact or inhalation of the vapors, has the potential for serious injury or fatality. The vapors are highly irritating to the eyes and respiratory tract, and exposure to vapor concentrations of 500-1000 ppm for periods of 30 to 60 minutes is dangerous to life. Bromine is not flammable, but it is a powerful oxidizing agent and can cause combustible materials to ignite on contact. Table 1 summarizes some of the physical and toxicological properties of bromine.

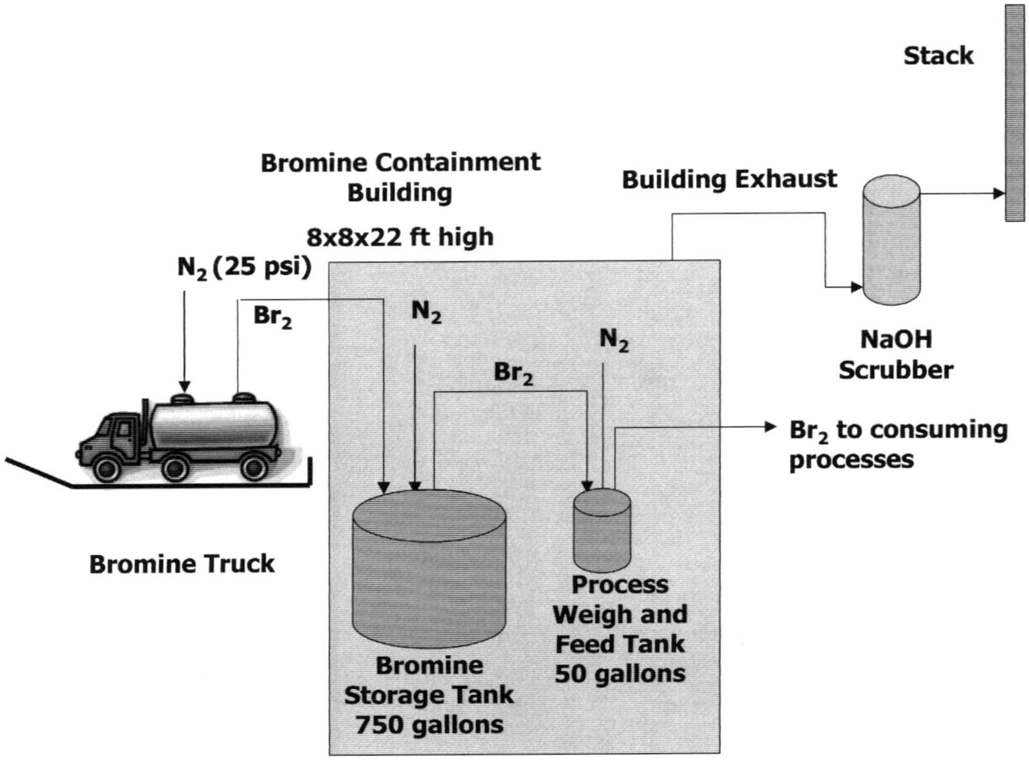

Figure 1: Existing Bromine Unloading and Storage Facility

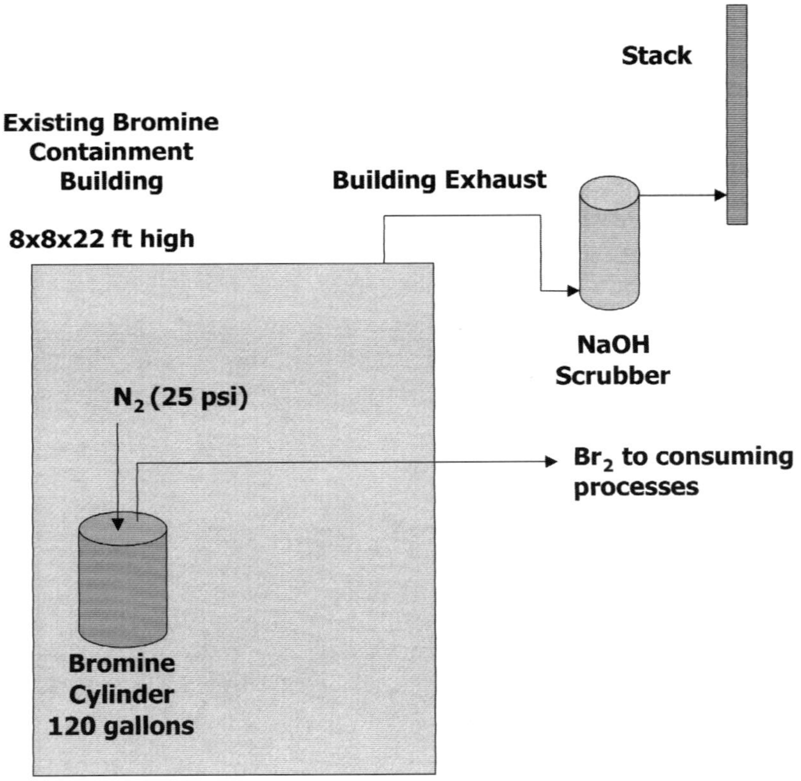

Figure 2: Proposed Bromine Cylinder Facility

Property	Value
Molecular Weight	160
Boiling Point	59 ºC
Freezing Point	– 7 ºC
Density	25.9 pounds/gallon at 77 ºF
Permissible Exposure Limit (PEL) (OSHA)	0.1 ppm
Emergency Response Planning Guideline (ERPG)[2]	
ERPG-1	0.2 ppm
ERPG-2	1 ppm
ERPG-3	5 ppm

Table 1: Properties of bromine [3]

For the QRA, the bromine toxicity dose-response relationships was expressed in the form of a "Probit" equation. Probit equations are generally derived from animal exposure data. The use of probit equations to describe the toxicity dose-response relationship for toxic gases is described by CCPS in the *Guidelines for Quantitative Risk Analysis*[3]. For this QRA, the bromine toxicity probit relationship recommended by the United States Coast Guard (1980)[4] was used:

$$\text{Probit} = A + B \ln (C^N t)$$

A, B, N – constants derived from experimental vapor exposure data

$A_{Bromine} = -9.04$

$B_{Bromine} = 0.92$

$N_{Bromine} = 2$

t – exposure time in minutes

C – exposure concentration in ppm

IV. Proposed Modifications

After many years of service, an inspection of the bromine storage tank indicated that significant repairs to the tank and its structural supports would be required in the near future for the tank to remain in service safely. The plant decided to evaluate other options for bromine supply as alternatives to repairing the existing bromine tank. Use of bromine from 120 gallon (3100 pound) cylinders was determined to be a viable alternative, one which had the potential to reduce the hazard to the surrounding community by reducing the on-site inventory of bromine – an application of the ISD principle "minimize". The inventory of bromine on site would be reduced from slightly greater than 15,000 pounds to a maximum of 6,200 pounds in two separate cylinders – one at the feed location and a second in a storage area. As an additional benefit, the reduced inventory of bromine is below the threshold for coverage for the EPA Risk Management Program (RMP) regulations (10,000 pounds for bromine)[5]. Since there were no other materials handled on the site which exceeded the RMP threshold quantities, this would mean that the site was no longer an RMP facility.

The proposed design for bromine cylinder handling is shown in Figure 2. The existing bromine storage tanks would be dismantled and removed from the bromine containment building. One bromine cylinder feed station would be installed inside the existing containment building, on a scale to monitor the bromine feed to the process. All cylinder connections and disconnections would be done inside the building, and all flexible hoses would be inside the containment building. Thus, operations with the highest likelihood of an error or failure causing a bromine leak would be inside the containment building, where any possible leak would be captured by the building ventilation system, and treated by the caustic scrubber prior to discharge to the elevated stack.

V. Consequence Analysis

Initial consequence modeling was completed to get an idea of the potential magnitude of the hazard reduction. Modeling of potential releases was done using the PHAST (version 6.4) chemical release consequence modeling software from DNV. All releases inside the bromine containment building are equivalent because bromine is a liquid at room temperature. Any leak or spill will be contained inside the building and the spilled material will cover the floor of the containment building. The building and air flow characteristics determine the evaporation rate of bromine from the spilled pool. For purposes of the "worst case" consequence modeling, the bromine containment building scrubber was assumed not to be working.

The consequence analysis estimated a maximum distance of nearly 9 km to the ERPG-2 concentration of bromine for worst case weather conditions (F atmospheric stability, 1.5 m/s wind speed) for an outdoor spill from the bromine truck into the bromine unloading containment area. This modeling took no credit for any emergency response actions, and assumed that the spill would continue long enough for the spilled bromine to cover the maximum available area in the containment, and that the spill would continue to evaporate for at least 1 hour.

For the proposed cylinder handling operation, all bromine operations would be done inside the containment building. The consequence modeling for all indoor spill or leak scenarios, assuming that the containment building scrubber was not working, estimated a distance to ERPG-2 of about ½ kilometer, a large reduction in hazard distance.

The consequence analysis results provide a quantitative measure of the ISD benefit of handling bromine in cylinders. While there is intuitively a benefit to reducing the inventory of bromine, and from eliminating all handling operations outside the containment building (except for moving the closed cylinders into the building), the consequence analysis provides a quantitative measurement of the reduction in potential hazard distance.

VI. Quantitative Risk Analysis

The cylinder bromine option reduces the potential consequences of a bromine release, but it is possible that the overall risk would not be reduced, at least to some potentially impacted populations. Feeding bromine from cylinders will require more frequent connection and disconnection of flexible hoses connecting the bromine cylinders to the process piping.

To confirm that the cylinder option would reduce risk, a QRA was completed to compare the existing operation and the proposed cylinder handling. Tables 2 and 3 summarize the release cases that were considered for the QRA, for the truck and cylinder delivery options respectively.

Modeling Case Number	Failure Description
1	Leak in pipe from truck to 750 gallon tank, inside building
	Leak from 750 gallon tank
	Failure of 750 gallon tank
	Leak in pipe from 750 gallon tank to 50 gallon tank
	Leak from 50 gallon tank
	Failure of 50 gallon tank
	Leak from pipe from 50 gallon tank to reactor, inside building
2	Catastrophic failure of tank truck at normal unloading conditions
3	Bromine truck overpressurized because of a fire in the unloading area or adjacent building – heat exposure and boiling of bromine, relief valve fails to operate
4	Bromine truck overpressurized because of a fire in the unloading area or adjacent building – heat exposure and boiling of bromine, relief valve operates properly resulting in a discharge, but not a rupture of the truck
5-7	Various size leaks in truck shell equivalent to 1 inch hole
8	Unloading hose failure with rapid response (10 minutes) by operator to shut off the flow.
9	Unloading hose failure with no response by operator to shut off the flow. The flow continues for one hour.

Table 2: Bromine release cases for existing truck delivery system

Modeling Case Number	Failure Description (Note – all failures are inside the containment building)
10	Unloading hose failure (2 hoses), or hose not properly connected
	Cylinder valve leak
	Catastrophic failure of the bromine cylinder (including dropping or mishandling)
	Failure of pipe, or leak in pipe from cylinder to reactor
	Leak directly from cylinder

Table 3: Bromine release cases for proposed cylinder delivery system

The frequency (likelihood) of the various release cases was estimated using published failure rate data, and Rohm and Haas Company internal data sources. Population distribution information for the plant and surrounding community were estimated based on the known plant workforce data, and the known population of the town, using aerial photographs to determine the residential areas of the town. Weather data (wind speed and direction, atmospheric stability class) was obtained from a nearby airport. Figure 3 shows the weather data used for the analysis.

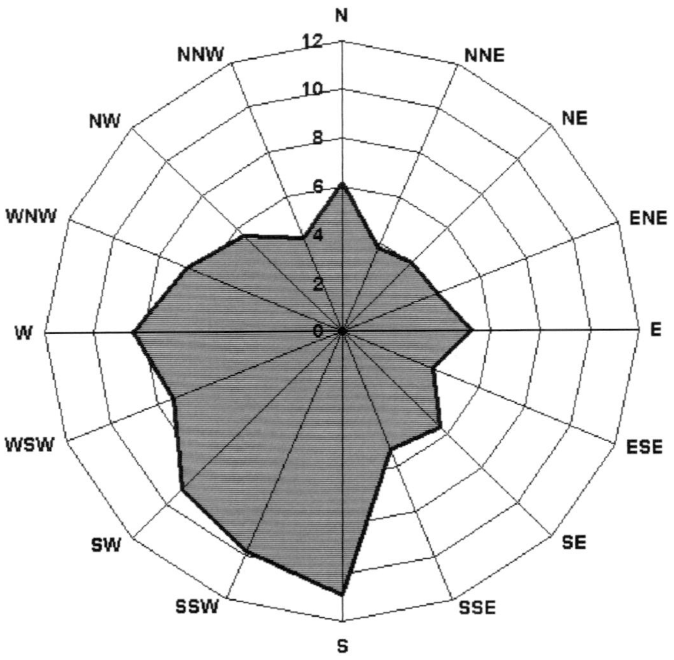

% of time wind is blowing from a given direction

Figure 3: Weather conditions used in QRA

The risk analysis consequence and risk calculations were done using the SAFETI 6.4 risk analysis computer program from DNV.

VII. Risk Analysis Results

Figure 4 shows the individual risk contour map for the existing (truck delivery) bromine supply facility. These results show that the existing facility does not meet Rohm and Haas Company risk guidelines[6, 7], which require that the estimated risk of fatality off site be less than 1 in 100,000 per year. The QRA results indicate that there is a small area of the community to the northwest of the plant where the estimated risk exceeds the guideline value. Therefore, Rohm and Haas standards require that system improvements to reduce risk to the guideline value be evaluated.

In this case, most of the risk for the existing truck delivery option arises from the outdoor operations – truck unloading. Any bromine leak or spill outside the containment building (in the truck unloading spill containment area) cannot be captured and mitigated by the bromine containment building ventilation, scrubber, and stack. Rather than looking at opportunities to improve the existing truck unloading operation, the plant evaluated the risk of the proposed cylinder bromine supply.

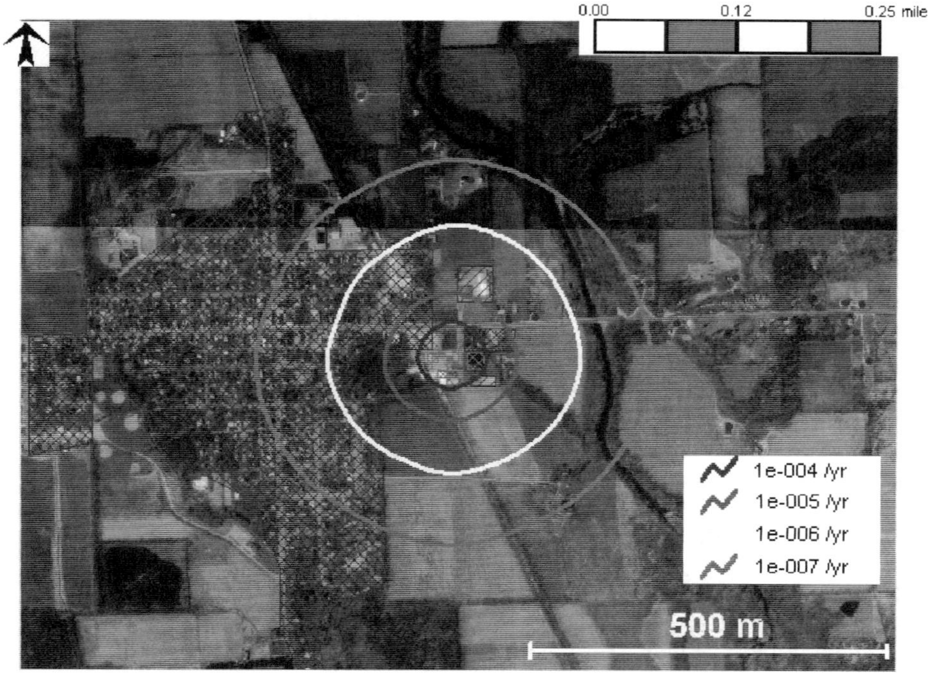

Figure 4: Individual risk contours for bromine tank truck delivery (crosshatching represents population areas)

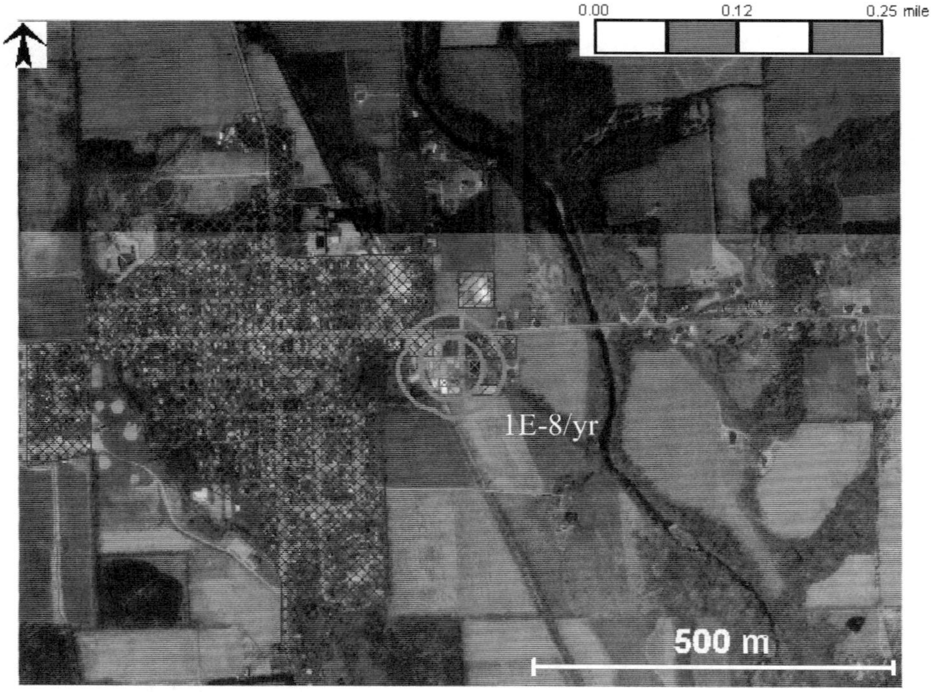

Figure 5: Individual risk contours for bromine cylinder delivery

Figure 5 shows the individual risk estimate results for the proposed bromine cylinder operation. The highest individual risk contour shown is 10^{-8} per year, or a risk of fatality of one in 100 million per year. Clearly the cylinder bromine supply option meets the corporate risk guidelines.

The risk of fatality of 1 in 100 million per year estimated for the cylinder operations should not be taken too literally. This is essentially background risk, and the risk of injury or fatality from events which are not taken into account in the analysis is likely to be much higher than the risks which were actually analyzed. For example, the analysis does not consider sabotage or terrorism, tornadoes or other severe weather, earthquakes, etc. The likelihood and consequences of these types of events is difficult to quantify and include in a risk analysis, and they will likely have other significant impacts beyond their effect on the bromine system.

In general, in a quantitative risk analysis we would not even present risk contours at the 10^{-8} per year level. However, we are showing the risk contours in this paper because they illustrate an interesting characteristic of elevated releases of heavy vapors such as bromine through a stack. The individual risk exceeds the 10^{-8} per year value in a doughnut shaped area surrounding the bromine containment building stack. Risk in the doughnut ring is higher than the risk inside the inner boundary of the ring, and also higher than the risk outside the outer boundary. This is because the low residual risk is a result of leaks inside the bromine containment building, which are discharged through the building stack if the caustic scrubber fails. The heavy bromine vapors discharged to the stack fall down to ground level, and reach the ground at a point downwind of the stack. From there they disperse in the atmosphere. Therefore the risk is lower close to the stack because the toxic vapors have not reached the ground at that location. As one moves away from the stack, one reaches an area where the vapor plume reaches the ground, and the risk is higher. Moving still further away, the vapors are diluted in the atmosphere, and the risk is reduced because the concentration of toxic vapors is lower. The maximum risk is in the area where the plume from the stack first reaches the ground, where the concentration of vapors is the highest. This phenomenon is illustrated in Figure 6.

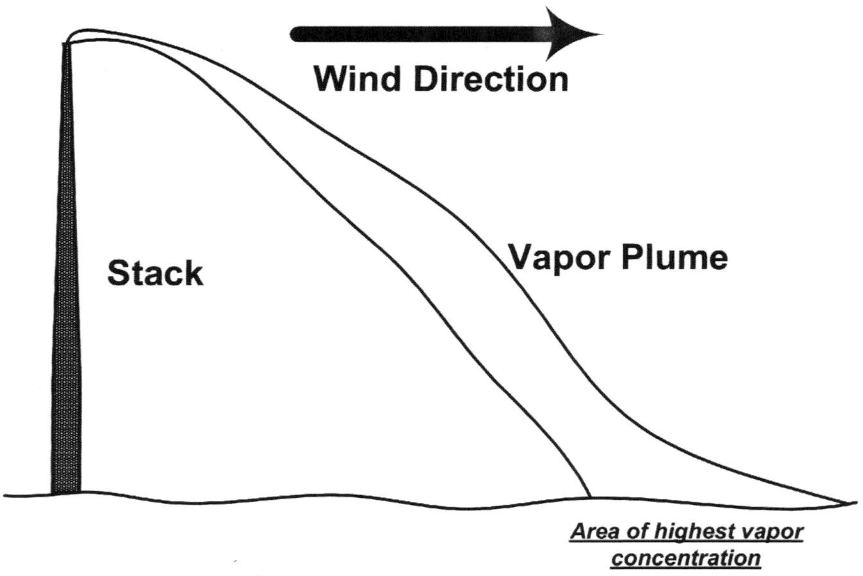

Figure 6: Bromine release from containment building stack

VIII. Conclusion and Action

Receiving bromine in cylinders was determined to be an economically viable option, particularly when considering the necessity for spending money to repair the existing system. The existing system would also require some type of modifications to the equipment or operating procedures to reduce the risk and meet the corporate risk guidelines. The plant decided to implement the supply of bromine in cylinders, an inherently safer supply option.

This case study illustrates that it is worth while to consider ISD for existing plants. In this case, the need for repairs to the existing bromine storage system prompted the plant to investigate potentially inherently safer alternatives. Rather than simply repairing the existing tank, plant personnel asked if there was a better way to supply bromine. Leak and spill consequence analysis, and quantitative risk analysis, are effective tools for understanding the ISD benefits of alternate designs.

In this case study, the changes to the bromine handling system reduced the inventory to a point where the plant is no longer covered by the EPA RMP regulations. This does not have a major impact on how the plant is operated – bromine, even in smaller quantities, is a very hazardous material and internal Rohm and Haas standards require process safety management systems which are essentially the same as those required by RMP. However, since the plant is no longer covered by RMP, a lot of paperwork and other regulatory burdens associated with the regulations are eliminated. It is also good for community relations to show that the plant has taken steps to reduce the hazards associated with handling of hazardous materials on the site.

References

[1] Carrithers, G. W., A. M. Dowell, and D. C. Hendershot. "It's Never Too Late for Inherent Safety." *International Conference and Workshop on Process Safety Management and Inherently Safer Processes*, October 8-11, 1996, Orlando, FL. New York: American Institute of Chemical Engineers. pp. 227-241, 1996.

[2] *Guidelines for Chemical Process Quantitative Risk Analysis.* 2nd Edition. New York: American Institute of Chemical Engineers, 2000.

[3] *Bromine: safety and handling*. West Lafayette, IN: Great Lakes Chemical Corporation. Not dated (downloaded from http://www.e1.greatlakes.com/corporate/prodserv/pdf/bromine_safety_guide.pdf, December 2004).

[4] *Guidelines for Consequence Analysis of Chemical Releases.* New York: American Institute of Chemical Engineers, 1999, p. 254.

[5] *Code of Federal Regulations.* 40 CFR 68, subpart B, Appendix A, Section 68.130. "List of Substances." Washington, DC: US Government Printing Office, June 20, 1996.

[6] Renshaw, F. M. "A Major Accident Prevention Program." *Plant/Operations Progress* **9**, 3 (July), pp. 194-197, 1990.

[7] Renshaw, F. M. "A Major Incident Prevention Program: Ten Years of Experience." *Process Safety Progress* **23**, 2 (June), pp. 155-162, 2004.

Continuous Process Improvement Toward a Safer Ester Amidation Process

Adrian David, Srinivas Tummala, Boguslaw Mudryk, Jing Liang, Lucius Rossano, Karen Tenhuisen, Simon Leung , David Kacsur

Bristol-Myers Squibb Pharmaceutical Research Institute
Process Research and Development
One Squibb Drive
New Brunswick, New Jersey 08903-0191

Key Words: ammonolysis, amidation

Prepared for Presentation at
American Institute of Chemical Engineers
2005 Spring National Meeting
Process Plant Safety Symposium
Atlanta, Georgia
April 11-13, 2005

Unpublished

December 2004

Abstract

The evolutionary development of a pilot-plant procedure for the conversion of an aromatic ester to the corresponding primary amide by ammonolysis using gaseous ammonia to a safer, more efficient process involving the reaction of the ester with formamide and sodium methoxide will be presented.

Ammonolysis is the most common industrial method for converting esters to unsubstituted amides. The reaction of an aromatic ester with gaseous ammonia was carried out in neat propylene glycol, because the transesterification to the propylene glycol ester greatly increased the reaction rate with ammonia. The ammonolysis reaction had to be run under pressure in order to increase the solubility of ammonia in propylene glycol at the reaction temperature. The reaction was slow and difficult to control. It also required a significant excess of ammonia, most of which was lost via venting through a scrubber after completion of the reaction. Ammonia was introduced more efficiently in a second-generation process by feeding liquid ammonia subsurface which improved the controllability but still required a large excess of ammonia. In both cases, stringent process and engineering controls were required to ensure the safety of the operation. Managing the heat released in the process and safe scrubbing of the excess ammonia were critical.

Following a continuous process improvement strategy, a simpler, faster and safer amidation procedure involving the reaction of the same aromatic ester with formamide and sodium methoxide was developed and successfully scaled up. The amidation reaction with formamide and sodium methoxide was carried out in a homogeneous solution at 40-50°C. The fast reaction kinetics allowed for reaction completion in less than two hours. No dosing was necessary as the heat of reaction was trivial.

Introduction

There are few options at the industrial scale for preparing aromatic primary amides from the corresponding esters without going through and intermediate derivatization. The most commonly used method is the direct ammonolysis – the reaction of the ester with anhydrous ammonia. The process is usually carried out in a two-phase system employing water-miscible alcohols as solvent[1,2], but there have also been recent reports of the use of supercritical ammonia as both solvent and reagent[3,4]. Both procedures require elevated pressure and the implementation of safety protocols for the handling of the large excess of ammonia.

This paper describes the engineering and safety issues associated with the scale-up of an ester ammonolysis reaction for the production of a novel pharmaceutical intermediate (Figure 1).

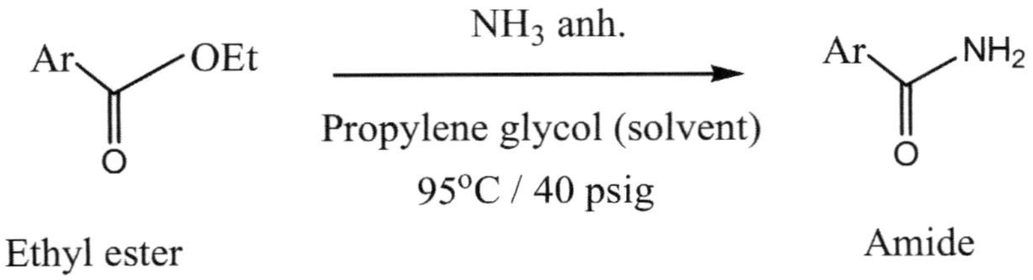

Figure 1. Ammonolysis of an aromatic ester.

In order to reduce the requirement for specialized equipment design for a multi-purpose pilot facility, a simpler and intrinsically safer process than the ammonolysis was evaluated. Thus, the development and scale-up of an amidation process based on the homogeneous reaction of the ester with formamide in the presence of sodium methoxide[5,6,7] is presented (Figure 2).

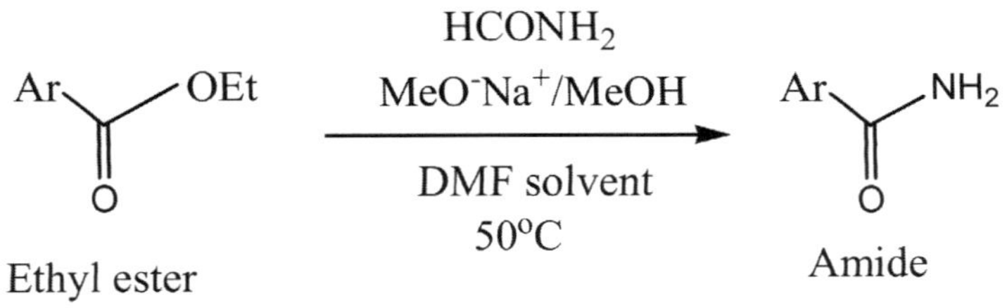

Figure 2. Amidation of an aromatic ester with formamide/sodium methoxide.

Experimental

The ammonolysis reaction was carried out in propylene glycol at 95°C and 40 psig of ammonia pressure. The first-generation process at the pilot plant scale (100 gallon) used 17 equivalents of gaseous ammonia fed subsurface to the ester/PG solution. The thermochemical analysis (RC-1) conducted during the development stage indicated that the gaseous ammonia had to be dosed in order to manage the heat released in the process. Thus, the overall heat liberated in the process was –652 kJ/mol of ester or –38.7 cal/g of reaction mixture, resulting in a calculated ΔT_{ad} = +57.4°C. A significant portion of the heat, -385 kJ/mol of ester was attributed to the heat of solubility of ammonia (gas) in the propylene glycol solvent. Controlling the reaction temperature solely by the dosing rate of the gaseous ammonia proved to be challenging at the pilot-plant scale. Initially, most of the gaseous NH_3 was added at <40°C to handle the heat of solubility, then the batch was heated to reaction temperature. This increased the pressure and reaction temperature rapidly, making the addition of the balance NH_3 and the control of pressure and temperature difficult. In subsequent batches, the gaseous ammonia was added at >80°C to enhance temperature and pressure control.

Ammonia was introduced more efficiently in a second-generation process by feeding liquid ammonia subsurface, and thus taking advantage of the latent heat of vaporization. Using liquid ammonia improved controllability, but did not significantly decrease the requirement for a large excess of the reagent. Both the gaseous and the liquid ammonia process were carried out in reactors with excellent agitation, suitable for gas-liquid reactions. The equipment diagram used at the pilot scale is presented in Figure 3.

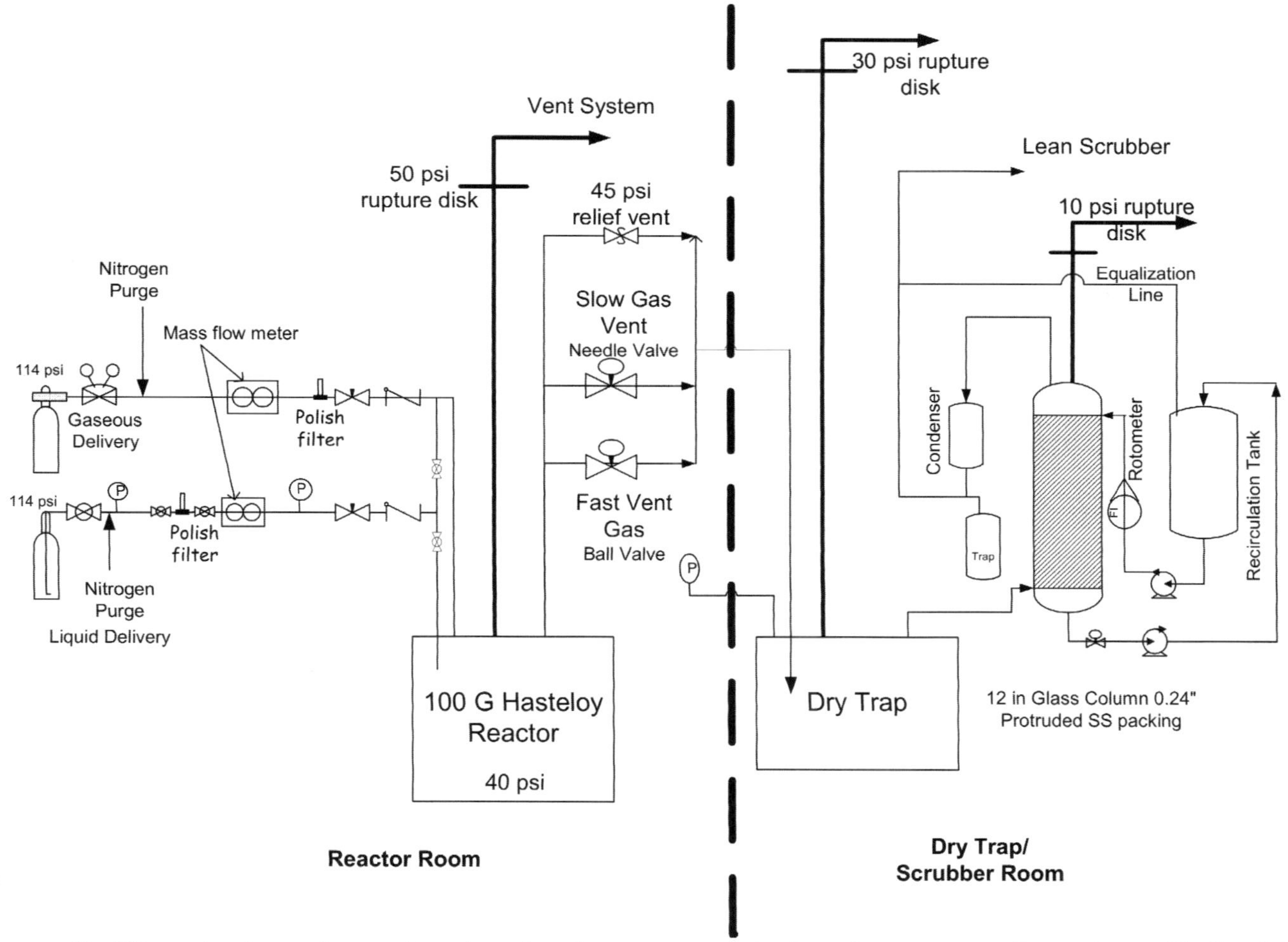

Figure 3. Pilot-plant equipment setup for the ammonolysis of an aromatic ester with liquid ammonia.

About 95% of the ammonia was charged in liquid form over about 2 hours at 85-95°C, with the remainder 5% being charged in gaseous form at 95°C over an additional 1-1.5 hours. The batch was then held for about 16 hours under a regulated ammonia atmosphere (40psig), bringing the total reaction time to ~20 hours. The pressure regulator placed at the end of the gas line constantly maintained the pressure of the reactor at 40 psi (Figure 3). When the reaction end-point was reached, the excess ammonia was vented to a containment system comprised of a dry trap and wet scrubber-column.

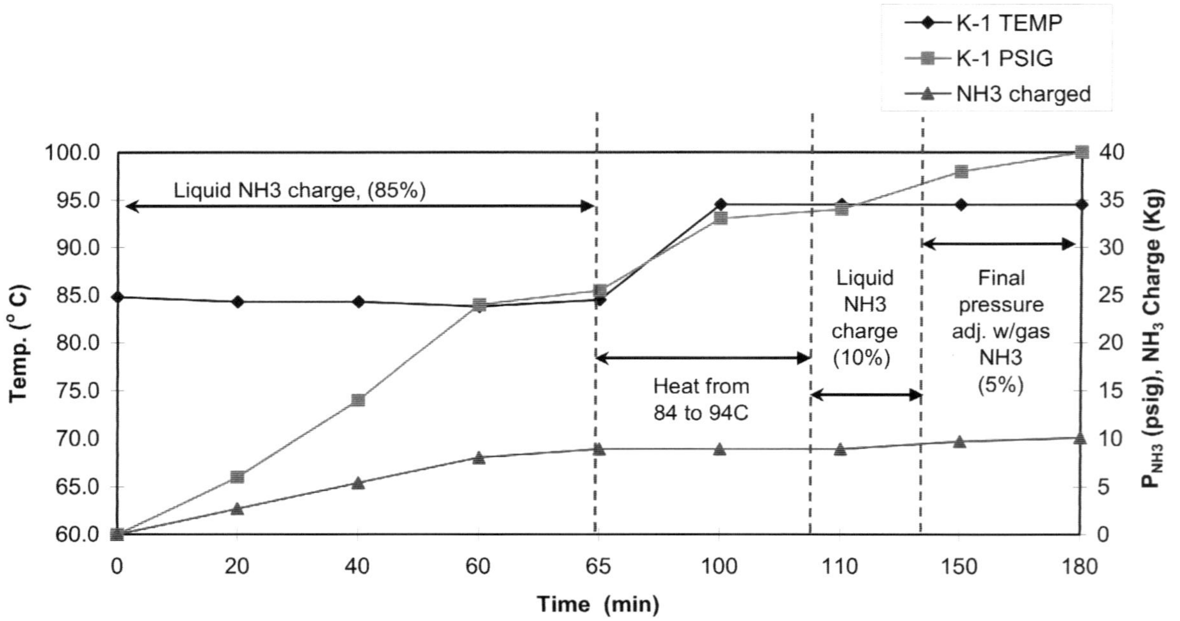

Figure 4. Ammonolysis reaction: Liquid ammonia charge profile.

The ammonolysis reactor was provided with a vented DoPak® system for discrete sampling. Nonetheless, in order to minimize operator exposure during sampling, the reaction was continuously monitored by FT-IR (Figure 5), with additional confirmation of the end-point by HPLC.

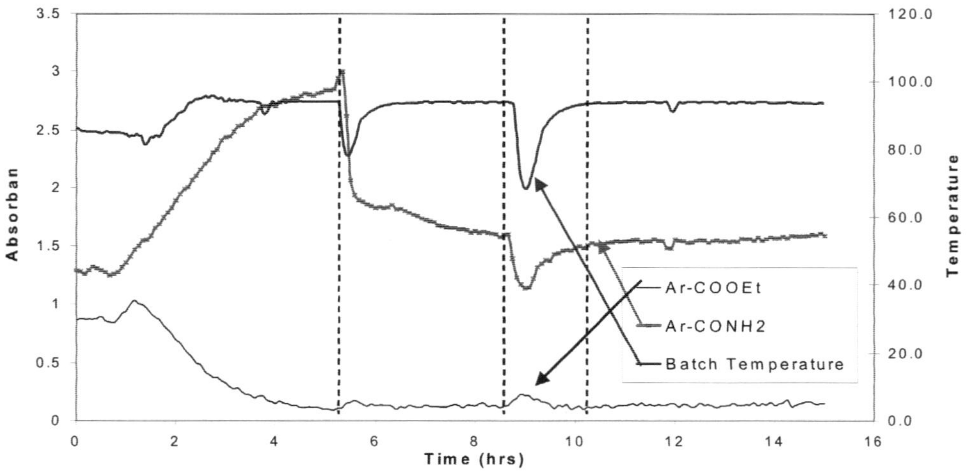

Figure 5. FT-IR monitoring of the ammonolysis reaction.

The challenges posed by the ammonolysis reaction prompted the development team to adopt a simpler faster and safer amidation procedure based on the chemistry illustrated in Figure 2. The postulated mechanism[5] for the reaction is shown in Figure 6.

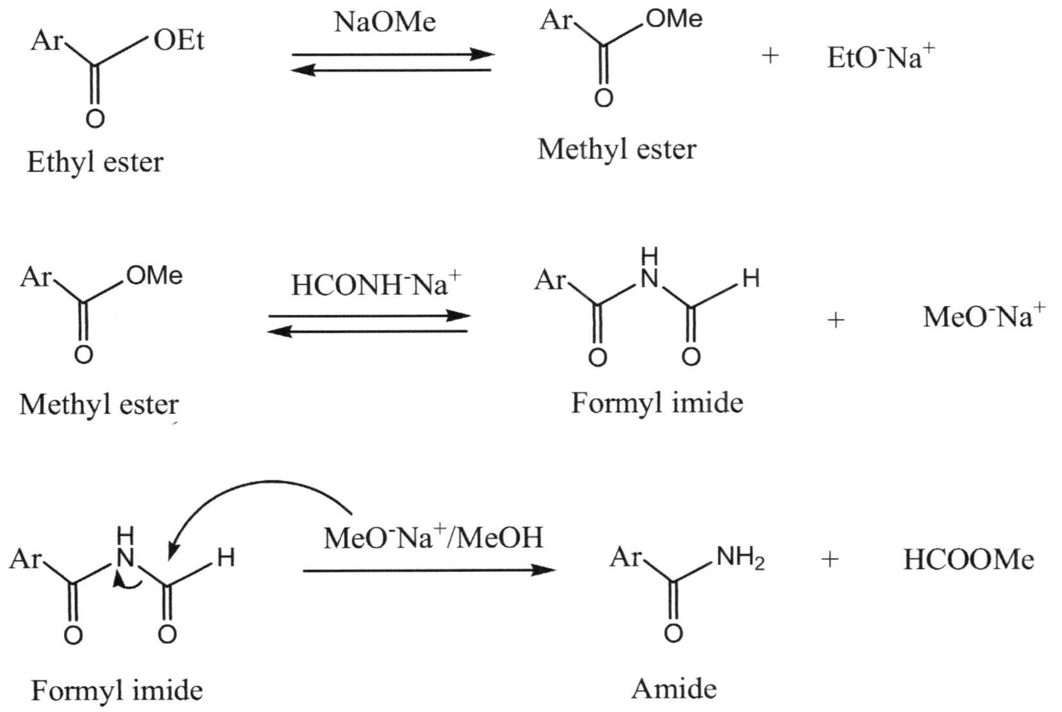

Figure 6. Mechanism of amidation reaction with formamide/sodium methoxide.

At the pilot plant scale, all reagents were mixed in N,N-dimethylformamide (DMF) solvent (Note: sodium methoxide was charged last, no dosing was necessary) and the conversion of the ester to the amide took less than two hours at 50°C in a homogeneous system. More than 95% of the conversion occurred in the first ten minutes, but reducing the concentration of the methyl ester intermediate to the 0.3% (relative area%) specification required another 1.5-2 hours (Figure 7). The RC-1 calorimetry evaluation showed that the heat of reaction was –8.77 kJ/mol of ethyl ester or –0.38 cal/g of reaction mixture, resulting in a calculated ΔT_{ad} of +0.7°C.

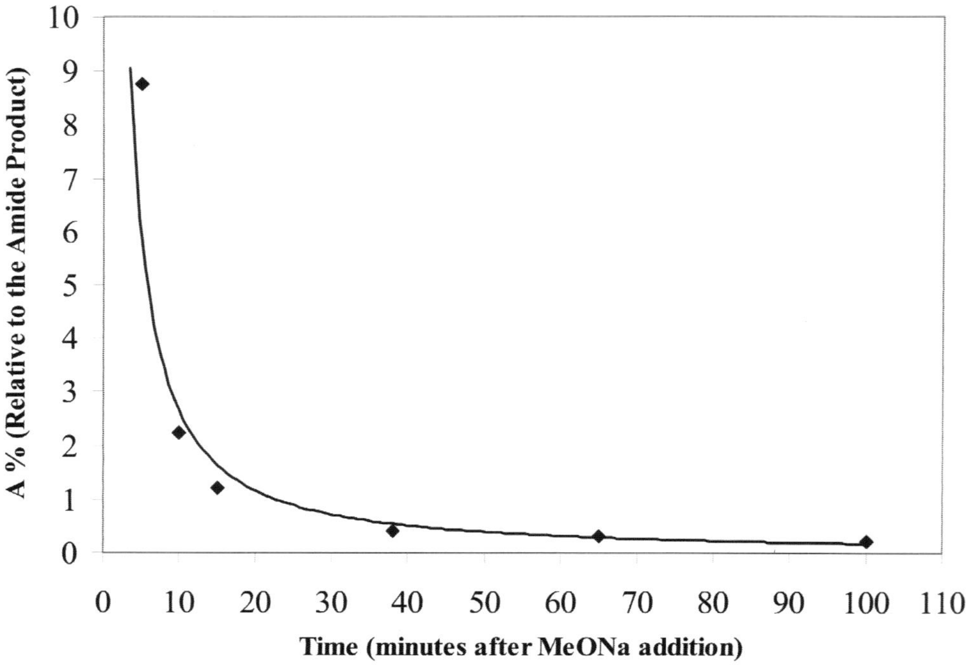

Figure 6. Consumption of the methyl ester intermediate during the amidation reaction with formamide/sodium methoxide.

Results and Discussion

In the ammonolysis process, the propylene glycol solvent also acts a mediator/catalyst via a transesterification reaction (Figure 7) that transforms the ethyl ester starting material into a propyl ester (two isomers) which reacts faster with anhydrous ammonia than the ethyl ester.

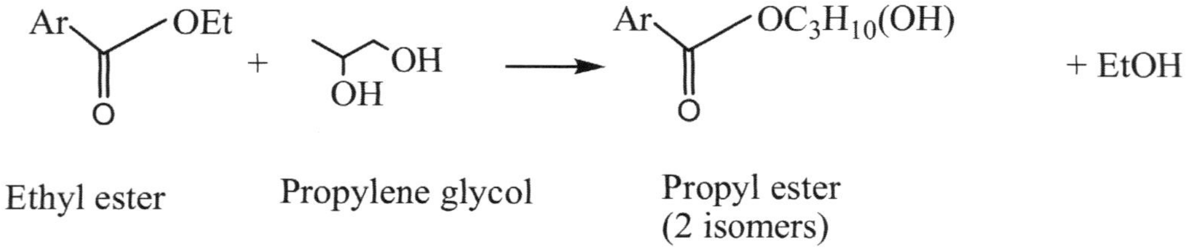

Figure 7. Transesterification of the ethyl ester with propylene glycol.

As specified earlier, the ammonolysis using liquid ammonia afforded better control of the reaction temperature. The main reason for the better control was the assistance from the heat of vaporization (+23.2 kJ/mol of ammonia) which counteracted the heat of solvation of ammonia in the PG solvent (-31.0 kJ/mol). The net heat for the addition of liquid ammonia and dissolving it in PG is of -7.8 kJ/mol of ammonia compared to -31.0 kJ/mol in the case of gaseous ammonia. Thus, the ΔT_{ad} for the overall process decreased from +57.4°C to about

+24°C upon replacing gaseous with liquid ammonia. The liquid ammonia process was a significant improvement over the gaseous ammonia procedure in terms of reaction control, but it did not address the issue of the excess ammonia requirement.

The amidation procedure based on the chemistry illustrated in Figures 2 and 6, on the other hand, did not require working with a hazardous gas under pressure, and the installation of additional equipment for purging the excess ammonia. The amidation reaction mixture is homogeneous throughout, and the process only requires a temperature of 50°C. Control by reagent dosing is not necessary as the heat of reaction is trivial, and the process is robust over a variety of mixing conditions. In addition, the amidation with formamide/sodium methoxide is much faster than the ammonolysis reaction (2 vs. 20 hours). Both the amidation with formamide/sodium methoxide and the ammonolysis processes gave yields of ~90% at similar volume efficiencies.

Conclusion

The reaction of esters with formamide/sodium methoxide was determined to be an intrinsically safer alternative for producing amides at the pilot plant scale than the widely used ammonolysis reaction of the ester with anhydrous ammonia under pressure. At similar volume efficiency and product yield, the amidation with formamide/methoxide has been proven to be more robust, easier to implement in the field, faster and less equipment intensive. In summary, the main advantages of the amidation procedure over the ammonolysis process are:

- Trivial temperature/pressure requirements.
- Minimal engineering calculations and design for venting, pressure relief and scrubbing.
- Fewer unit operations and overall reduction in equipment and instrumentation requirements.
- Use of general purpose equipment.
- No scrubber waste.
- Significant improvement of cycle time.
- Homogeneous reaction
- Energy savings (utilities).
- Intrinsically safer operation.

Acknowledgments

The authors would like to acknowledge the contributions from Frank Okuniewicz, Joe Tang and Alice Yang.

References

1. Arcelli, A., Porzi, G., and Sandri, S., *Tetrahedron*, **52**, 4141 (1996).
2. Gordon M., Miller, J.G., and Day, A.R., *J. Am. Chem. Soc.*, **71**, 1245 (1949).
3. Wang, S., and Kienzle, F., *Ind. Eng. Chem. Res.,* **39**, 4487 (2000).
4. Wang, S., Karpf, M., Kienzle, F., *J.Supercrit. Fluids*, **15**, 157 (1999).
5. Allred, E.L., and Hurwitz, M.D., *J.Org.Chem.*, **30**, 2376 (1965)
6. Jagdman, G.E. et al, *Synth. Commun.*, **20**, 1203 (1990)
7. Deo, N.M., Crooks, P.A., *Synth. Commun.*, **25**, 691 (1995)

Preventing Incidents at Newly Acquired Facilities – Implementation of Lessons Learned

Brian R. Dunbobbin, C. Lorn Paxton, Glen A. Peters
Air Products & Chemicals Inc.
Allentown, PA 18051 USA
dunbobbr@apci.com

Martin A. Dennehy
Air Products plc.
Walton on Thames,
England KT12 4RZ
dennehma@apci.com

The large number of acquisitions taking place in the chemical industry has increased the need to efficiently assess the process safety risks at such facilities so that they meet the requirements of the acquiring company. This can be difficult if the technology issues are not fully disclosed during the "Due Diligence" phase of an acquisition.

This paper describes a tiered approach to ensuring that the acquiring company identifies as quickly as possible key process safety risks so that employee, community and company interests are adequately protected. In particular, three aspects of the work process for identifying process safety risks at an acquired facility are shown to be important. The first is the method used to identify process safety risks during the "Due Diligence" phase. This involves the selection of appropriate team members, and development of comprehensive checklists that the acquisition team can use to identify potential risks at an early stage. The second is the ability of the acquiring company to focus its own technical and process safety experience on the technologies being acquired; this experience is often in the form of standards that are used in the design and operation of similar facilities. Whether or not to retrofit these standards becomes an important issue when the newly acquired facility is operated by the new company. Finally, a Process Safety Assessment work process, with defined protocols, has been developed so that senior Process Safety, Operations and Engineering personnel can assess the potential process safety risks at a facility quickly and consistently in the first few months of operation following an acquisition.

1.0 Introduction

Following acquisition of a chemical facility it is important that the acquiring company quickly understand the potential hazards and risks that the facility may present. This ensures that employee, community and company interests are adequately protected. In particular there are three issues that should be understood by the acquiring company:

1) Are there major hazards present in the facility that could cause a serious incident?
2) Are the management systems at the facility adequate to ensure safe operation? Are these management systems compatible with existing company systems?
3) Do the standards and practices utilized at the newly acquired facility meet those of the acquiring company?

Unfortunately getting answers to these questions can be a difficult task, especially if the technology issues are not fully disclosed during the pre-purchase, or initial due diligence, phase of an acquisition.

Environmental issues are often easily identified during the pre-purchase phase of an acquisition since regulations normally require disclosure and public reporting. Process safety issues on the other hand are often tied closely with the technology of the processes being acquired and the selling company has little incentive to give full disclosure of technology "trade secrets" prior to finalization of the sale. The process safety hazards and risks may not be fully identified until the new owner has taken over the facility and reviewed the details of the technology and the hazards that may be present. It can then be an expensive task to bring the acquired facility into compliance with the process safety standards of the new owner.

Air Products utilizes a tiered approach to ensure that potential key process safety risks are identified as quickly as possible. In particular, three aspects of the work process for identifying process safety risks at an acquired facility are shown to be important:

1. **Pre-Purchase Assessment (sometimes referred to as the due diligence phase).** A consistent work process and methodology must be used to identify process safety risks during the pre-purchase phase. This involves the selection of appropriate pre-purchase assessment team members, including experienced process safety practitioners, and development of comprehensive checklists that the acquisition team can use to identify potential risks at an early stage. This checklist must be efficiently prioritized so that it can be used quickly in the normally brief pre-purchase period. At this stage of the acquisition a transition plan is also put in place. This transition plan identifies key elements that must be in place on the first day of ownership.
2. **Technology Issues.** The ability to focus technical and process safety experience on the technologies being acquired; this experience is often in the form of standards that are used in the design and operation of similar facilities. Whether or not to retrofit these standards becomes an important issue which can often be complicated by the fact that the newly acquired facility safety record may be excellent, but operates to different standards and work processes.
3. **Post Purchase Assessment.** A process safety assessment work process, with defined protocols that build upon the checklists used in pre-purchase phase, has

been developed so experienced process safety, operations and engineering personnel can assess the potential process safety risks at a facility quickly and consistently in the first few months of operation following an acquisition.

In the following sections of this paper we will describe in detail the major elements of each of these three aspects of the work process.

2.0 Pre-Purchase Assessment Phase of an Acquisition

After the business areas have determined that a potential acquisition is feasible, a small team is usually formed to review documentation and visit the potential acquisition candidate for a pre-purchase assessment to identify potential environmental, health and safety impacts. This is often done in secret by a small, two or three person team, so employees and stakeholders are not alarmed at the prospect of a potential sale of their facility. During the usually brief site visits, sometimes lasting for only several hours, the team will usually walk around the facility and get an overview of the technology, equipment and operating capabilities of the facility. It is normally not meant to give a detailed assessment of the technology since this often remains confidential until later in the acquisition process. Obtaining as much documentation as possible prior to the site visit can be extremely helpful. Documentation normally needed at this early stage includes plot plans, permits, lists of chemicals on-site, chemical inventories, etc.

It is at this stage of a potential acquisition that the acquiring company would like to identify key issues that could impact the economic viability of the acquisition. Within the process safety area these would include major hazards that could cause serious incidents and the possible costs necessary to provide an acceptable level of risk. The pre-purchase assessment team also develops a transition plan that summarizes actions for management and identifies items that must be in place on day one of the transfer of ownership.

It is extremely important that the team contains representatives from the correct functional areas. We will not get into the full team composition here, but rather we will concentrate on the need for adequate experienced process safety representation. This process safety representative must have a level of experience and prestige so representations on the safety of the facility are heard by senior management.

To assist the process safety member of the pre-purchase assessment team we have developed a checklist that can be followed to ensure that the "right" questions are asked both prior to and during the site visit.

The major sections of the checklist include elements of any good process safety management system and are given below:

1. Determination of Regulatory Applicability
2. Process Safety Information
3. Process Hazard Reviews
4. Operating Procedures
5. Training*
6. Pre-startup Safety Reviews (PSSR)*
7. Hot Work*

8. Management of Change (MOC)*
9. Mechanical Integrity
10. Incident Investigation
11. Emergency Planning and Response Procedures
12. Compliance Audits (Internal & External)
13. Trade Secrets*
14. Facility Public Risks
15. Transportation Risk
16. Site Security
17. Safety Systems and Standards*
18. Technology Specific Items*

Items in the checklist marked by an asterisk are often not addressed fully, if at all, during this pre-purchase phase due to time constraints and a lack of information, therefore these are usually treated as low priority items and are only addressed if time allows. In particular, the last two items, Safety Systems and Standards, and Technology Specific Items, may be extremely difficult to assess at this stage if the technology is new to the acquiring company. Also, the site visit assessment team is **often limited to two or three team members** and having all the necessary technology skills on the team can be difficult. All of the items not completely covered in this pre-purchase assessment will be reviewed in detail as part of the transition plan and during the post purchase assessment.

Each section of the checklist contains detailed questions that can be answered by the process safety pre-purchase assessment team member after discussing the questions with facility management.

In the Mechanical Integrity section of the checklist for instance the following would be addressed:

a. *Determine whether there is a mechanical integrity system/procedures for the following equipment:*
   - *Relief devices*
   - *Major process equipment like reactors and distillation columns*
   - *Storage tanks*
   - *Piping*
   - *Pumps and compressors*
   - *Safety instrumented systems (critical interlocks)*

b. *Determine if initial thickness baselines for tanks have been completed.*

c. *Determine the current plant test or inspection frequencies for this equipment.*

d. *Review previous and current year maintenance budgets to identify any significant planned expenditures for equipment replacement.*

e. *Review MI test records for large equipment (reactors, distillation columns, large rotating machinery, storage tanks, etc.) to determine if significant useful life is still available and whether the need for replacement is imminent.*

f. *Identify what organization performs the MI inspections (contractor, corporate group, etc.)*

g. *Determine expenditures for MI inspections for the previous year and budget allocation for current year.*

Answers to these questions give good insight into how well maintained the facility is, how close to its expected life the equipment is, and whether or not relief devices have been identified and serviced. If a facility cannot produce a list of relief devices, it is extremely unlikely that relief device testing and inspection is being performed on a routine basis.

After completing as much of the checklist as possible, the pre-purchase assessment team will be in a better position to assess what is known and what is not known about a facility in respect to its hazards and risks. Unfortunately not all of the information required will be freely given to the team during these initial phases of the acquisition. However, one has to determine if there is enough information available to assess the potential costs to perform any necessary upgrades to bring a facility into compliance with the safety requirements of the acquiring corporation as a whole.

The types of problems that can show up during pre-purchase assessment include:

1) Facility siting issues caused by local population centers being too close to the storage of hazardous materials.
2) Chemical reactivity problems with materials being used at the facility e.g. pump seal fluid compatibility with main process fluids.
3) Inadequate documentation on the maintenance and design of relief devices.
4) Lack of process safety information and corresponding inadequate hazard reviews.

Some of these issues are readily handled by putting a contingency on the value of the facility. It is the responsibility of the pre-purchase assessment team to recommend to management which issues require significant contingencies.

2.1 Transition Plan

The major element of the transition plan is the identification of items that must be in place on day one of the transfer of ownership. These may include:

1) Emergency response issues such as emergency contact numbers on documentation. Sometimes these will involve contracting with the selling company to continue providing these services until the acquiring company personnel are adequately trained.

2) Regulatory notifications such as US EPA RMP or European Seveso ownership and registration.

3.0 Technology & Standards Evaluation

During and after the pre-purchase phase of an acquisition it is important to focus technical and process safety experience on the technologies being acquired. This experience is often in the form of standards that are used in the design and operation of similar facilities. For example Air Products has many standards covering the safe operation of air separation plants. When we acquire such a technology, we can quickly assess whether or not the facility in question conforms to our standards.

Whether or not to retrofit these standards onto a newly acquired facility becomes an important issue, but is not always straight forward. Just because a facility operates to different standards does not mean that it is unsafe.

To assess whether or not to retrofit standards onto facilities we utilize a work process that ensures that senior process safety, technology and operational input is given. A list of standards that have required retrofitting in the past has been developed and this is given to the process safety assessment teams when looking at new acquisitions. When questions arise related to the need to retrofit standards, senior management is consulted to ensure that adequate technical resources are dedicated to answering the question.

4.0 Process Safety Assessment

Shortly after the transfer of the assets, a process safety assessment is performed by experienced process safety, operations and engineering personnel to assess the potential process safety risks at a facility in detail. The assessment will generate an integration plan that outlines how the facility will meet the acquiring company safety and technology standards and how the facility plans to resolve any potential safety risks. This plan should be agreed to by all parties. This assessment typically takes about 2-3 days for a large facility and utilizes an assessment protocol that ensures consistency. This assessment protocol is similar to the pre-purchase checklist referred to earlier. It contains the same elements as the pre-purchase checklist, but since more time is available to conduct this assessment, it has more in-depth questions and more guidance for the assessors.

Figure 1 is an extract of questions from the hazard review element of the protocol. These protocol questions range from the basic, "Has a hazard review been conducted for the processes at the site?" to "Are sufficient funds budgeted to correct the deficiencies and complete the action items?" and "Has a system been established to promptly address the team's findings and recommendations?" Note that if documentation cannot be found to support each question then it is assumed that the work hasn't been done. Guidance notes are also included to assist the process safety assessor in the task. These are important since they help to ensure consistency between assessors.

The work process for the process safety assessments consists of three distinct phases:

1) Assessment Pre-Planning
2) On-Site Activities
3) Post Assessment Follow-Up

A brief look at each of these areas gives the major considerations for each phase:

4.1 Assessment Pre-Planning

This phase is probably the most significant since it includes communication with the site, team selection and definition of the protocols to be used. It sets the ground rules for the assessment and ensures that all required documentation is available for review by the assessment team. Several weeks prior to the site visit the team leader will identify the appropriate personnel to conduct the assessment. Key employees will include experienced process safety specialists, operations and engineering personnel. Ideally the people selected should have experience and knowledge of the processes being reviewed at the acquired facility. For example, we utilize persons knowledgeable in air separation technology (operations and engineering) when conducting an assessment of an air separation facility. These knowledgeable persons can quickly ascertain potential design and operations hazards of the acquired facility based on a review of company engineering standards.

Communication with the site will allow the site to understand exactly what will be reviewed, when it will occur, and who will be participating. As part of pre-planning, the team leader will prepare a draft daily schedule showing specific areas the team will review and the morning or afternoon the specific area will be reviewed. This schedule is sent to the site with a request that names of appropriate site people are assigned to review a particular area with the assessor on a particular day. This allows the site manager to ensure sufficient resources are available during the assessment for an adequate review, and gives the assessor a point of contact once on-site. The site is also sent a copy of the Process Safety Assessment Protocol questions so they can begin gathering the documentation required to answer the protocol questions.

This pre-planning phase has been shown to be extremely important since it ensures an efficient on-site assessment process, facility buy-in with the assessment process and leads to true team building once on-site. This team building is important since the assessment is not intended as a simple audit of the site. It is intended as a mechanism to identify potential process safety issues and to develop an integration plan that the newly acquired facility can practically achieve using resources from the site and the acquiring company.

4.2 On-Site Activities

Having communicated the intent of the assessment during the pre-planning phase, the first thing that the assessment team does after arriving onsite is to conduct an opening meeting with the site team. This opening meeting allows the team leader to introduce the assessment team, communicate the activities to be completed on-site, as well as the post assessment follow-up activities. The team leader also reviews the daily schedule agreed upon prior to coming onsite to ensure the appropriate people are still available and no emergencies, either plant or people wise, will cause interference with the assessment.

At the completion of the opening meeting, the assessment team takes a site tour to familiarize the team with site layout and conditions. Once the tour is completed each individual from the assessment team meets with the assigned site contact to begin the assessment. At the end of each day the assessment team meets with the site team to communicate the areas reviewed, and discuss findings and items that may need more detailed follow-up. This daily debriefing is very important. It informs the site of the actions identified during the assessment, and it allows for no surprises at the closing meeting.

At the completion of the on-site assessment, the assessment team prepares written actions and communicates these actions to the site team and facility management at a closing conference. The site should have previously been informed of all actions at the daily site debriefings, and only agreement on the wording of each item is usually necessary.

While performing this assessment one quickly determines the adequacy of the process safety management systems in place at the facility and a determination can then be made as to areas that require attention. In some cases this can mean implementing, for example, a whole new mechanical integrity program, while in other cases only small adjustments may be necessary to bring the facility into compliance with corporate requirements. It is important that the assessment team contain experienced process safety and technology members since they have to be able to assess the hazards and recommend to management whether or not a facility is safe to operate.

4.3 Post-Assessment Follow-up and Integration Plan

Within 30 days of the on-site assessment, facility management, in conjunction with the assessment team, prepares an integration plan that addresses assessment actions and puts forth an action plan for resolving them. The assessment team reviews the plan and recommends modifications as necessary to ensure that it is adequate to closeout all of the assessment findings. Once finalized, the integration plan is entered into a database for follow-up until all items are completed.

An extract from an integration plan is shown in Figure 2. Each item has a corresponding action plan, with the responsible person identified. The plan should also contain an expected and an actual completion date.

Where necessary, a plan for gathering necessary process safety information and for performing hazard reviews should be part of the integration plan. These hazard reviews are performed in addition to the process safety assessments since a hazard review is much more detailed and can take a significant amount of time to prepare and execute, especially if a facility has little in the way of process safety information documented.

5.0 Conclusion

A multi-tier approach to assessing the hazards and risks associated with newly acquired facilities has been used successfully to ensure that employee, community and company interests are quickly addressed and adequately protected.

Adequate process safety representation on the teams working on the early stages of new acquisitions is essential and a checklist outlining key issues to be addressed simplifies the process and ensures consistency. This checklist has to be concise since the pre-purchase phase of an acquisition has to be completed in a very short time period by a small multi-functional team. Identification of all potential hazards and risks is not always possible in these early stages due to confidentiality concerns, but early identification of potential issues can be very beneficial and ensure a smooth transition of ownership. A transition plan, outlining items that need to be in place on day one of transfer of ownership, is prepared by the pre-purchase team if the acquisition is finalized.

Where the technology is known, and existing standards exist within the acquiring company, a determination of whether or not to retrofit these standards onto the newly acquired facility has to be made. This can be a difficult decision since the existing operating procedures and design may be perfectly safe. A list of standards that should be considered for retrofit should be developed for such technologies so that process safety and technology management can determine the retrofit requirements. In general, process safety standards that are not technology specific are usually retrofitted.

After taking over operation of the facility, a more detailed process safety assessment should be performed to obtain a thorough evaluation of the hazards and risks associated with a facility. Since more time is available to conduct this post purchase assessment than is available for the pre-purchase assessment, an assessment protocol containing more guidance and detail is used as a tool to ensure thoroughness. An integration plan is developed showing how and when the newly acquired facility will comply with corporate requirements. The team performing this assessment should be experienced and senior enough to ensure that management gets a good picture of the true risks to the corporation.

Figure 1 - Excerpt from the Process Safety Assessment Protocol

	PROCESS SAFETY ASSESSMENT for ACQUISITIONS/JOINT VENTURES **Hazard Reviews**	
PSM-3	Has a hazard review been conducted for the processes at the site? Assessor Guidance: • Assessor should verify a list of the various processes exists for the site and confirm that a hazard review has been conducted for the specific process.	Y *N* NA ***DK***
PSM-4	Determine which of the methodologies listed below were used for the hazard reviews; • What-if/Checklist; • Hazard and Operability Study (HAZOP); • Failure Mode and Effects Analysis (FMEA); • Fault Tree Analysis; or • An appropriate, equivalent methodology.	Y *N* NA ***DK***
PSM-5	Does the hazard review address the areas listed below (Assessor should assess the quality of the hazard review)? a) The hazards of the process a) The identification of any previous incident which had a likely potential for catastrophic consequences in the workplace b) Engineering and administrative controls applicable to the hazards and their interrelationships, such as appropriate application of detection methodologies to provide early warning of release c) Consequences of failure of engineering and administrative controls d) Facility siting e) Human factors Assessor Guidance: • Confirm that the hazard review documentation includes a discussion of facility siting, human factors, and previous incidents. • Assessor should review the hazard review to assess the quality of the hazard review. • Facility siting means considering the location of covered processes within the plant property. Facility siting issues can include location of vessels containing HHCs and their proximity to other equipment, control rooms, maintenance shops, and administration buildings • Human factors involve issues and techniques with the functional relationship between human operators and engineered systems. Human factor considerations include potential human error causes of accidents; examining the location of and access to critical safety instruments, alarms, and equipment; or reviewing critical procedures used by operators and maintenance personnel	Y *N* NA ***DK***

PSM-6	Was the hazard review performed by a team with expertise in engineering and process operations, including at least one employee with experience and knowledge specific to the process and one member knowledgeable in the specific hazard review methodology used? <u>Assessor Guidance:</u> • Hazard review must identify through documentation who these individuals are. Also, determine what organization(s) performed the hazard reviews (plant only, support from corporate, contractor, etc.)	Y ***N*** NA ***DK***
PSM-7	Has a system been established to promptly address the team's findings and recommendations? Does <u>each</u> resolved hazard review recommendation have documentation to support resolution of the items? <u>Assessor Guidance:</u> • Assessor should determine if any high-risk action items are not complete. How many action items are beyond their due date?	Y ***N*** NA ***DK***
PSM-8	Has a written schedule been established of when the findings and recommendations are to be resolved? <u>Assessor Guidance:</u> • Each item requiring resolution must have a planned completion date and individual responsible for implementing the corrective action. A single completion date for the entire hazard review does not qualify as a "system" to address the recommendations as required above.	Y ***N*** NA ***DK***
PSM-9	Are sufficient funds budgeted to correct the deficiencies and complete the action items? <u>Assessor Guidance:</u> • The assessor should review hazard review recommendations and identify any items that will require significant capital to implement	Y ***N*** NA ***DK***
PSM-10	Are the hazard reviews updated and revalidated periodically by a qualified team? (Assessor should record the date of the last hazard review). <u>Assessor Guidance</u>: • US PSM regulations require a 5-year revalidation frequency for covered sites. • Air Products requires all sites receive a minimum 10-year revalidation frequency when not governed by regulation.	Y ***N*** NA ***DK***
PSM-11	Has the site identified fatal accident rate (FAR) events and estimated capital required to upgrade systems to meet this target, as appropriate?	Y ***N*** NA ***DK***

Figure 2 – Extract from an Integration Plan

Action	ACTION PLAN With Responsible Person	Estimated Comp. Date	Actual Comp. Date	STATUS
A complete P&ID was not available nor a hazard review for Process A. • ***Develop P&ID's for Process A.*** • ***Conduct hazard review for Process A.***	• Complete P&ID's update for Process A (Facility Manager) • Ensure that adequate process safety information is available for the hazard review (Facility Manager) • Complete the hazard review for Process A. (Facility Manager)	Mar. 31 2004 June 30 2004 Sept 31 2004		

Paper 141b

Organizational Factors That Influence Safety

Behavioral Science Technology, Inc.
417 Bryant Circle
Ojai, CA 93023
bstojai@bstsolutions.com

Prepared for Presentation at
American Institute of Chemical Engineers
2005 Spring National Meeting
Process Plant Safety Symposium
Atlanta, Georgia
April 11-13, 2005

UNPUBLISHED

Introduction

For the past 20 years the author and his associates have worked with over 1,600 organizations to improve safety performance. In the course of that work, we have discovered that organizations optimize their efforts when they provide and support strong roles at each of three core roles: the front-line employee, supervisory and senior manager levels. These three roles, when appropriately engaged, make up a cord of three strands that is not easily broken.

Of these three areas, the leadership role has proved to be both the most impactful for safety improvement, as well as the most complex to understand and influence. One striking example of this is seen in the study of sites using the same improvement methodology. Even when compared by industry, site size, and systems, some organizations achieve steady improvement while others consistently struggle. More than any other factor, the quality of the organization's leadership, in particular its influence on the organizational culture, determined the level of success achieved.

While it is generally recognized that leadership is important to performance, the "how" of that leadership is often debated. What makes a leader effective at influencing and improving safety? What are the qualities of such a leader? What are the key tasks that a leader must perform in order to generate desired results?

While this paper will focus on these questions as it pertains to senior leaders, the principles discussed are applicable to any level in the organization.

How Leadership Influences Safety

How is it that safety leadership assures performance improvement?

The primary activity of safety initiatives, whether at the site or corporate level, is to reduce the amount of *exposure* that occurs in the workplace. While not all exposure is equal in terms of severity potential, all incidents result from exposure to hazards and reducing exposure is the primary mechanism of safety improvement. Exposure arises from management and work system problems and the pressures of the organizational climate. These lead to hazards in the workplace and worker exposure. Leaders directly and indirectly impact systems and climate, and thereby hazards and exposure.

In order to understand these connections, the following sections present findings from ongoing research that answer some of the critical questions surrounding safety leadership.

The Four Elements of Successful Safety Leadership

At its most basic, leadership is about establishing a direction and helping people to move in that direction. Practically speaking, leadership comes down to two tasks:

- Getting subordinates to do the right work the right way, and
- Maintaining a successful relationship with the people doing that work.

In many organizations, particularly in the area of safety, the two tasks appear to be at odds with each other. Oftentimes leaders believe that doing one well means sacrificing the other. While there is undeniably a tension that must be balanced between these two tasks, we have found that the leaders in organizations who perform both of these tasks well are able to maintain a healthy and appropriate balance. And these leaders are more effective at fostering high levels of safety performance.

The successful realization of this balance, both on the interpersonal level and on a level that impacts the whole organization, starts with who the leader is and what is important to him or her. It radiates out from there in how the leader goes about influencing others concerning what's important to the organization. Then it expresses itself in what things the leader does day-to-day and especially in how he or she does them. And it ends by shaping nine elements of the organization's culture and safety climate, elements that play out directly in safety performance.

Personality & Values

The leader's personality and values are at the core of who the leader is, and consequently at the core of his or her effectiveness as a safety leader. These two elements strongly influence what the leader will tend to focus on and the likelihood that the leader will favor or use the most effective influencing style for safety and the best safety leadership practices.

How the Leader Influences

Leaders differ in their style of influencing others. The influence style describes how the leader interacts with subordinates to enlist their energy in an enterprise. Research shows that a transformational leadership style is characteristic of the most effective safety leaders.

What Leaders Do

The leader enacts his or her personality, values and leadership style in daily practices. These are the day-to-day behaviors through which the leader effectively (or not) influences the vigor of the safety climate and builds strong safety systems.

Organizational Culture

The safety climate is the readily measurable, more accessible and more rapidly changed aspects of the organizational safety culture. It is the place where the leader has leverage to impact the culture, or "how we do things around here." Research indicates that there are specific aspects of the organization's culture and safety climate that are predictive of high safety performance.

Personality & Values

At the very core of who a leader is, and consequently how he or she acts and responds as a leader, are personality and values. Studies have identified a strong link between a leader's personality and many dimensions of employee performance. Personality describes the stable attributes of the person on five dimensions, known in the research literature as the *Big Five.* The Big Five dimensions are:

- **Extroversion** - involves warmth towards others, outgoingness, assertiveness, level of activity, seeking excitement, optimism and positive emotions.

- **Agreeableness -** involves trusting others, being straightforward, consideration for others, compliance with standards, modesty, and sympathy.

- **Conscientiousness** - involves a sense of competence, orderliness, sense of responsibility, need to achieve, self-discipline, and deliberateness.

- **Emotional Stability** - involves confidence in self and others, an upbeat approach to challenges, and the ability to handle stress well.

- **Openness to Experience** - involves a sense of curiosity, an exploratory approach to challenges and an imaginative mindset.

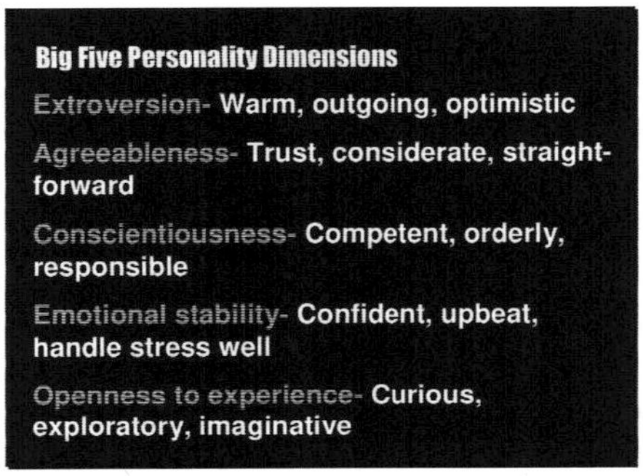

Exhibit 1. Big Five Personality Dimensions.

In addition, the leader's values need to be compatible with those required for the job. In our experience, the most successful safety leaders have a high value for service to others, problem solving and quality as well as a low tolerance for exposing others to risk. The makeup of a leader's personalities and values dispose them to their influence style, and the likelihood that they will engage in the practices identified as optimal for producing a high-performing organizational safety culture.

While a personality is well-formed by adulthood and its Big Five dimensions are stable over time, the important point is that leaders gain flexibility and breadth of options through insight into their own values and personality structure. Effective leaders understand their values and personalities and are able to develop highly effective ways for bringing their assets to the forefront and offsetting their liabilities.

How the Leader Influences

There are a number of influence styles that leaders use. The research shows that leaders who have a strong transformational leadership style typically have groups that perform better in various ways, including safety outcomes.

Transformational leadership has four dimensions. The first is **charisma**. Does the leader provide vision and a sense of mission, instill pride, gain respect and trust and increase optimism? The second is **inspiration**, which is sometimes grouped with charisma. Essentially, it defines whether the leader acts as a model, communicates a vision, sets high standards, and uses symbols to focus efforts. The third is **individual attention**. Does the leader coach, mentor, provide feedback, link the individuals' needs to the organization's mission, and provide personal attention? The fourth dimension is **intellectual challenge**. Does the leader provide subordinates with a flow of challenging new ideas aimed at rethinking old ways of doing things, challenge dysfunctional paradigms, promote rationality and careful problem solving?

Interestingly, leaders who have high levels of transformational leadership are not dependent on their bosses to place a high priority on safety. Their safety best practices are strong whether or not there is an external emphasis on safety. It is likely that this is related to who the leader is: a transformational leader is more likely to demonstrate a value for the well being of his or her subordinates, and this motivation to protect employees comes from a different—and more interior—place than organizational authority. Another finding related to transformational leadership is that its relationship to safety outcomes is entirely mediated by preventative action. In other words, it's not just the leader's influence style that matters, but also what the leader does: supervisors with strong relationships with workers (transformational style) talk and listen to them and take action about safety (preventative action) and that leads to lower injury rates.

What a Leader Does: Best Practices

What a leader does and how he or she does it is a manifestation of the leader's personality, values influence style and level of insight or understanding of his or her own personality. The leader's typical behaviors or practices in turn strongly influence the organizational safety climate and ultimately the culture.

The author and his colleagues have identified several best safety leadership practices. In our work we noted that certain management and leadership practices recur across those organizations with high-performing cultures. Comparing that experience with existing literature on leadership influences on safety and organizational culture the author and his colleagues came to the conclusion that there are at least eight definable leadership practices connected to the development and support of a high-performing safety culture. These are Vision, Credibility, Collaboration, Feedback and Recognition, Accountability, Communication, Values Safety, and Action Oriented.

Vision – The effective leader has identified the strategic role of safety performance to his or her organization, is able to "see" what such performance excellence would look like, and can convey that vision in a compelling way to the organization.

Credibility – The effective leader is credible to other people in the organization, is willing to admit his or her mistakes to others, "goes to bat" for direct reports and the interests of the group, and gives honest information about safety even it if is not well received.

Collaboration – The effective leader works well with other people, promotes cooperation and collaboration in safety, actively seeks input from people on issues that affect them, and encourages others to implement their decisions and solutions for improving safety.

Feedback & Recognition – The effective leader is good at providing feedback and recognizing people for their accomplishments. This person publicly recognizes the contributions of others, uses praise more often than criticism, gives positive feedback and recognition for good performance, and finds ways to celebrate accomplishments in safety.

Accountability – The effective leader gives people a fair appraisal of the efforts and results in safety, clearly communicates people's roles in the safety effort, and fosters the sense that every person is responsible for the level of safety in their organizational unit.

Communication – The effective leader is a great communicator. He or she encourages people to give honest and complete information about safety even if the information is unfavorable. This leader keeps people informed about the big picture in safety, and communicates frequently and effectively up, down, and across the organization

Action-Orientation – The effective leader is proactive rather than reactive in addressing safety issues. This leader gives timely, considered responses for safety concerns, demonstrates a sense of personal urgency and energy to achieve safety results, and demonstrates a performance-driven focus by delivering results with speed and excellence.

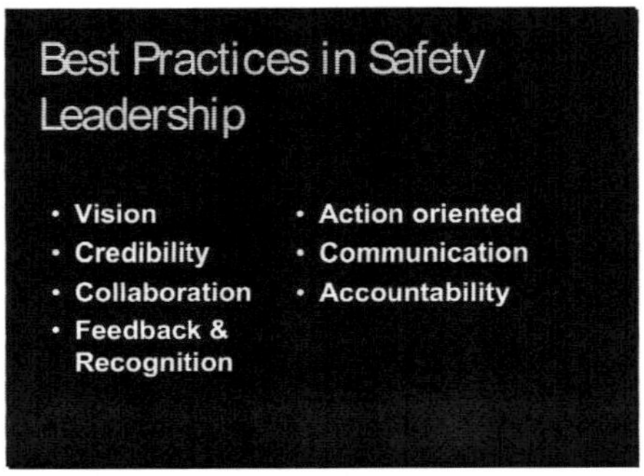

Exhibit 2. Best practices in Safety Leadership.

The Organizational Culture and Safety Climate the Leader Creates

Our experience working with organizations interested in performance improvement, and subsequent ongoing research, have highlighted the fact that far from being an ambiguous facet of organizational life, organizational culture and safety climate are definable and measurable in very practical terms. High-performing organizations consistently show high trust, good communication, management credibility, and organizational value for safety. Low-performing organizations tend to show the opposite. We have identified nine characteristics of organizational culture that are predictive of successful safety outcomes. These are:

1. Teamwork–the effectiveness of workgroups in meeting targets and deadlines. It is clear how the leadership best practice of collaboration and feedback can build teamwork. Transformational leadership can build teamwork when the transformational leader directs attention to the team and uses it to develop and implement safety solutions.
2. Workgroup relations–the degree to which coworkers respect each other. Transformational leadership instills pride, gains respect and trust, increases optimism among members of a workgroup and collaborates practices increasing the likelihood that team members will reciprocate these things with each other.
3. Procedural justice–the level that workers rate the fairness of first-level supervisors. When transformational leadership is strong in a leader, the individual attention is shared across subordinates, not focused on particular favorites, which supports this procedural justice. But procedural justice also requires action-oriented practices: systems and procedures need to be aligned to visibly support fairness.
4. Perceived organizational support–the level to which employees feel the organization is concerned for their overall well being. This factor in particular requires the transformational style and follow-up action oriented practices. Part of the definition of transformational leadership is individual attention, which influences Perceived Organizational Support only when the leader actually does things to demonstrate the concern for employees well being and interests.
5. Leader-member exchange–the strength of the two-way relationship that workers feel they have with their supervisors. Strong perceptions of leader-member exchange require that the leader follow-up his or her individual attention, intellectual challenge, and inspiring words with behaviors and practices, for example actually going to bat for employees when needed. This requires not only action oriented practices, but those needed for good communications and credibility.
6. Management credibility–the perception of consistency and fairness of management in dealing with workers. Part of the definition of transformational leadership is that the leader gains respect and trust of employees, and acts as a model. This dovetails with leadership best practices needed for credibility. When the majority of leaders are successful in this, overall management credibility will be improved.
7. Organizational value for safety–this is often called 'safety climate', the perceived level of the organization's commitment to safety. Interestingly, this factor matters less if the leader has the right values. This means that even when the organization itself is not supporting safety at a high level, the individual transformational leader can effectively support it. Conversely, if the leader's style does not include strong transformational elements, the leader is more dependent on the organization's support for safety performance.
8. Upward communication–the adequacy of upward messages about safety. Listening to subordinates is an action that needs to accompany transformational leadership for it to be effective. But when both are in place, upward communication is positively impacted.
9. Approaching others–the probability that workers will speak to each other about safety issues. In a feedback-rich environment, modeled by the leader, people are more likely to approach each other.

Companies with cultures that are high in these nine characteristics tend to be more successful in initiatives they undertake, generate change more rapidly, and achieve higher performance overall in critical business functions than companies that score low on these factors.

Conclusion

Organizational culture and safety climate are strongly influenced by the quality and type of leadership enacted throughout the organization and this in turn impacts safety performance. Leaders can gain insight into their own values and personality structures and how these play out in their influence style and practices. Understanding how these factors interact enables leaders to use their own behavior as a central leverage point for safety. It shows them what to do to build strong safety performance.

142a Keynote Paper Holder

Scott Berger

Keynote Paper Holder

The 27 March 2003 Billy-Berclau Accident: A Technical And Organisational Investigation

Jean-christophe Lecoze, Nicolas Dechy, Samantha Lim,
Emmanuel Leprette, Ruddy Branka
INERIS (Institut National de l'Environnement Industriel et des Risques)
Parc Alata, 60550 Verneuil-en-Halatte, FRANCE
Tel: 33 3 44 55 66 77, ineris@ineris.fr

Abstract

Following a dynamite explosion in a pyrotechnic plant in Billy Berclau, France, which caused the death of 4 persons and injured 13 persons, the Environmental Ministry appointed INERIS (National Institute For Environmental Safety) to carry out an investigation. The purpose of the investigation was to consider technical and organisational issues related to the origins of the accident as well as the damages of the accident. This paper will introduce the results of the investigation with a focus on its organisational side.

The organisational requirement within that investigation along with the technical dimension is a first in France in the chemical industry and was enforced partly by the SEVESO II European Union directive context. The context of the Toulouse disaster played its role also. Indeed this directive requires the implementation of a safety management system, dedicated to major hazard prevention for certain plants, according to their risk level (based partly on the quantity of hazardous materials stored). The plant where the accident occurred was under that regulation and had to implement such a management system. The need for an understanding of the management system's underlying root causes in that accident was therefore high.

INERIS has been developing research programs on the organisational sides of major hazard prevention for few years and could therefore use its expertise in that field for the investigation.

In a first part of the paper, the accident will be presented, through a chronology of the events, damages and through the technical causes that have been identified and retained as most probable during the investigation. This first part is a preliminary step for understanding the case itself and to get into the second part of the paper.

The second part will discuss the organisational investigation, from a methodological point of view and then from the findings. The methodological aspect will question the ability to get into the complexity of the organisational life. Organisations are complex systems that need to be looked at with appropriate lenses. These lenses are influenced by the investigator experience and background and it is part of the methodology to address this issue. The level of opening of the organisational investigation is another strong methodological issue. As organisations are complex systems, they interact at all levels, and their understanding implies opening the system to many individuals involved inside and outside the boundaries of the organisation. That is another important issue of the investigation. The outcomes of the investigations are presented according to different levels of the system and are discussed.

1. Introduction And Context Of The Investigation

Following a dynamite explosion in a pyrotechnic plant in Billy Berclau, France, which caused the death of 4 persons and injured 13 persons, the Environmental Ministry appointed INERIS (National Institute For Environmental Safety) to carry out an investigation. The purpose of the investigation was to consider technical and organisational issues related to the origins of the accident as well as the damages of the accident. This paper proposes to describe some related issues around the investigation, with a focus on its organisational side.

1.1 Context Of The Investigation

A dynamite explosion, in a pyrotechnic plant in Billy Berclau, France, occurred on 27 march 2003 at 6h16, and caused the death of 4 persons and injured 13 persons. Fires resulting from the explosion were quickly extinguished by emergency services.

After the event, several investigations were launched by Authorities:

- the litigation inquiry carried out by the Police and the Justice,
- the Labor Ministry carried out its investigation,
- the investigation of the company Nitrochimie S.A. that merged with the Health and Safety committee: their focus was to find out the causes of the explosion,
- the administrative inquiry of IGE-CGM-IGAS (Inspection Générale de l'Environnement, Conseil Général des Mines and Inspection Générale des Affaires Sociales) set up by four Ministries: Environment, Defence, Labor and Industry; the investigation had to focus on circumstances of the accident, causes (technical, human and organisational), safety measures, recommendations that could lead to regulation changes, and should use expertise of INERIS and IPE (Inspection des Poudres et Explosifs); the report would be made publicly available (on the web in April 2004),
- the investigation of INERIS, launched by the Environment Ministry, focusing on technical and organisational dimensions of the accident.

The general objective of administrative inquiries carried out by IGE, IGAS and CGM, that are organisations within their Ministries, is to focus on the circumstances (in case of accidents, pollution,...) and on the local application of the law by companies and local control authorities. The investigation of INERIS brought (final report was issued in September 2003) the parallel inquiry of the IGE-CGM-IGAS, a technical and expertise point of view, independent of the other investigations (litigation, company...). It was also used as a support for the local control authority (DRIRE) to require implementation of safety measures from the company to get authorization to re-start.

This investigation occurred in a broader context following the Toulouse disaster that occurred the 21st September 2001. As a reminder, several investigations were launched by stakeholders (a few more than in this case), and provided numerous findings, lessons and recommendations for changes (Dechy et al, 2004). Disasters are usually seen as catalyst for changes and the Environment Ministry was designing a new law on industrial risk prevention (Law of 2003-699 of 30th of July 2003) and discussing it with stakeholders at the time of this event.

After Toulouse disaster, other decisions of the Environment Ministry were taken to upgrade the existing funding for carrying research at our Institute on the learning from experience, accident investigation methodologies and the integration of human and organisational factors into the major hazard prevention. This investigation was the first one to use most of these developments. As a reminder, INERIS has activities for certifying explosives and hazardous materials (for transportation) in France and EU.

1.2 Objectives, Contents And Management Of The Investigation

The objectives of the investigation were set up by the risk bureau on chemical risk (SEI/BRTICP) and the BARPI (that runs the ARIA database on industrial accidents) within the Environment Ministry and were accepted by the scientific director of INERIS.

The objectives were to:

- estimate the effects of the explosion (inside the site and in the environment), its strength, its eventual domino effects, the differences with the safety studies scenarios,
- find the origin, the circumstances and the possible technical and organisational root causes that should be investigated; for the technical causes, the nature, quantities, reactions of material involved, interactions with the process have to be investigated; organisational factors should be investigated also with regards to the Seveso II regulation and its safety management system requirement,
- formulate technical recommendations on process, workshop design and organisational recommendations such as cleaning of units, risk analysis, learning from experience, etc.

To fulfil these requirements, INERIS management appointed a multidisciplinary team of 5 main investigators (the Institute has 150 people working in accidental risk prevention) that were competent in the field of pyrotechnic process, the pyrotechnic and Seveso II regulations and its safety management requirements, the inspection process, technical investigation, materials sensitivity testing, explosion and pressure waves effects, human and organisational factors. The reports were reviewed by two top level managers (technical director of the Accidental Risk Division and the technical director of the Certifying Division) and approved by the scientific director.

The first visit consisted in collecting facts, photos of the damages, material samples, and documents and were followed, during 3 months, by several interviews of process operators, frontline managers, maintenance technicians, safety managers, top management of plant, top management of the group and control authorities.

2. The Context Of The Accident

2.1 The Company

The company, Nitrochimie S.A., is owned by a French group (SAEPC, Société Anonyme d'Explosifs et de Produits Chimiques) quoted to the stock exchange market of Paris and by the Davey Bickford Smith English group. The turnover was 120 M€ in 2001. The company Nitrochimie S.A. owns two sites in France: the Billy-Berclau site in north of France that produces explosives such as dynamites, blasting gelatine, powder dynamites, and ammonium nitrate fuel oil (ANFO) with 80 to 90 people, and another site (with the Research and Development laboratories) that manufactures blasting emulsions, expansive cement, and monomethylamine nitrate.

The plant is on a site of 750 000 m², on Billy-Berclau and two other cities and its first neighbours are at 700 m. The plant was created a hundred years ago in 1904 to produce nitroglycerine based explosives for the coal mines nearby. It was destroyed during WWI and rebuilt in 1920 with two areas for different manufacturing purposes: one for ANFO and one for dynamites. In 1991, the nitrate activity was stopped and the number of employees was reduced from 170 to 117. Ten years later, in 2003, they were 84 permanent workers and 4 to 6 interim workers.

The explosives manufacturing rates per year were roughly 6 500 tons of dynamites, 8 500 tons of ANFO and requires the consumption of 4 000 tons of sulfonitric mix, 240 tons of DinitroToluène, 10 700 tons of AN and 550 tons of light fuel. These manufacturing explosives to be sold on the EU market needed to be approved according to EU Directives specifications. INERIS does certify explosives in the EU and since 1997 certified those of Nitrochime S.A. and the last time was in May 2001 for 15 different dynamites and 2 ANFO.

2.2 The Safety Related Regulations

These manufacturing activities were under two main type of safety-related regulations:

- The labor regulation and its pyrotechnic regulation: a specific regulation from the labor Ministry was set up in 1979 (Decree n°79-846) and implemented with positive results measured by reduced rates of accident. It required specific prevention measures to be implemented. An emphasis of the investigation was on a comparison between observed effects and estimated effects from the safety cases,
- The environment regulations and its Seveso II regulation (96/82/EC) transposed in 2000 in France: indeed this Directive requires the implementation of a safety management system, dedicated to major hazard prevention for certain plants, according to their risk level (based on the quantity of hazardous materials stored). The plant where the accident occurred was under that regulation for 3 years and had to implement such a management system. The need for an understanding of the management system underlying root causes in that accident was therefore high.

2.3 The Process

There exist several kinds of dynamites that differ in rate of nitroglycerine or nitroglycol. To manufacture dynamite gelatine, four main steps are briefly described:

- Nitration (in a workshop, continuous process) of a mix of glycerine/glycol by a mix of nitric and sulfuric acid to get nitroglyceroglycol (NCG)
- Paste or gelatine manufacturing: it can be done in two different ways: an automated mixing bath holder called Tellex operated at distance in a safe place - or by batch in a tank called Guedu where NCG and adsorbents are poured by operators and the blast gelatine is removed after it has been automatically mixed. The Tellex produces 3 tanks in a row (that are made of polyethylene) for a total amount of 135 kg of dynamite gelatine. These tanks are equipped with caps and 4 wheels. The Guedu produces 3 tanks for a total amount of 120 kg. Then the tanks are driven on a trolley to an intermediate workshop storage n°49 that has a storage capacity of 1 300 kg. From this storage, tanks are automatically driven following a cable, to the processing workshop n°50 to manufacture cartridges of dynamites. These workshops are all separated by barricades of 5 to 6 meters high.
- The automated processing of gelatine to dynamite cartridges at workshop n°50: The inlet of the tanks into the feeder (see figure 1) is automated when the process is started. After the extrusion of the gelatine into the die (photo 1 and 2) and its cut at the roller, the gelatine is rolled in a wax paper before introducing it in the cartridge. The cartridge, processed in the Rollex machine (rate of 900 kg/hour), are transferred on a conveyor to be packed in boxes in a neighbouring workshop n°52. This neighbouring workshop has the video surveillance reporting screen (and recorder) for operators to monitor the process once it is started.

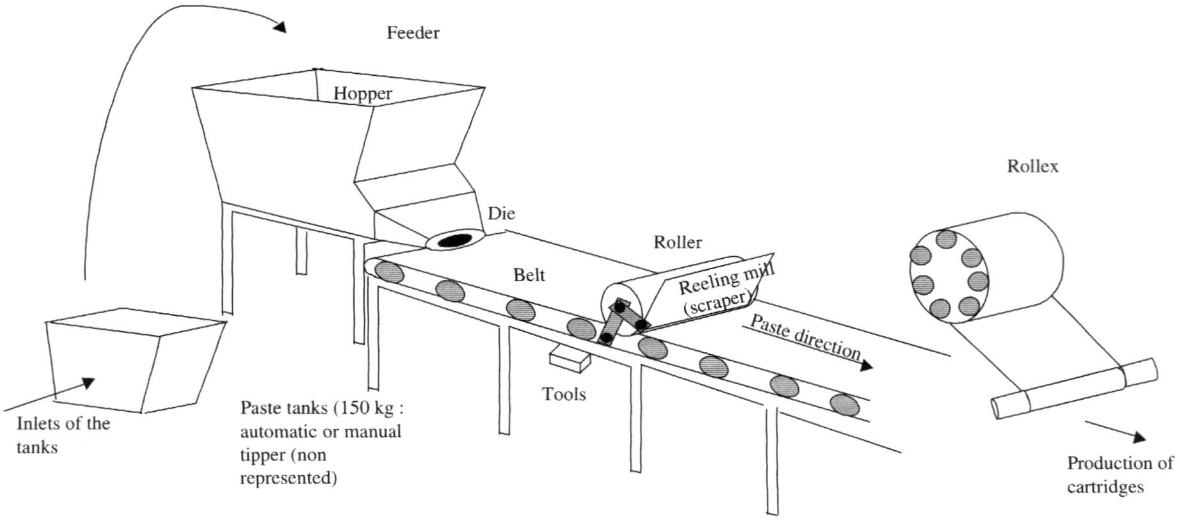

Figure 1: The Rollex Machine And The Feeder

Photo 1: View Of The Feeder	**Photo 2**: Inside The Feeder (Post-Accident Experiments By The Company)

2.4 Organisation To Operate The Process Involved In The Explosion

Several people are involved on site in the operation and management of the high risk process of dynamites cartridges manufacturing. The key actors were:

- the operator of the Rollex machine: its role is required on the Rollex at the starting of the process; to prepare the automated processing state, its actions are to pour the tanks into the feeder, realising the checks and mechanical adjustments, verifying the size of the cartridge, the level of the conveyor belt, the level of the roller before moving to workshop n°52 to monitoring the process on the video screen; the starting of the automated process is known as a critical operation for the operator's life due to the vicinity with the paste processing into the mechanical machine;
- the assistant of the operator of the Rollex, that helps but does not make the adjustments
- the cleaner of the pyrotechnics waste for the all site,
- a mechanical technician of the maintenance team,
- 3 frontline managers and 1 production manager

The 3 frontline managers and the production manager are responsible for scheduling the production on a monthly basis and then on a weekly basis. The shifts are from 5h-12h and to 12h-19h or from 7h-15h.

2.5 A Summary Of The Chronology

In the days before the accident, a decision was taken to recycle a specific blast gelatine (27 tons of Supedopex) that was stored in UK after the Irish subsidiary decided to cancel the production of these explosives. The gelatine would be recycled in the process for the manufacturing of a specific type of dynamite called Dynaroc 6 after a testing phase. These tests started at the beginning of March 2003.

Because it is not a current product in Billy-Berclau, a new recipe had to be created. The results then have to be approved by the internal laboratory to confirm the quality and safety specifications. The production started three days before the accident.

In the following chronology, there are two frontline managers (A and B): A is responsible for the nitration and mixing process steps and B is responsible for the processing of gelatine to cartridge manufacturing.

The day before the accident, some problems of homogeneity of the mix of Superdopex gelatine and other reactants had delayed the production. The frontline manager B decided to start the shift work at 5 a.m. instead of 7 a.m. to process stocks of paste. In parallel, some testings for a new dynamite called Frangex, have resulted in 8 boxes of 20 kg to be recycled in the production. The operator of the Rollex and its frontline manager B, are unaware of these testings.

So, the day of the accident, the rollex operator arrived at 5h15 in the workshop n°50 (notice that some steps of the chronology have been identified with the video recorder). He looked for gelatine to be processed into the pyrotechnic area and met some operators. From 5h20 to 5h30, he started the rollex machine, made the adjustments, processed some gelatine and packed two boxes (50 kg of cartridges were found after on the workshop 52). There were two other operators scheduled but they were not there that day: one did not wake up, and another switched with a colleague and was not aware of the schedule change. During a half hour some discussions and actions involving the rollex operator and the frontline managers are performed to find two other operators.

After being informed that he will be helped, the rollex operator came back at 6 a.m. to start the process. He faced problems with processing the gelatine through the die. It required him to do a lot of operations (mechanical, cleaning) on the feeder, the plasticating screw, the die, the belt, the roller, etc.

A few minutes before the accident, the assistant joined him. A few seconds before it, the cleaner came into the workshop and took the pyrotechnic paper garbage. He was supposed to start at 7 a.m. but showed up early that day, and with the agreement of frontline manager B, was authorised to start his work of unit cleaning. Also the mechanical technician for maintenance, for some operating reasons (he had forgotten a tool) was walking and passing in front of the tunnel linking workshop n°49 and n°50 when the explosion occurred at 6h16'34.

3. The Effects Of The Accident

3.1 Damages To Structures

The explosion produced extensive damages in the plant and in the nearby village of Billy-Berclau. The methodology of data collection and damage assessment was the same as followed for the investigation of the Toulouse accident by INERIS for IGE (Mouilleau and Dechy, 2002). Some principles of theories on shock wave propagation (Wright J.K., 1961, and Baker W.E. et al, 1983) were used for damages assessment. As a reminder, this damages investigations involves uncertainties and methodological biases that lead, for example, to several TNT equivalent estimates for the explosion of Toulouse (Dechy and Mouilleau, 2004).

Workshop n°50, where the explosion took place, was totally destroyed (photo 3).

Photo 3: Workshop 50, where the explosion occurred $\Delta P > 100$ kPa

Photo 4: Crater in Workshop 50 $\Delta P > 100$ kPa

Photo 5: Damages in workshop 49 (30 m from Workshop 50) - $\Delta P = 10 - 15$ kPa

Photo 6: Damage to a building, 370 m from Workshop 50. (dislodged tiles and broken panes) - $\Delta P = 1\text{-}2$ kPa

The heavy Rollex parcelling machine was thrown against a barricade, barricades were partially collapsed, and two small craters (about 1 m diameter) were dug in the concrete at two locations were dynamite charges were stored on the ground (photo 4). Nearby workshops within about 100 m from the explosion suffered extensive damages (photo 5), but mainly to light structures (roofs, sidings, wood beams). Barricades isolating each workshop from the others proved once again to be an efficient protection safety barrier against the explosion. In the far field, minor damages to light structures were observed (photo 6).

Most buildings in the plant suffered from dislodged tiles and asbestos-cement panes, broken windows and panes. Similar effects were noticed in Billy-Berclau village, about 1000 m distant from the explosion epicentre.

It was not mentioned any domino effect. The whole dynamite charge present in workshop n°50 exploded, but the explosion did not propagate to nearby workshops where high quantities of explosives were also stored.

3.2 Analysis Of The Damages

Overpressure estimates were made on 34 damage points related to typical overpressure level or ranges available in the literature (Mouilleau and Dechy, 2002; Cozzani and Salzano, 2004). The ranges of estimated overpressures recorded are plotted versus distance from the explosion epicentre (figure 2).

The screen effect provided by barricades between workshops was taken into account in the assessment methodology, according to the following rules recommended by the French Pyrotechnic Regulations (figure 3):

- within a distance twice the height of the barricade, the overpressure level at ground is divided by 4 to 6, compared with the overpressure that would be measured without barricade.
- At a distance between twice and four times the height of the barricade, the overpressure level at ground is divided by 2 to 4, compared with the overpressure that would be measured without barricade.

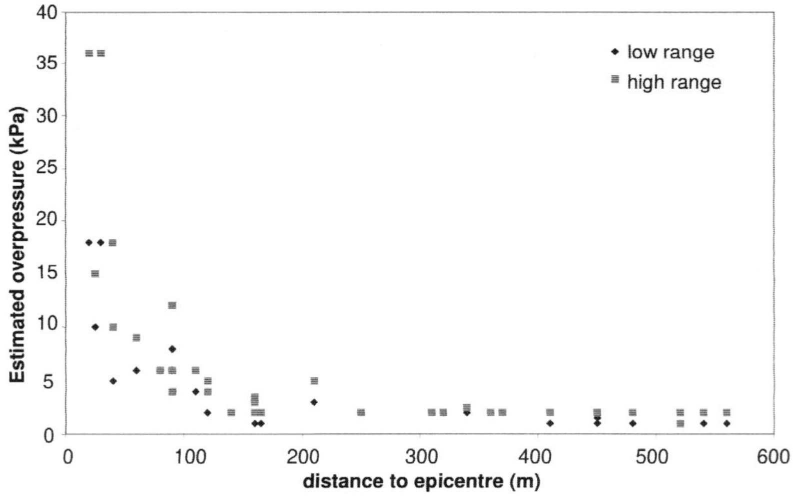

Figure 2: Overpressure Estimations Vs Distance To The Epicentre

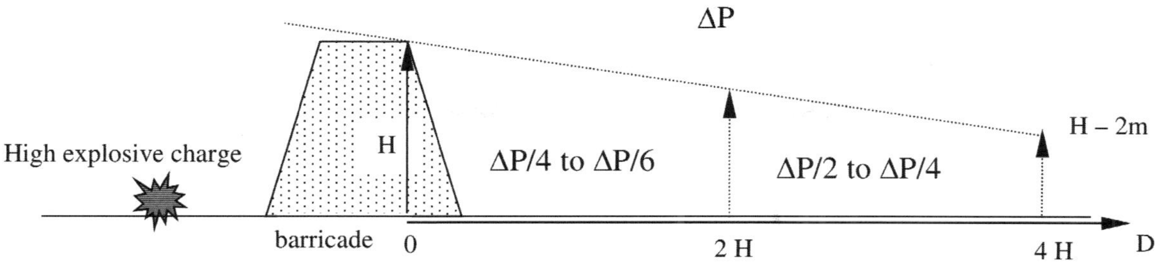

Figure 3: Screen Effect Of A Barricade (Adapted From The French Pyrotechnic Regulations)

As one would expect, the estimates show a decrease in the overpressures the greater the distance measured from the epicentre. In the pyrotechnic area, the shock wave propagation was clearly disturbed by reflection and diffraction phenomena in crossing the barricades, that complicates the damage interpretation (see Wenzel A.B. and Bessey R.L., 1969). After correction by the screen effect, a mean overpressure level from 5 to 15 kPa seems plausible. Outside the pyrotechnic area (beyond 250 m from workshop n°50), the overpressure level seemed about 1 to 2 kPa.

From the very first analysis of the damage points, it has been deduced that the TNT equivalent of this explosion turned out to be between 70 to 130 kg. These values correspond to the medians of the weight values calculated from the overpressure estimates respectively on the low side and on the top side. Considering an energetic equivalent of 1.1 kg TNT for 1 kg dynamite, that would provide a dynamite charge from 60 to 120 kg. However, a discrepancy was noticed in TNT equivalents from inside and outside the pyrotechnic area. The analysis of the inside damage points would suggest a dynamite charge from 50 to 100 kg, whereas the outside damage points would give 100 to 300 kg.

This disparity should be explained by considering:

- in the pyrotechnic area, the overpressure wave disturbance caused by the barricades;
- in the far field, the difficulty to quote accurately damage to light structures, and the possible disturbance by external phenomena, such as the wind speed and the wave reflection by the lower atmospheric layers (Grant et al, 1967, and Reed, J.W., 1973).

Those estimates can be compared with the real explosive charge present in workshop n°50 when the explosion occurred (figure 4). This has been assessed to 580 kg, distributed into:

- two dynamite tanks on hold at the entrance of Workshop 50: 260 kg,
- dynamite in the feeder: 150 kg,
- one tank discharge near the belt: 90 kg,
- 4 boxes of Frangex dynamite: 80 kg

The first explosion has probably occurred in the feeder containing 150 kg of dynamite. Applying the French Pyrotechnic Regulations rules of calculation, the detonation in the feeder would have been transmitted by pressure wave to other close dynamite charges within a distance of no more than:

$$D = 0.5.Q^{1/3}$$

where Q is the explosive mass.

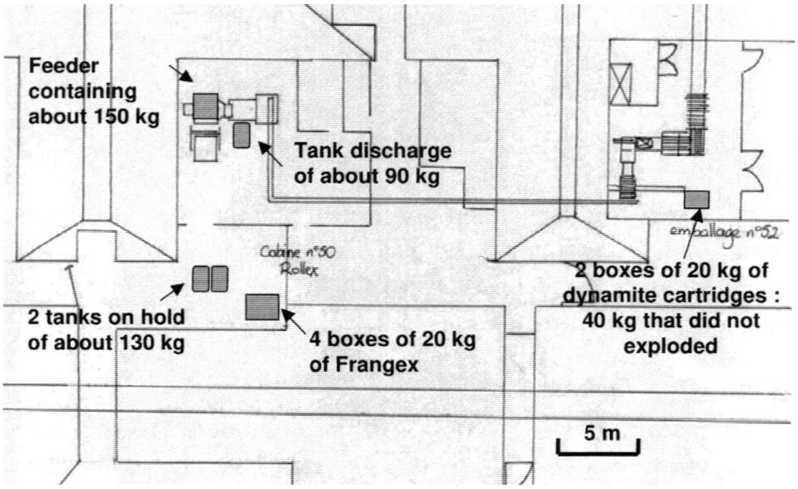

Figure 4: Gelatine Quantities At The Moment Of The Explosion

Thus, it could have been transmitted by pressure wave to the tank discharge closed to the feeder. Regarding the other charges distant from about 10 m from the feeder, the detonation would have rather been propagated by missile projection. Barricades must have prevented the explosion transmission by missile projection from workshop n°50 to nearby workshops.

However, the real total dynamite amount in workshop n°50 turns out to be quite larger than the one estimated by the damage analysis. This gap could be explained considering that several successive detonations generated independent overpressure peaks which did not cumulate. Many witnesses mentioned a “long, dull noise” which could emphasise this assumption.

It could be admitted that two main charges were located in Workshop 50:

- 240 kg near the feeder (feeder + tank discharge)
- 340 kg at the entrance (2 tanks + 4 Frangex dynamites boxes).

If the explosion was transmitted by a missile impact from the first charge to the second one, the speed of transmission would have been about 1000 m/s, meaning that the second charge would have detonated about 10 ms after the first one. As the overpressure pulse of a detonation has a typical period of 1 ms, the global pressure wave signal would have been a succession of independent overpressure peaks that would not have cumulated, at least in the near field. Therefore, the overpressure damages would have been caused by the detonation of the biggest charge (340 kg) rather than the total charge (580 kg).

One can also consider that not two but four independent dynamite charges, as identified on figure 3, generated four successive detonations and four independent overpressure peaks. Thus, the overpressure damages in the near field would have been caused by the detonation of 150 kg of dynamite in the feeder. This hypothesis would agree with the TNT equivalent estimate of 50 – 100 kg from the near field damages interpretations.

Beyond the first barricades, very complex waves diffraction and reflection phenomena should have occurred, and pressure signal should have partially recombined. That could explain the discrepancy between TNT equivalent estimates from the damage analyses in the pyrotechnic area and in the far field.

As a conclusion, it seems that the overpressure effects were not generated by the massive explosion of the whole dynamite charge in workshop n°50, but were most probably caused by several successive and independent detonations whose overpressure signals might have partially recombined in the far field, according to shock waves propagation theories.

3.3 Effects To People

Fourteen accident at work notifications were registered by the company during the days following the explosion. Casualties are located on figure 5. Among them are:

- 4 deaths,
- 2 minor injuries by blasting onto the ground,
- 1 minor injury to ears,
- 1 minor injury by ceiling fall,
- 5 psychological shocks,
- 1 disease,

Notice that five people located within 100 m from the explosion epicentre were not injured.
The four deceased people were located in workshop n°50 and for one of them, in the communicating tunnel between workshops n°49 and n°50. Other people located in the nearby workshops suffered minor injuries, highlighting the protection efficiency against pressure waves offered by barricades between workshops.

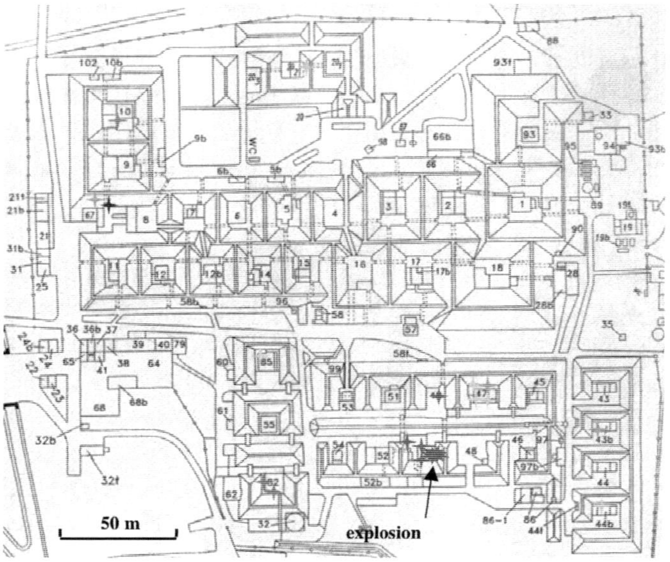

Figure 5: Victims location

death, injury by blasting onto the ground, injury to ears injury by ceiling fall, psychological shock, disease, no injury

4. The Technical And Direct Causes Investigation

The technical analysis of the investigation was performed by using an analytical tool (fault tree) to identify in a structured way the probable causes of the explosion of the workshop n°50. The analysis with a fault tree is an inductive-type method used to determine the technical causes and some human actions leading to an accident.

The building of this fault tree was based on:

- the information gathered during the interviews,
- the tests on the samples of the explosive that was packed and stored in the workshop n°52 at the time of the accident,
- on the analysis of the damage caused by the explosion,
- on the expertise of INERIS in its knowledge about the phenomena of explosives explosion.

The possible location of the explosion had to be determined as a starting point in order to be relevant in our assumption in the remaining analysis. Thus, based on data from the damage plotting, the chronology and the video recorded file, the feeder was identified as the most probable location of the explosion.

Then, the investigation focused on the nature of the blasting material in the feeder. The material in hand here is obviously dynamite but many questions arose (see chronology):

- What kind of dynamite was it ?

- Did the dynamite meet the required technical specifications ?
- Or was it more sensitive than the required technical specifications ?

About the type of dynamite, there were two possibilities: the Dynaroc 6 made on a regular mode at the Tellex or the Dynaroc 6 made with the Superdopex gelatine on a manual mode.

Two charges were put into the hopper: the first one was some Dynaroc 6 made of Superdopex gelatine produced the day before the accident. But, the origin of the second container is still unknown. Because of the sequence of events, it certainly came from one of the storage buildings, but the inventory made after the explosion did not help much to draw clear conclusions.

Concerning the suitability of the dynamite to the technical specifications, the tests carried out at INERIS on the samples collected from the company, if one considers them as representative of what was in the feeder at the time of the explosion, helped to conclude that the explosive was adequate and not more sensitive to shock or to friction than the dynamite usually produced by Nitrochimie S.A..

Yet, even if the explosive was adequate, its sensitivity can be questioned as, for example, the gelatine had to remain a long time (one night) in the feeder and the die. The gelatine was hard to process (see chronology) that had an effect on the exposure of operator to risk. Many factors may have increased the sensitivity of dynamite:

- the formulation of the gelatine can be inadequate,
- the viscosity of the gelatine can be off-specification,
- the absorbents can have a high acidity rate,
- exudation can occur,
- the dynamite can be unstable,
- the gelatine can have an inadequate homogeneity.

Consequently, except the issue of acidity, all the factors were kept for different reasons. For example, an exudation could have occurred in the die due to the long stay of the paste in it. Besides, even if it is difficult to set a relation between an inadequate homogeneity of the gelatine and its level of explosibility, this factor cannot be turned down.

There is a minimum amount of energy required to trigger the explosion and thus knowing the paste sensitivity appears to be of high interest. Some of the ignition sources could have been present but not powerful enough. Several ignition sources were identified:

- heat action by an external fire,
- electrical, electrostatic arc,
- external pressure wave,
- missile impacts,

- mechanical action from the process or something else.

The approach consisted in making an inventory for each type of ignition source of the situations or circumstances where they could have appeared given the equipment, the labor organization, the task performance itself in the workshop, etc.

According to the available data, only the mechanical action was kept as probable. Among the possible causes of a mechanical ignition, there were possibilities of friction due to different metallic parts between the container, the feeder, the hopper or the tipper or the friction of a foreign matter on the feeder. This last option was kept as the most probable.

Because not enough evidence was collected, many options are still conceivable. The foreign matter could have been introduced:

- in the hopper by the operator, notably when he threw a ball of gelatine (that fell into or close to a tool box placed along the Rollex belt) to the feeder a few seconds before the explosion,
- in the hopper by the fall of an object of the tank when unloaded
- in the hopper by the fall of an object from above the feeder
- in the tank during the manufacturing of the dynamite, the transportation of tank on the trolley to the workshop n°50 or in the workshop n°50
- in the hopper before the paste was loaded in.

As a preliminary conclusion of that step of the investigation, there are still some uncertainties about the exact technical causes of the explosion. It would have been interesting to continue the tests on the gelatine used at the moment of the explosion to clarify the link between its properties (viscosity, exudation, ...) and its sensitivity to explosion. Besides, there is still this uncertainty about the origin of the second tank.

The most probable scenario is the presence of a foreign matter in the feeder either brought inadvertently by an operator or due to the fall of objects from the container or the hopper. But the timing and the circumstances where it could have been introduced are still unspecified. The internal company investigation concluded also that the most probable scenario is a foreign matter in the feeder.

However, this was not the only issue as a matter of a fact that four people died in the explosion, introducing the risk factor of the exposure of operators. Limiting exposure is a key thing in the pyrotechnic industry. In that accident, the circulation issue appeared therefore as a key element of the understanding of the severity of the human losses. The cleaner, the maintenance operator as well as the assistant of the main operator were not supposed to be where they were in these particular circumstances, at the time of the explosion.

With this most probable technical scenario assumption and the circulation problem, we could get into the issue of why the organization was not able to prevent this scenario. Safety barriers and controls to prevent this scenario and its severity, which was a known risk by the company, were analysed in their organizational context.

5. The Organisational Investigation

Three main points are discussed in the following part: the methodological aspects of the organisational investigation and the background models, the outcomes and a discussion around them.

5.1 Methodological Issues

5.1.1 Rationale For Investigating Organisations

Organisations are complex systems. They imply a high number of interactions, at all levels, from the shop floor to the board committee but also the outside of the organisation, with individuals located in its environment (regulators, economic partners). These interactions are rich; they involve the nature of the social world. This social world is made of situation of power between people, of cultural dimensions, of social construction of reality, of norms, of values, of rule. Organisations also can be seen through several lenses like the financial one, the human resources one, the economical one, the institutional one, the historical one, the societal one... Organisations are therefore a subject of research for various disciplines like economy, politics, sociology, anthropology, psychology, psychosociology, all of them applying different filters. This makes the organisational life a quite dense subject of study (J.Rasmussen and I.Svedung (1997, 2000) suggested a representation of the socio-technical system where many of these features are introduced and linked).

Qualitative studies investigating on large organisations but even on small organisations deal with the problem of collecting and analysing data for interpretation, and also deal with the ability to consider various possible angles and dimensions, across levels.

However, with investigations, one helping thing is that the focus following an accident is given. Indeed, the accident sequence reveals some inadequacies that are the starting point of the investigation. The aim of the investigator is therefore to give a context to the events. This implies to go back into the history of the organisation in order to better describe in which circumstances the decisions and actions of the individuals were taken. The idea of going back in history is to find out more about the conditions that "*shaped the landscape in which the accident unfold itself*" (J.Rasmussen I. Svedung, 2000*)* further back from the close events, like human errors and technical failure leading to the accident.

The purpose of an organisational investigation could then be said to describe the events with a broad scope, to bring some explanation (defining some causal or rather conditional links between the events and a context) and sometimes to build a predictive model from it. The latter has not been performed so far many times. It is the work of researchers, social scientists, who conceptualised further their interpretation of the organisational dimensions, applying social science techniques of theorising from grounded data, and single cases.

What is actually performed is represented in this figure n°6.

As said, it is nevertheless not such a common thing to theorise from accidents. Thus, when we go through the existing literature on the organisational side of accidents, it shows that there are actually several types of possible approaches for investigating.

We classified them in three types (JC Lecoze 2005).

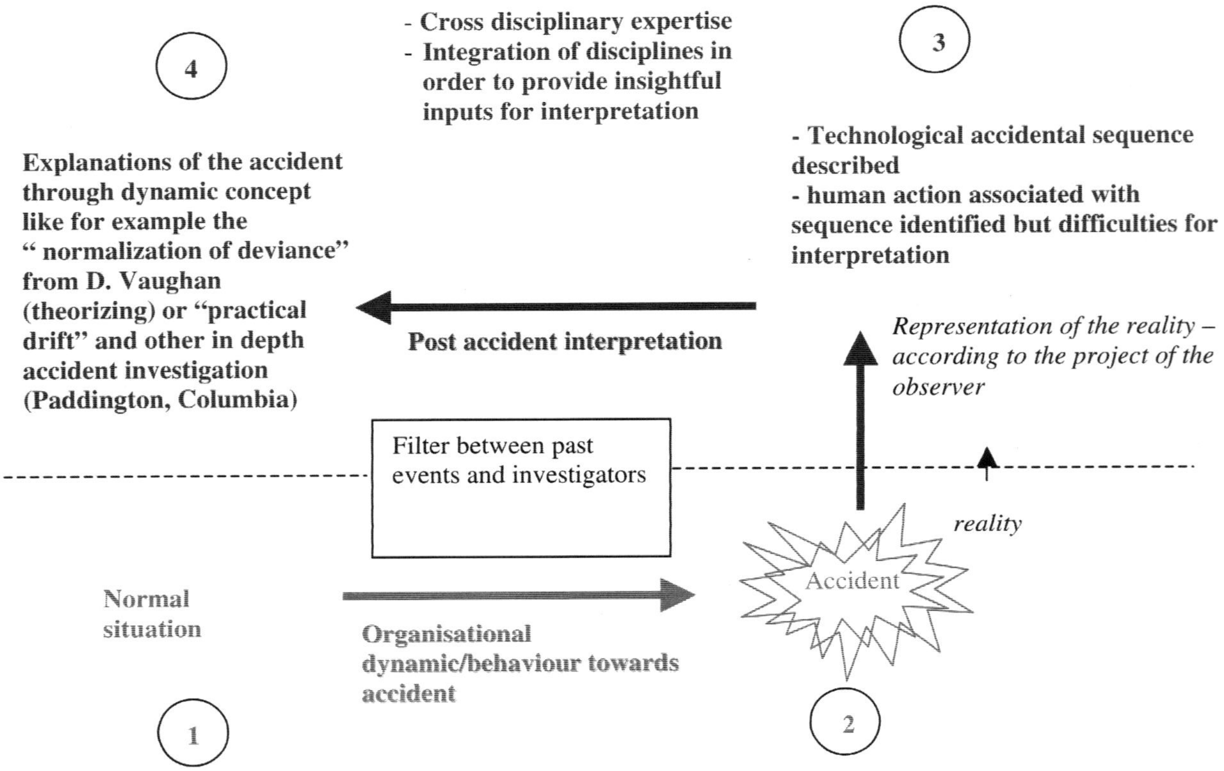

Figure 6: Purpose And General Process Of Investigating (Lecoze Et Al 2002)

5.1.2 Three different approaches of investigations

First, these works already mentioned with research and theorizing perspectives (i.e D.Vaughan, 1996, S.Snook, 2001), secondly the commissions set up after disasters (i.e: Columbia, 2003, Ladbroke grove, 2000) and then the structured root causes analysis (i.e MORT, STAMP).

All of these three types must face the bias of forcing the data.

The "forcing the data" concept comes from one of the difficulty of the social science theorizing process. In social science, theories can be grounded, generated and inferred from the data collected through fieldwork. These are not to be generated purely by a logico-deductive approach. However, in front of the richness of the organisational life, theories can be validated through strategy consisting in forcing the data to fit the theory. This is a symptom that also could be called the "data that should fit the model" rather than the "model that should fit the data" approach (from Glaser and Strauss, 1967).

The three types of investigating techniques deal with it differently.

Pre defined models, in structured root causes analysis, apply a "data that should fit the model" approach but are made for systematic recommendation purposes and are useful in that sense. Commission and research work are more descriptive and explanatory, and do not specifically rely on pre defined model. However, they also introduce underlying models and theories to bring some light on the data extracted from the investigation.

These three types involve different relationship with the underlying models that they use for treating the data, but also different time constraints, techniques, resources (number of people, competences, expertise etc), scope (micro-meso-macro levels) and objectives (describing, explaining, making recommendation).

5.2 Methodology Used For The Case

5.2.1 The Underlying Models

In the investigation performed for the Billy Berclau accident, the approach was based on an investigating process that was half way between a structured root cause analysis (i.e MORT, implying focused questions and pre-defined safety management system) and a more descriptive approach based on open interviews and openness of the system with micro-meso-macro perspective as well as a historical approach.

The investigation report made clear in a introductory part that several organisational dimensions would be investigated, with references to works in the field, like the High Reliability Organisation (i.e Roberts 1990, Rochlin 1996, 1999, Weick, 1987), Normal Accident theory (C.Perrow, 1999), Incubation period (B.Turner, 1978), Normalisation of deviance (D.Vaughan, 1996), Latent and active failure (Reason, 1990, 1997) as well a the idea of organisational behaviour towards accidents (Rasmussen and Svedung, 2000).

Also, it has been stated that some dimensions were investigated and articulated with the help of a type of matrix as suggested by Glandon and Waring (1998). This matrix contains the following dimensions:

Risk contexts
Organisational environments
Economies and market
Public policy, legislation and regulation
Social and political climate

	Technology
	History, operating territories and conditions
Human Factors	
	Culture
	Power relations, political processes and decision making
	Perception, cognition and meanings of success
Formal coping arrangements	
	Risk management
	Risk assessment
	Management systems
	Approaches to change

Table 1: Thematic table linking theoretical aspects of risk with empirical studies (A.Waring and A.I. Glendon, 1998)

All of the dimensions can be investigated with related theories in the background, models or works in the safety field (like in perception, power in companies, safety management systems etc).

The idea was not to validate any of the frameworks suggested by any theories and concepts but rather to explicitly mention that they were in the mind of the investigator when collecting and analysing the data. In her article dealing with the heuristic of case analysis (1992), D. Vaughan has debated this point of the pre-existence of models in mind when investigating and theorizing.

The importance of the references and underlying models in organisational investigation is high. It has been quite rightly stated that an investigator could find what he/she wanted to find in accidents.

5.2.2 The link between technical and organisational factors

The technical anchoring of the organisational investigation has been performed through the barriers and controls revealed through the technical sequence. Twenty three barriers and controls were identified based on the most probable scenario retained. The barriers from B1 to B12 were aimed at decreasing the likelihood of foreign object presence in the feeder, B13-B16 were designed for preventing the explosion of the feeder in case of a foreign object friction and the B17-B23 decreased the severity of the human loss in case of accident. Each of the barriers were investigated in detailed through interview and then replaced in a broader context.

To illustrate it in a simple way, the investigation has been performed like the next figure n°7 is suggesting it (adapted from a representation used for MORT, Johnson, 1980 and mixed with Waring and Glendon, see table 1).

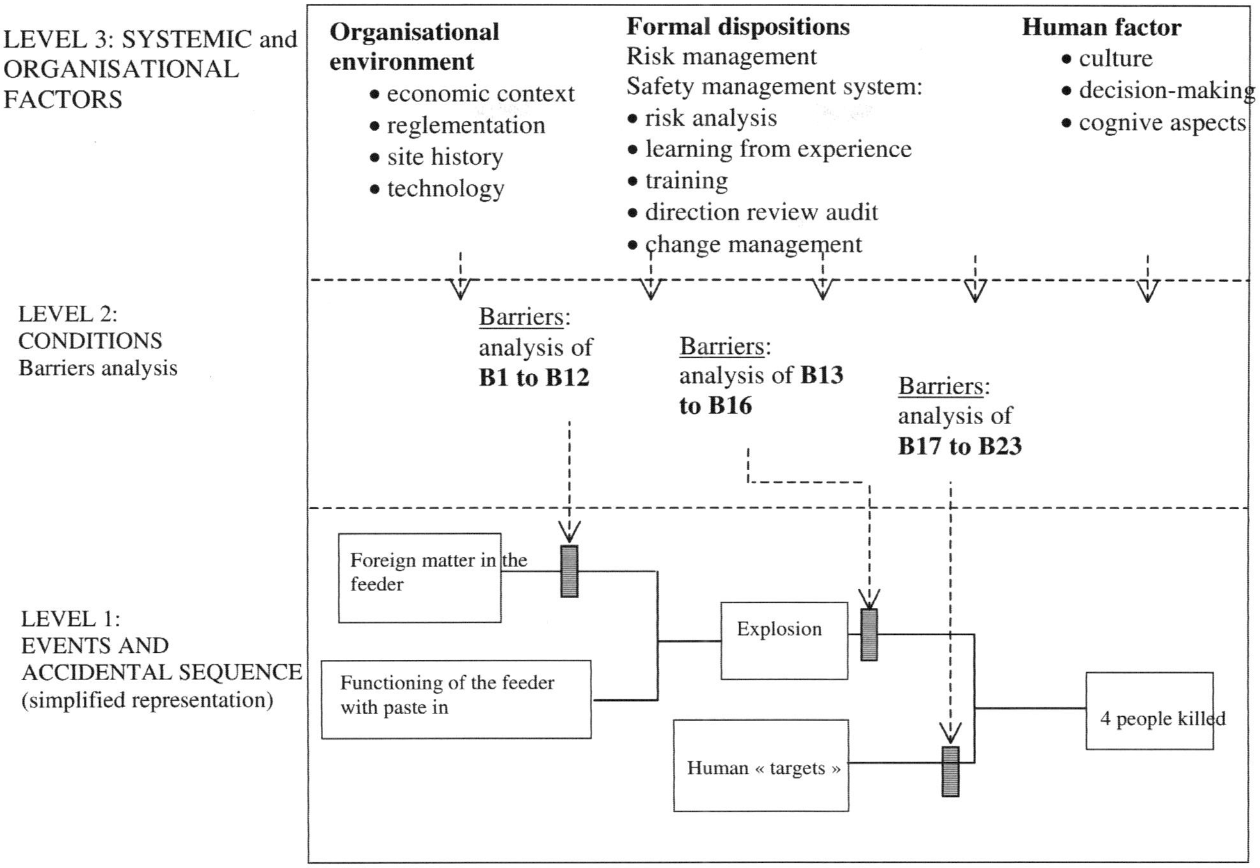

Figure 7: Link Between Technical And Organizational Factors In The Billy Berclau Case

The first level is the accident sequence with the inadequacies of the barriers and controls, the second deals with the conditions for these barriers and control not to fulfil their functions and the third level, which brings some insights on the overall context influencing the conditions.

In fact, this investigating process shares some similarities with what has been described for example in Fortune and Peters (1995:110):

- Pre-analysis
- Identification of significant failures (similar to level 1)
- System selection (similar to level 2)
- System modelling (similar to level 2)
- Comparisons (similar to level 2 and level 3)
- Further analysis (similar to level 3)
- Synthesis

The investigation consisted in a series of 30 interviews (70% for the levels 1 and 2, 30% for the level 3) with people from the shop floor to the board committee at the corporate level. An extensive investigation focusing on the documents provided by the company describing quality and safety management systems as well as safety cases, and the records of the activity of learning from experience, safety meeting, risk analysis, etc was also performed. The use of a book about the culture and history of the company celebrating a century of existence was used and very helpful for getting to know better the company and its history.

5.3 Outcomes Of The Investigation, Lessons To Learn

A certain numbers of findings made are quite consistent with some already mentioned influencing organisational factors, in other investigations like in A. Hopkins, (2001) but also general lessons learnt throughout accident (Llory et al., 2004) and were similar to accident models mentioned earlier in this paper. Ninety recommendations were made for the company to improve the level of safety. They were made at several levels of the system, from the technical level to the management and regulatory level.

Here is a selection and discussion of relevant aspects that were identified.

Some lessons at the level of the authorities and regulations (risk contexts Public policy, legislation and regulation, in regard with table 1):

The regulations are increasingly dense for industries, as much from their contents as from their overlapping. The pyrotechnical site was liable to many regulations concerning safety under the pyrotechnics regulation depending on the health and safety regulations but also under the environmental regulation (the classified installations in France and Seveso II Directive in EU) dealing with major hazards. The site was classified high threshold concerning the transposition of the SEVESO II directive. In this case, there is an overlapping of two regulations which however both deal with the prevention of major hazard.

These regulations have nevertheless basic differences. One of these difference relates to prescriptive nature for one (the law on pyrotechnics safety) and the other by objectives (in particular the SMS within the framework of the application of the SEVESO II directive).

Consequently, a time for integrating the new regulation was required. This adaptive time is all the more important than the culture of the industry has been shaped by the regulations applied for more than two decades. In the case of pyrotechny, the decree of 1979, very detailed and seen as an effective regulation, which allowed notable improvements of safety on sites at the time of its implementation, had somehow played this shaping role.

Another one of these differences relates to the risk assessment practices and calculations of safety distances, which is not without causing administrative issues. Indeed, the two regulations have different thresholds, different classifications. This brings confusion and a difficulty of treatment of the safety reports and a difficulty in meeting regulations requirements. Concretely, at the level of the plant, this also resulted in an important volume of administrative work which has been regularly growing since the nineties with the increase of the environmental regulation and requirements. This risk context took part in diverting the safety department from its operational activities.

The diversity of the regulations implies also the diversity of controls. This can be a good thing, but can also be a bad thing when the links between the parts have consequences for the consistency of the whole. In particular, it is necessary to take into account that this company faced variations in terms of number, objectives and quality of inspection. An inspectorate like the IPE (Inspection des Poudres et Explosives - Explosive and Powder Inspectorate) has the means and experience for understanding well the operations of the installations and can bring a useful and constructive help for risk management, while keeping a control and regulative function. This is therefore, not a surprise to find clearly stated in the regulation the possibility for the assistance for in site inspection and safety reports by the IPE for the DRIRE and the labour inspectorate (in charge of work safety regulation as well labour regulations). Consequently, in case of numerous updating of safety reports, the company has different strategies available for dealing with the submission of the reports, to one of the control authorities. In that case, this precise situation occurred and revealed consequently several non-conformities.

This general context of regulation overlapping with sometimes different objectives and contents, with multiple control authorities, creates a complex regulatory situation for which the company must organise its resources in order to cope specifically with it. These remarks do not, of course, justify non-conformity, but remind about the sometimes dense and complex regulatory contexts that organizations are facing, and shed some light on the difficulties that they meet through their attempt to comply fully with them. The pyrotechnical site faced a situation from this point of view for which a great part of its safety operational activity was diverted to treat these administrative tasks involved.

Authorities in France work actively on this issue of overlapping regulations.

At the level of the organization of the company and site (human factors and formal coping arrangement, in regards to table 1)

- ✓ *Changes in organizations*

Changes in organizations must be done with a careful attention given to the potential impact of these on the safety practices at the operational level. Usually, the impact of changes is approached for technical issues but much less in terms of organizational issues. It is clear that the impact can't be measured the same way. The impact of organizational changes is clearly not easy to assess, but some measures should be taken before and after. In this accident, important changes were done at the level of the site top management and at the level of the technical corporate department. These two changes impacted on the operational safety of the site. Indeed, following the changes, the safety department was not able anymore to play its safety role against the production pressure, as it had been working for years. Thus, it is believed that any changes that imply an evolution in the way the system sustains its level of safety through a balance between safety against the natural production pressures, increases the risk of operating outside the boundaries of safe practices. The evolution that was identified through the investigation in this case concerned the way information was available at the level of the board committee. This information was important for high levels in order to hear about the level of safety at the operational level. This situation was not without reminding the "trickle down effect" described by D.Vaughan about the Challenger accident (1997) when she describes the effect of political, economical and organisational arrangements made at the top of the system that slowly and gradually spread across the lower levels.

✓ *The cultural dimension*

Culture is seen as an important concept today for explaining how companies operate. However it is still a scientific concept that is strongly debated. For the purpose of the investigation, it appeared however relevant for some aspects. We retained about culture that, according to E.Schein, (1992), it is:

" *a pattern of shared basic assumptions that the group learned as it solved problems of external adaptation and internal integration, that has worked well enough to be considered valid and, therefore, to be taught to new members as the correct way to perceive, think, and feel in relations to this problems"*

The history of the company is hundred years old and the site is one of the oldest. A long tradition of operators has been working on the site, a dynasty as they would call them (a term that was found in the document celebrating the century of the company). This long history, through the events and the technology used, shaped the practices and maintained a tradition in the way the blasting gelatine was handled. Despite the new technology, requiring less manual handling of the paste, some remaining practices described as "pratique à l'ancienne" equivalent for "old time practices" were still in use. They consisted in regularly manipulating the paste with bare hands. This kind of old practices interacting with more recent practices required by the new technology can interact inappropriately with safety principles introduced with the technology. In the accident, the video revealed a behavior of the operator taking a piece of gelatine that fell on a tool box and putting it directly back in the feeder. This is not without reminding an old way of dealing with the paste.

It was also found interesting to talk about safety culture in our case as it is today a widely shared concept (or rather assumption) in the research and practice though also still very debated. There is again no consensual definition among scientists. Here, for practical purposes, we can retain the HSE (UK Health and Safety Executive) definition about safety culture as being:

"*The product of individual and group values, attitudes, competencies, and patterns of behaviour that determine the commitment to and the style and proficiency of, an organisation's Health and Safety programmes. Organisations with a positive safety culture are characterised by communications founded on mutual trust, by shared perceptions of the importance of safety, and by confidence in the efficacy of preventive measures*"

As a guidance through the concept of culture the HSE gives these items related to the dimensions of safety cultures for which some are relevant in this accident case:

- **Organizational commitment and communication**
- **Line management commitment**
- Risk taking behaviour and some contributory influences

- Personnel role
- Workmates' influence
- Competence
- Supervisors' role
- Permit to work systems
- **Reporting of accidents and near misses**

In bold the dimensions that appeared illustrative, as items representative of aspects of safety in the site. The line management commitment and organisational commitment appeared through the interview and documents analysis to be less than adequate to maintain the activities against natural and expected drifts of safety practices (this point is highlighted below). However, the concept of culture is not used further here as it tends to give a deterministic vision of the organisation. A vision which implies that culture creates the conditions for people to behave the way they behave although individuals have also a strategic and adaptive dimension, with their own intentions, that are never fully determined by outside constraints (it is one of the difficulties for predicting organisational behaviour as a whole).

✓ *Safety management (formal coping arrangements in regards with table 1)*

From a more structural point of view of the organisation, it was clear that the SMS required by SEVESO II directive was currently being under implementation, but had some shortcomings in the way it was thought and designed. A context for this difficulty to fully apply the safety management principles has already been stressed with the existence of a prescriptive regulation.

Some are the shortcomings pointed out during the investigation:

- the principle of defences in depth (also called the barrier and control approach) extracted from the risk analysis was not fully applied and only limited to a small number of defences. This strategy was implemented in order to focus only on the last defence, called the ultimate defence principle. The idea for the company was not to identify too many defences which is very heavy afterwards in terms of management, maintenance, inspecting and auditing. Moreover only technical defences were included and not behavioural ones (like the circulation issue). It is a problem not to make barriers and controls more explicit for a risk control perspective. This was a serious flaw in the way the system could have been thought. Also, it appeared that the risk analysis sessions were not opened to operators. This can be surprising as it is a widely known practice within industries but was eventually less surprising in a prescriptive regulatory context where everything was already codified and written, ready to be applied.

- The learning from experience is a key activity that must help operators and managers to think and work on the conditions leading to the incidents. These conditions can be technical as well as organisational. In this case, the organisational dimensions of the incident were not so much analysed, as the safety department was no longer involved into the learning sessions following the recent changes in the management of the plantAS a consequence, there was more focus more on technical rather than organisational issues. A long list of warnings concerning foreign objects in the last 3 years triggered some concerns about the cleaning within the installations. The management had therefore planned to improve cleaning practices.

- The training was also a concern for the company who had to manage in a flexible manner the production for coping with the low visibility of the market. This implied recruitment of temporary workers, which created the problem of the following up on training. Training must also be used as a way of reminding people about safety principles, to prevent the work routine from becoming unsafe, according to the law. The company had not been very cautious on the follow-up for all operators on the refresher sessions.

- This aspect of routine introduces the idea of dynamic drifting (as for example described in S.Snook, 2000). This consists in shifting at the level of the operators from a rule based approach to a more task based approach, where the individual can locally cross the safe boundaries of the system (J.Rasmussen, 1997), and therefore can create risk, but also being exposed to risks. Micro deviations can install in the organisation with no impact on the production as a whole but create a kind of deviance within the system in terms of safety (D.Vaughan, 1996, 1999). As dynamic and evolving, it is of no surprise to discover that operational practices defined by rules, confronted with multiple dimensional and sometimes conflicting aspects, are moving towards adaptive tasks. It appeared, with retrospect, that what was observed in this accident, had obvious common points with what J. Rasmussen (1997) described:

- "In any well designed work system, numerous precautions are taken to protect the actors against occupational risk and the system against major accidents, using a 'defence-in-depth' design strategy. One basic problem is that in such a system having functionally redundant protective defences, a local violation of one of the defenses has no immediate, visible effect and then may not be observed in action. In this situation, the boundary of safe behaviour of one particular actor depends on the possible violation of defenses by other actors. Therefore, in systems designed *according to the defence-in-depth strategy, the defenses are likely to degenerate systematically through time, when pressure toward cost-effectiveness is dominating."*

Although the pressure towards cost effectiveness didn't show to be so important through the analysis of the economical constraints of the company, an internal pressure for production was however clearly stated through the interviews. In the accident the various drifts around the rules involved the shift schedule that was modified without front line managers knowing about it but also the early start of the installations without the required number of people, the issue of circulation in the pyrotechnic area and the cleaning of installations. They all appeared, with hindsight all to be outside the rules defining the way things should have been performed. However, they couldn't be seen as individual and isolated mistakes of operators in a traditional sense, but had to be seen as a general behaviour moving outside expected practices and safe boundaries, and involving all levels within the organisation, that allowed this situation to occur.

K.Weick (1987) gave an insightful comment about the problem of drifting and crossing the safe boundaries. He suggests an interesting explanation for the reasons why safe boundaries are crossed. It also had some quite strong echoes with our case, particularly with the issue of circulation, with the installations cleaning but also the specific morning situation:

"Reliability is invisible in the sense that reliable outcomes are constant, which means there is nothing to pay attention to. Operators see nothing and seeing nothing presume that nothing is happening. If nothing is happening and they continue to act the way they have been, nothing will continue to happen. This diagnosis is deceptive and misleading because dynamic inputs create stable outcomes"

Operators regularly do things to cope with situations in order to be able to make their way out of conflicting goals but also unexpected situations, and therefore can cross, without totally noticing it, the boundaries of safe operations. In this accident for example, without any adaptation and with a strict respect of the rules, the operations wouldn't have started that day as two operators were missing at the beginning of the shift.

Deviance or drifts can sometimes be sustained in the long run, as the normal(ized) practices, in some specific organisational, economical and historical contexts. Companies should acknowledge this situation and work on these micro adaptations through the design of organisational means where people can talk about them and find way to overcome the risky ones. These kind of dynamic aspects have clearly to be addressed specifically with an appropriate organisational support. This would need to be experienced and developed in the future, for safer operations.

- The safety indicators related to occupational health and safety (LTI for Lost Time Injury) were very good. The company had a past records of no accidents for four years. This led to the conclusion, following the accident, that these types of indicators do not make clear how good a company manages the major hazard prevention. This is also very consistent with what has been identified by A. Hopkins in his investigation of the Longford plant gas accident (2001). Indicators of these types can be misleading and give a wrong picture of the safety state of a plant, in terms of major hazard prevention. Efforts should therefore be made to indicate the level of safety through the concept of barriers for example, where specific links with the major hazards are possible. A quality type of approach for indicators can be used for that purpose, which could be seen maybe as a bit expensive but that is believed to be worthwhile implementing. The limits of such systems should however also be stated, and complemented for example with softer approach where people can deal with deviation from expected practices which have no direct and measurable impact on safety, and are sometimes necessary in light of conflicting or constraining goals.

- The last point discussed here from this case is the usefulness of cross sites auditing. In the company, this practice had been gradually decreased throughout the years. It seemed that this decrease has revealed to be detrimental to the level of safety. It is believed, that quite consistently with some ideas derived from the normalisation of deviance concept, that external auditing from competent individual is a great help for getting out and pointing out established practices at odds with what is being done outside or known outside as being safe. Control authorities can of course fulfil this mission but internal cross auditing between plants make the process of exchanging maybe sometimes easier, and more open to discussion.

6. Conclusion

The multidisciplinary approach developed in this accident investigation covered a wide range of issues from technical to organisational ones. Damages analysis, technical cause analysis, material tests, chronology, barriers analysis and organisational investigation all gave complementary insights, in order to better understand the accident. The damages analysis, chronology and material test supported the technical cause analysis and the barriers approach helped to create a link with the organisation. There are of course many aspects that could be developed further. But the time spent on the case helped to get deep enough to generate insights.

The methodological approach developed for the organisational investigation has been described as half way between a type of structured root cause analysis and a more descriptive and contextual approach looking for conditional links between facts, using the existing literature on the subject as the background models for guiding the interpretation. The methodology has the potential for being applied again in other accident investigations with organisational dimension, but will be also practicable for smaller investigations.

The findings about the organisational side of this accident have some similarities with in depth investigations dealing with this level of analysis. Some of them have been described in the outcomes discussed in this paper.

7. References

Baker W.E., Cox P.A., Westin P.S., Kulesz J.J., Strehlow R.A., 1983: Explosion hazards and Evaluation, Elsevier, New York.

Columbia Accident Investigation Board, 2003, Vols. II-VI of Final Report.

Cozzani and Salzano, 2004, The quatitative assessment of domino effects caused by overpressure Part I, Probit models, Journal of Hazardous Materials A107 (2004) pp 67-80.

Cullen, Lord. 2000 The Ladbroke Grove Rail Inquiry Part 1 and 2 reports, HSE books, her majesty stationery Office, Norwich, U.K.

Dechy N. and Mouilleau, Y. 2004, « Damages of the Toulouse disaster, 21st september 2001 », Proceedings of the 11th International Symposium Loss Prevention 2004, Praha, 31 May - 3 June 2004.

Dechy N., Bourdeaux T., Ayrault N., Kordek M.-A., Le Coze J.-C., 2004, First lessons of the Toulouse ammonium nitrate disaster, 21st september 2001, AZF Plant, France, Journal of Hazardous Materials 111 - July 2004, (special issue on JRC-ESReDA seminar on Safety accident investigation, Petten, the Netherlands, 12-13 May 2003)

Fortune J. Peters G. (1995) Learning from failure, the system approach. John Wiley and Sons.

Glaser B. G. Strauss A. L. (1967) The discovery of grounded theory: strategies for qualitative research. Aldine de Gruyter.

Grant, R.L., Murphy J.N., Bowser M.L., 1967: Effect of weather on sound transmission from explosive shots, Bureau of Mines Report of Investigation 6921

Hopkins A. 2000. Lessons from Longford: the Esso Gas Plant Explosion. Sydney. CCH Australia.

Lecoze J-C., Lim S., Intégration des aspects organisationnels dans le retour d'expérience: l'accident majeur, un phénomène complexe à étudier, pour le Ministère de l'Ecologie et du Développement Durable, 2002, downloadable on http://www.ineris.fr

Lecoze J-C, 2005, Organisation and disaster: from lessons learnt to theorising, forthcoming

Llory.M, Dien Y, Montmayeul R. (2004) Organisational investigation methodology and lessons learned. Journal of Hazardous Materials 111 – July 2004

Mouilleau Y. and Dechy N., 2002. « Initial analysis of the damage observed in Toulouse after the accident that occured on 21st of september on the AZF site of the Grande Paroisse company », *International ESMG Symposium, Nürnberg, Germany 8-10 October 2002, on Process safety and industrial explosion protection*

Perrow C. (1999) Normal accident theory, living with high risk technology, second edition

Rasmussen, J. (1997) Risk management in a dynamic society: a modelling problem, in safety science vol 27 n°2/3, p 183-213.

Rasmussen J. Svedung I. (2000). Proactive risk management in a dynamic society. Swedish Rescue Service Agency. Karlstad.

Reason.J. L'erreur humaine. Presse Universitaire de France. Collection Le Travail Humain. 1993.

Reason J. 1997. Managing the risk of organisation accident. Ashgate.

Reed, J.W., 1973: Distant Blast Predictions for Explosions, Minutes of the Fifteenth Explosives Safety Seminar, Dpt of Defense Explosives Safety Board, Washington D.C., vol. II, pp. 1403-1424

Roberts K.H. (1990) Some characteristics of one type of high reliability in organisation. Organisation Science. 1(2). 160-176.

Rochlin G.I. (1996) Reliable organisations: Present research and future directions, journal of contingencies and crisis management, volume 4, june.

Rochlin G.I. (1999) Safe operation as a social construct. Ergonomics. Volume 42. N° 11. 1549-1560.

Schein.E, Organisational culture and leadership, second edition, Jossey-bass, 1992.

Snook, S.A. (2000) Friendly fire, the accidental shootdown of US black hawks over northern Iraq, Princeton university press.

Turner B. Pidgeon N. (1997) Made-man disaster: the failure of foresight. Second edition, Butterworth-Heinmann.

Vaughan D. (1992) Theory Elaboration: The Heuristics of Case Analysis, in Charles Ragin and Howard S. Becker (eds.), What is a Case?: Exploring the Foundations of Social Inquiry. Cambridge University Press.

Vaughan D. (1996) The Challenger launch decision: risky technology, culture and deviance at NASA, University of Chicago Press, Chicago.

Vaughan, D. (1997) The trickle down effect: policy decisions, risky work and the challenger tragedy, California Management Review. vol 39. n°2.

Weick K. (1987) "Culture as a source of high reliability" , California Review of Management.

Waring A, Glendon A.I. (1998) Managing risk, critical issues for survival and success into the 21st century. Thompson Learning.

Wenzel A.B., Bessey R.L., 1969: Barricaded and Unbarricaded Blast Measurements, Contract n° DAHC04-69-C-0028, Subcontract 1-OU-431, Southwest Research Institute, San Antonio, Texas

Wright J.K., 1961: Shock tubes, John Wiley & Sons Inc., New York

Reports of INERIS on the Billy-Berclau accident (to download on http://www.ineris.fr)

R. Branka, N. Dechy, J.-C. Lecoze, E. Leprette, S.Lim, Intervention de l'INERIS après le sinistre survenu le 27 mars 2003 sur le site de Billy-Berclau de la Société Nitrochimie, Rapport final, Septembre 2004, 30 pages. (*global summary report*)

S. Lim, J.-C. Lecoze, Chronologie de l'accident du 27 mars 2003 survenu à Billy-Berclau sur le site de Nitrochimie, Rapport Annexe 1, MEDD/DPPR, août 2003, 15 pages. (*chronology*)

R. Branka, Examen de l'explosif en cause dans l'accident du 27 mars 2003 survenu à Billy-Berclau sur le site de Nitrochimie, Rapport Annexe 2, MEDD/DPPR, Juin 2003, 8 pages. (*material tests*)

N. Dechy, E. Leprette, Analyse des dommages observés à Billy-Berclau (62) après le sinistre du 27 mars 2003 survenu sur le site de la société Nitrochimie, Rapport Annexe 3, MEDD/DPPR, Août 2003, 63 pages. (*Damages investigation*)

S. Lim, J.-C. Lecoze, Détermination de l'arbre des causes de l'accident du 27 mars 2003 survenu à Billy-Berclau sur le site de Nitrochimie, Rapport Annexe 4, MEDD/DPPR, Août 2003, 24 pages. (*technical analysis*)

J.-C. Lecoze, S. Lim, Analyse organisationnelle de l'accident du 27 mars 2003 survenu à Billy-Berclau sur le site de Nitrochimie, Rapport Annexe 5, MEDD/DPPR, Août 2003, 133 pages. (*organisational investigation*)

Reports of IGE-IGAS-CGM on the Billy-Berclau accident (to download on http://www.ladocumentationfrancaise.fr/brp/notices/044000226.shtml)

Y. Lecointe, J.-P. Chassine, H. Legrand, P. Monsel, Accident survenu à l'usine Nitrochimie de Billy-Berclau (Pas-de-Calais), Inspection Générale de l'Environnement, Conseil général des Mines, Inspection Générale des Affaires Sociales, Mars 2004.

Paper TN014 - 3

When Risk Becomes Reality:
Formosa Plastics' Response to a Plant Explosion

Robert Thibault
Formosa Plastics
9 Peach Tree Hill Road
Livingston, NJ 07039
rthibault@fpcusa.com

Prepared for Presentation at
American Institute of Chemical Engineers
2005 Spring National Meeting
39th Annual Loss Prevention Symposium
Atlanta, Georgia
April 11-13, 2005

UNPUBLISHED

When Risk Becomes Reality: Formosa Plastics' Response to a Plant Explosion

Robert Thibault
Formosa Plastics
9 Peach Tree Hill Road
Livingston, NJ 07039
rthibault@fpcusa.com

Abstract

At 10:45 p.m. Central Time on Friday, April 23, 2004, an explosion ripped through Formosa Plastics' polyvinyl chloride plant in Illiopolis, IL. The explosion effectively destroyed the plant, leaving five people dead, several others injured, caused the evacuation of thousands of people for two square miles and generated headlines in media around the world.

Formosa's response to the incident has been called a "textbook example of how to handle a crisis." This presentation will outline the steps Formosa Plastics took before, during and after the crisis to help reduce the impact on victims and their families, the community, the company and its employees and customers.

Handling Chemicals in Small Containers

Albert Ness
Rohm and Haas
3100 State Road
Croydon, PA, 19021
aness@rohmhaas.com

Richard Gibson
Rohm and Haas
4800 State Route 12
Elma, WA. 98541
rgibson@rohmhaas.com

Prepared for Presentation at
American Institute of Chemical Engineers
2005 Spring National Meeting
39th Annual Loss Prevention Symposium
Atlanta, Georgia
April 11-13, 2005

January, 2005

UNPUBLISHED

Handling Chemicals in Small Containers

Albert Ness
Rohm and Haas
3100 State Road
Croydon, PA, 19021
aness@rohmhaas.com

Richard Gibson
Rohm and Haas
4800 State Route 12
Elma, WA. 98541
rgibson@rohmhaas.com

Abstract

This paper describes three incidents involving buckets or other small, open containers that have occurred at Rohm and Haas facilities. In one, a reaction between Sodium and water propelled a bucket with considerable force, resulting in an injury. In another, a fire occurred when the lid a plastic container containing a liquid peroxide became contaminated during filling and was blown off. In the last, a solid peroxide and several different initiators and catalysts were staged on a pallet for use in batch reactors. A fire was caused by mixing of small spills from the containers on or under the pallet used to hold them. The fire caused severe damage and temporary shutdown of a reactor train.

1. Introduction

Trevor Kletz once wrote that, in a chemical plant, buckets are more dangerous than a book of matches![1] This paper describes three incidents involving buckets or other small, open containers that have occurred at Rohm and Haas facilities.

In the first, an operator was injured when a bucket previously containing sodium had not been handled and cleaned by the normal procedures. The bucket was then exposed to water from a heavy rain, and the resulting reaction propelled the bucket into the operator's leg with considerable force.

In the second, a fire occurred when the lid of a plastic container containing a Cumene Hydroperoxide (CHP) was blown off. A contaminated container was used, resulting in a reaction of the contaminant with CHP.

In the third, a fire started on or under a pallet used to stage small containers of several different initiators and catalysts. Over time small spills of a solid peroxide and other materials from the containers accumulated on a cable tray located under the pallet and the resulting mixture finally caught fire. The fire caused severe damage to electrical and instrumentation wiring, resulting in a temporary shutdown of a reactor train.

1 Kletz, Trevor, "Improving Chemical Engineering Practices; a new look at old myths of the chemical industry" second edition, Hemisphere Publishing Corporation, New York, 1990.

2. Bucket Propelled by Sodium - Water Reaction

2.1 Incident Description

At a plant that uses sodium, small pieces of waste metallic sodium are generated during line removals and clean-outs, and each time an empty sodium rail car is disconnected. The waste sodium pieces were kept in the sodium handling building in a stainless steal bucket filled with mineral oil until enough pieces were generated for reclamation. The reclamation process involved dissolving the pieces of metallic sodium in clean, dry methanol, with a water analysis of less than one percent. This operation was done in a stainless steel bucket. When the sodium had fully reacted in the methanol, as determined by the absence of sodium "chunks", the mix was transferred from the bucket to a sample recovery pot.

Earlier in the week of the incident, an operator successfully dissolved sodium in methanol in two other buckets, and noticed nothing unusual. On the day of the incident, the operator came across a bucket that contained 2-3" of a sludge material. He asked co-workers about the material and what to do with it. The co-workers were not sure where the sludge material in the bucket came from so they collectively recommended adding methanol to ensure that there was no sodium remaining in the sludge. The operator added methanol to the bucket, placed a lid on it, and placed it on a steam-out pad. About 6-7 hours later the operator returned to the bucket and assumed that, if there was any sodium in the sludge, it would have reacted and transferred the contents of the bucket into the recovery pot. After completing the transfer, the operator noticed there was still residual material left in the bucket. Thinking that all the sodium had reacted, he proceeded to an effluent building to dump the sludge into an effluent pit and wash the bucket out. The operator did not see any sign of reaction when the sludge from the bucket was dumped into the pit. The operator placed the bucket down on the walkway and went to get a hot water hose to rinse the bucket out. At this time an extremely heavy rain started, igniting the contents of the bucket. While attempting to cover the bucket, he knocked it over and its contents spilled onto the wet pad. The resulting reaction propelled the bucket into the operator's leg, resulting in an injury.

2.2 Findings

The investigation found that the bucket still contained sodium even after a 7 hour hold with methanol to recover the sodium, because the sodium was under a layer of dolomite and did not contact the methanol.

The sodium - dolomite mixture was left over from an earlier maintenance activity, and the bucket did not get properly labeled.

Although the operators discussed what to do with the "unknown" material, an adequate EHS management of change review was not conducted for this unusual situation.

2.3 Actions Taken

The sodium reclamation procedure was halted immediately.

The plant contacted the sodium supplier for advice on reclamation procedures, and discovered that the supplier sent their waste to a hazardous waste treatment facility. After that, the plant also started to do this and stopped doing the reclamation process.

For the possibility that some sodium reclamation may have to be done, the procedure was revised to include the following learning's: ensure all containers are properly labeled, do not perform the procedure outdoors when it is raining, ensure all the sodium chunks are dissolved by stirring until no more chunks are left, and, if any unusual situation arises, ensure that an EHS review is conducted.

Several training sessions were held on how to properly label the buckets, how to prepare the sodium waste for disposal and on the importance of management of change and EHS reviews.

3. Container Fire due to Contamination

3.1 Incident Description

This incident involved two operating areas in a plant which we will call Area 1 and Area 2 for convenience. Area 1 supplies an intermediate product to Area 2. Cumene Hydroperoxide (CHP) is a raw material for the intermediate. Storing and handling of CHP, as well as some other small charges, are done by Area 2. The procedure is to have the operators from Area 2 weigh up the small charge of CHP into a labeled plastic container with a hand pump and transfer the containers to Area 1 when needed.

At about 8:00 PM, the Area 1 operator located the plastic containers labeled for CHP use and contacted the Area 2 operator to fill the containers for a batch to be run early the next morning. The hand pump normally used was broken, so another one was located and washed out with water from a flush drum. The containers were covered and stored at Area 2. At approximately 11:15 PM, one of the CHP containers was found smoking. Pressure had blown the top off the container. The fire brigade was called and directed water onto all of the containers. The containers and the drum from which they were pumped were all moved to a safe storage area outside and the top bungs loosened to safely vent any possible reaction. No further reactions occurred.

3.2 Findings

The Area 2 operator noted that he had observed dark debris on the outside of one of the containers, and that it looked like Ferrous Sulfate. After the containers were filled, the operator also noted that the contents of the container looked darker than the other containers, but believed it was an effect from the debris on the outside of the container.

The MSDS notes that the CHP is incompatible with acids, oxidizing agents, reducing agents, oxidation catalysts, copper, lead and iron. Additionally, when CHP and Ferrous Sulfate are entered into the NOAA *Chemical Reactivity Worksheet*[2], it returns the entries; "Explosive when mixed with oxidizing substances", "Heat generated from chemical reaction may initiate explosion", "May cause fire" and "Heat generation by chemical reaction, may cause pressurization" for the combination.

The investigation also found that when the handling of CHP containers by the two areas began, clear responsibility for maintaining the containers was not established.

The investigators concluded that the most probable cause was contamination with Ferrous Sulfate from the outside of the container. A possible cause was contamination from residual chemicals in the flush drum, which is used to flush lances for the drum pump used for monomers, but the investigators considered this less likely.

3.3 Actions Taken

Clean replacement containers were obtained.

A dedicated storage area for the empty and full CHP containers that protects them from contamination from other chemicals was established.

A new pump dedicated to CHP service was obtained.

The operator training was updated and repeated to ensure all operators have an understanding of the hazards of CHP and to emphasize that containers for storing chemicals should be free of contamination inside and out.

The process operating procedures were updated to establish clear ownership of the procedural steps in making this product, including the maintaining and cleaning of the containers.

The process operating procedures were updated to limit the time frame for staging containers of CHP.

4. Fire from Oxidizer Spills

4.1 Incident Description

Small charges for several batch processes were weighed into small containers and staged near the reactors on a pallet. Among the materials handled in these containers were solid organic peroxide, a persulfate and other small charges. Figure 1 shows a re-creation of the staged pallets.

A fire occurred in the process building on the pallet of oxidizers containing the peroxide and persulfate. The sprinkler system activated, but its effectiveness was limited because a cable tray was between the sprinklers and the pallet. The fire caused extensive damage to electrical power,

---

[2] *Chemical Reactivity Worksheet*, U. S. National Oceanic and Atmospheric Administration, http://response.restoration.noaa.gov/chemaids/react.html

control and instrumentation wiring, resulting in considerable downtime in a reactor train. Figure 2 shows the actual pallet after the fire.

Figure 1: Re-Creation of Pallet with Pre-Weighed Materials in Process Area

4.2 Findings

The investigation revealed that, in the past, there had been spills of the different materials because buckets tipped over and cracks had been reported in the buckets. The floor was an open grate and a cable tray was located underneath the staging area, creating a location for the spills to collect because it was difficult to clean up.

The investigators concluded that the most likely cause of the fire was a spill of the peroxide from a bucket onto a cable tray below the grating which contained other, incompatible materials, particularly some activators that had spilled over time. The peroxide reacted with other materials starting a fire.

The location of the pallet near cable trays and junction boxes led to the extensive damage.

Contributing to the incident was a lack of detailed knowledge of how the specific materials would react if mixed. There was a general understanding that the materials should not be mixed, but not in-depth knowledge of what would happen. This led to a practice of staging incompatible materials needed for several batches on the same pallet.

Figure 2: Remains of pallet after fire

4.3 Actions Taken

All chemical residues on pipes and equipment below the grating on the floor of the process building were cleaned up.

All existing containers for pre weighed materials were inspected and replaced as needed. Procedures were established to inspect all containers within one year and adjust the frequency of the inspection or implement a replacement policy based on the results.

Separate storage areas for of catalysts and activators, away from critical equipment and potential contamination sources were set up.

Separate, dedicated, color coded pallets for storing incompatible materials were obtained and used.

Spill containment was provided for all pre-weighed, packaged materials staged on the process floor.

Limits on the amount of materials that could be pre weighed and staged on the process floor were established.

The peroxide containers were replaced with new, color-coded buckets with lids with gaskets.

5. Conclusions and Recommendations

These incidents have several things in common:

- Open buckets or other small containers were used for handling reactive chemicals.
- Contact occurred between the material in the containers and other, incompatible material(s) through spills or contamination.
- The corrective actions rely heavily on administrative procedures.

The incidents show that whenever small containers or buckets are used to handle small charges, the potential for contamination or inadvertent mixing exists. To prevent these kinds of incidents, look for ways to avoid use of small containers. If this is not possible:

- Determine the potential hazards of contamination. Constructing a reactivity matrix is a good way to do this (see the reference to the Chemical Reactivity Worksheet for a good source for beginning to construct a reactivity matrix).
- Provide training to personnel on the potential hazards of contamination and ignition of the materials being handled.
- Provide separate areas for weighing and staging incompatible materials such as activators and initiators.
- Use dedicated containers and handling equipment, such as pumps and pallets.
- Clearly label the containers, and consider color coding them.
- Provide lids for containers.
- Establish procedures for cleaning and inspecting of containers.